Interplay of Creativity and Giftedness in

[redacted]

ADVANCES IN CREATIVITY AND GIFTEDNESS

Volume 9

Advances in Creativity and Gifted Education (ADVA) is the first internationally established book series that focuses exclusively on the constructs of creativity and giftedness as pertaining to the psychology, philosophy, pedagogy and ecology of talent development across the milieus of family, school, institutions and society. ADVA strives to synthesize both domain specific and domain general efforts at developing creativity, giftedness and talent. The books in the series are international in scope and include the efforts of researchers, clinicians and practitioners across the globe.

Interplay of Creativity and Giftedness in Science

Edited by

Melissa K. Demetrikopoulos
Institute for Biomedical Philosophy, USA

and

John L. Pecore
University of West Florida, USA
Institute for Biomedical Philosophy, USA

SENSE PUBLISHERS
ROTTERDAM/BOSTON/TAIPEI

A C.I.P. record for this book is available from the Library of Congress.

ISBN: 978-94-6300-161-8 (paperback)
ISBN: 978-94-6300-162-5 (hardback)
ISBN: 978-94-6300-163-2 (e-book)

Published by: Sense Publishers,
P.O. Box 21858,
3001 AW Rotterdam,
The Netherlands
https://www.sensepublishers.com/

Printed on acid-free paper

TABLE OF CONTENTS

JOHN L. PECORE AND MELISSA K. DEMETRIKOPOULOS

INTRODUCTION FOR THE INTERPLAY BETWEEN CREATIVITY AND GIFTEDNESS IN SCIENCE

INTRODUCTION

Benjamin Bloom's Taxonomy of Student Learning attempts to classify the behaviors associated with learning into levels (Bloom et al., 1956). Bloom's original levels of learning included three lower levels (knowledge, comprehension, and application), and three higher levels (analysis, synthesis, and evaluation). This taxonomy spurred the development of the concept of higher level thinking skills, which are so important to educational practices to this day. The updated levels of Blooms Taxonomy are: Remember, Understand, Apply, Analyze, Evaluate, and Create (Anderson & Krathwohl, 2001). Thus, Create is considered to be the pinnacle of learning and so might be postulated to be best developed among gifted learners. Unfortunately, much of our current educational and testing constructs focus on the very basic foundation of learning Remember/Knowledge, which focuses on having students memorizing facts. The remaining components of the hierarchy are often neglected even for our most capable students. Thus, science instruction often focuses on content knowledge with students being expected to memorize numerous facts and formulas that relate to the natural world.

In many instances, talented and gifted (TAG) education also focuses on this lowest level of learning and merely consists of either having gifted learners process greater quantities of content (enrichment) or having them memorize the various fact and formulas at a younger age (acceleration). This approach is particularly problematic for science since the facts of science are in constant flux such that much of what students are taught during their precollege years will no longer be scientifically accepted by the time they either enter or complete college. For example, many of us learned that there were nine planets in our solar system and that single cell animals were the simplest animals. However now we are taught that there are eight planets in the solar system since Pluto was reclassified as a plutoid or dwarf planet based on its size and location in space; and the concept of single celled animals is nonexistent due to the fact that there is a separate Kingdom for all single celled eukaryotes.

While the content of science is constantly evolving, the general process skills of science have remained unchanged. Additionally, the mastery of these process skills, especially the higher level process skills, is necessary for success in the sciences and often defines truly great scientists. The ability of scientists to ask the right questions (Create), to discern the way to investigate these questions (Analyze), and to evaluate their findings (Evaluate) is what sets them apart from the technically competent bench scientist who can carefully follow a set of experimental procedures. The idea that creativity is important to the successful scientist was eloquently described by Albert Einstein (1931). "I believe in intuition and inspiration. Imagination is more important than knowledge. For knowledge is limited, whereas imagination embraces the entire world, stimulating progress, giving birth to evolution. It is, strictly speaking, a real factor in scientific research." Thus, it is critical for gifted learners to be taught how to create new postulates, to think logically and to reason rather than to be taught strategies to memorize a set of potentially irrelevant facts.

CHAPTER SYNOPSES

This book contains twenty chapters divided into five sections, which explore the interplay between creativity and giftedness in science. The first section titled "Historical Perspective" includes three chapters, and provide a historical context for the book. In "Historical Contribution of Creativity to Development of Gifted Science Education in Formal and Informal Learning Environments," Lynne Bailey and colleagues trace the history of gifted science education. In the next chapter, "Importance of Creative Thinking for Paradigm Shifts that Foster Scientific Advances," Mihyeon Kim examines the many aspects of creativity (i.e. imagination, intuition, insight and inspiration) that sustain the work of scientists, like Einstein, to progress in revolutionary scientific discoveries that result in paradigm shifts. Trudi Gaines, Jennifer Mesa, and John Pecore in "Twentieth Century Scientists who Exemplify the Interplay of Creativity and Giftedness" present the biographies of Luis Walter Alvarez, Barbara McClintock, and Peter Dennis Mitchell; three Nobel scientists who epitomize creative giftedness, and through the telling of their life experiences reaffirm the role of the environment in fostering creativity.

Section 2, which includes three chapters, is titled "Encouraging Scientific Creativity in gifted Learners." First, Erin Peters-Burton and Lisa Martin-Hansen in "Implications of Gifted Student Selection Techniques for Scientific Creativity" address issues with current metrics to identify scientifically creative children and the need for tests that better recognize scientific creativity when identifying gifted students. The next chapter, "Efficacy of Creativity Training for Gifted Science Students," by Anthony Washington and Lori Andersen discuss strategies for increasing gifted science students' engagement and developing creative skills. Then, Angela Stott and Paul Hobden in "A Belief System at the Core of Learning Science: A Case Study of a Critical and Creative Gifted Learner" present the belief system central to the self-regulated learning for science of André, a gifted learner, and describes his observed cognitive strategies motivated by critical and creative thinking.

The third section, "Developing Gifted and Creative Learners in Science Education Classrooms," contains four chapters that provide strategies for cultivating creativity in gifted learners. In the first chapter, "Mind your P's and E's: Developing Creativity in the Science Classroom," Claire Hughes and Timothy Goodale present a model that integrates science content instruction with the development of creativity. Next, Lina Soares advocates for providing hands-on and minds-on investigative inquiry experiences to foster curiosity and stimulate creative ideas in "Sciencing: Creative, Scientific Learning in the Constructivist Classroom." In "Quantifying the Effects of Personalized Assessment Tasks in Secondary Science Teaching," Adele Schmidt examines a chemistry programs impact of teaching and learning strategies on fostering personal engagement and creative thinking while upholding foundational knowledge. Then, Andrea Foster in "Fostering Creativity in Science Classrooms: Lessons Learned from a Brigadier General" highlights the significance of school experiences and the need for early identification of scientific talents to developing and nurturing gifted children.

Six chapters are included in Section 4, "Science, Creativity and Giftedness in Real World Contexts with Diverse Learners," which features a variety of settings for investigating creativity and science in diverse gifted learners. First, Tang Wee Teo, Jia Qian Woo, and Lay Kuan Loh in "Affordances in School Science Research: Narratives from Two Singapore Specialized Science School Students" examine, through the concept of affordances, the science research experiences of two high school students. Next, Jerry Everhart in "The Geography of Giftedness: Growing Scientists in Rural Areas" addresses three issues (space, readiness, and plan-of-action) that impact the transition of gifted rural students to studying science at university. In "Identifying Gifted and Creative Future Scientists who are Linguistically and Culturally Diverse," Maria Arreguín-Anderson and colleagues suggest changing the practice and ideology in the field of gifted education in order to open spaces in the science fields to the linguistically and culturally diverse gifted learner. Then, Corin Goodwin and Mika Gustavson in "Science, Creativity and the Real World: Lessons Learned from the U.S. Homeschool Community" advocate for valuing homeschool education as an existing resource for learning about fostering creativity in science. The next chapter is "Scientific Creativity within the Rules: Suggestions for Teaching Science to Gifted Children with Autism." Here Lauren Madden and Kristin Dell'Armo provide two examples of successful adults as examples for nurturing scientific creativity through classroom environments for students on the autism spectrum. Then Sally Carson, Steve Cutler, and Victoria Rosin in "Creatures, Costumes, Cryptic Creations: Integrating Creativity in a Secondary Science Gifted Program in Marine Science" present findings from New Zealand's Talent Development Imitative for integrating creativity in a residential science program.

The final five chapters in Section 5, "Approaches for Fostering Scientific Creativity in Gifted Learners," highlight innovative programs that promote creativity in gifted science education. In "Use of Analogy and Comparative Thinking in Scientific Creativity and Gifted Education," Audrey Rule and Benjamin Olsen discuss the applications of analogies as an essential component of science and gifted education. Then Keith Taber in "'Chemical Reactions Are Like Hell Because…': Asking Gifted Science Learners to be Creative in a Curriculum Context that Encourages Convergent Thinking" shares the use of a science analogy game to encourage creativity in secondary science teaching. In "Fostering Creativity Using Robotics Among Students in STEM Fields to Reverse the Creativity Crisis," Kyung Hee Kim and Steve Coxon discuss the creativity crisis and the potential for robotics programs to foster creativity. In "Attracting Dynamos: How Problem Based Science Opens Doors and Creates Opportunities," Mary Lightbody and Lisa Huelskamp discuss the use of problem based learning computer simulations to both attract gifted students to science and meet the needs of tomorrows science leaders. Finally, in "Developing a Rebel With a Cause through Creative Risk-Taking in Gifted Students," Carrie Rainwater and Nancy Wittner discuss ways for gifted students to gain confidence in taking risks associated with creative approaches to scientific problem-solving.

REFERENCES

Anderson, L. W., & Krathwohl, D. R. (Eds.). (2001). *A taxonomy for learning, teaching and assessing: A revision of Bloom's taxonomy of educational objectives: Complete edition.* New York, NY: Longman.

Bloom, B. S., Engelhart, M. D., Furst, E. J., Hill, W. H., & Krathwohl, D. R. (1956). *Taxonomy of educational objectives: The classification of educational goals; Handbook I: Cognitive domain.* New York, NY: Longmans, Green.

Einstein, A. (1931). *Cosmic religion: With other opinions and aphorisms.* New York, NY: Covici-Friede.

John L. Pecore
Department of Teacher Education and Educational Leadership
University of West Florida
USA
and
Division of Program Development and Assessment
Institute for Biomedical Philosophy
USA

Melissa K. Demetrikopoulos
Division of Program Development and Assessment
Institute for Biomedical Philosophy
USA

SECTION 1

HISTORICAL PERSPECTIVE

LYNNE M. BAILEY, LEE G. MORRIS, WESLEY D. THOMPSON, STEPHEN B. FELDMAN AND MELISSA K. DEMETRIKOPOULOS

1. HISTORICAL CONTRIBUTION OF CREATIVITY TO DEVELOPMENT OF GIFTED SCIENCE EDUCATION IN FORMAL AND INFORMAL LEARNING ENVIRONMENTS

INTRODUCTION

The needs of gifted learners are often not met in post-industrial revolution classrooms which were modeled after the work day in an industrial setting. This setting is structured with a foreman (teacher) and workers (students) who switch from task to task in response to a bell which is tolled by the factory owner (principal). In this model, students generally have their own individual work station (desk) and work independently, but in unison, on the same task. Students who are able to complete their assignments more quickly than others are expected to wait patiently until the other students catch up. Teachers' efforts tend to be focused on students who are struggling to complete the assignment since they are tasked with ensuring that all students meet a minimum performance standard and can be passed onto the next grade. Because their educational needs are not being met in this structure, gifted learners often turn to informal learning environments such as zoos and science centers to partially satisfy their science education interests. Additionally, some schools provide additional outlets for students that are not part of the regularly structured curriculum such as science fairs and science clubs which could also partially satisfy these students' science education interests. These supplemental avenues of science education differ from the standard elements of science education in that they are more flexible and thus facilitate the integration of the students' creativity with their science education.

REGULATION OF SCIENCE EDUCATION IN THE UNITED STATES POSITIVELY AND NEGATIVELY IMPACTS GIFTED STUDENTS

Jolly (2009) provided an historical review of the impact of federal funding efforts on the development of gifted Science, Technology, Engineering and Mathematics (STEM) education. STEM education initiatives in the United States can be traced back to West Point in 1802, as West Point graduates were partly responsible for the

M. K. Demetrikopoulos & J. L. Pecore (Eds.), Interplay of Creativity and Giftedness in Science, 3–13.

design and building of much of the physical infrastructure required for America's expansion. An important contribution to the advancement of STEM education was the Morrill Act of 1862 which greatly expanded STEM efforts at the College and University level by supporting the formal study of science, agriculture, mechanical arts, and engineering and ultimately led to the development of organized research in university programs (Butz et al., 2004). Nearly a century later, the National Defense Education Act was passed in 1958 in response to Russia's launch of Sputnik in 1957 which served as a call for immediate attention to the development of future U.S. STEM professionals. During this point in US history, a great effort was made to expand funding for promising STEM students. This reform effort in the 1950's and 1960's encouraged collaboration between science researchers and classroom teachers in an effort to cultivate America's underdeveloped population of future scientists (Dow, 1997; Flattau et al., 2006; Jolly, 2009). This was a time that embraced creativity within science programs for gifted students at the elementary level that focused on independent projects and the study of science phenomena occurring in everyday experiences. Efforts in high schools focused on having gifted students learn advanced STEM concepts at earlier ages and included dual enrolment, separate schools of science, and accelerated curriculum (Havinghurst, Stivers, & DeHaan, 1955; Anderson, 1961; Wiszowaty, 1961). During that golden era of gifted science education, the field benefited by support through Title III (Financial Assistance for Strengthening Science, Mathematics, and Modern Foreign Language Instruction) which provided STEM education funds to states, and Title V (Guidance Counseling and Testing: Identification and Encouragement of Able Students) which provided the funding to define and redefine how gifted students were identified. While the performance on intelligence tests has remained an important, if not primary, factor in the identification of giftedness, the concept of multiple talents as well as the importance of creativity began to take shape in the late 1950's and 1960's. The inclusion of creativity in defining gifted learners and as a factor in gifted science education was federally recognized in 1972 (Marland, 1972). The Civil Rights era of the 1960's and 1970's redirected the national focus from promoting our brightest STEM students to addressing the needs of a variety of other student groups, including students at the other end of the IQ continuum as well as students who were lacking appropriate STEM opportunities. While these new efforts were necessary and important, they came at the expense of programs for gifted learners which were often defined as elitist and thus not in keeping with popular ideas of equality. More recently, the No Child Left Behind legislation in 2001 further emphasized addressing the needs of struggling students and reduced the importance of meeting the needs of the gifted learner. However, both the National Academy of Science and the National Academy of Medicine, along with the Academic Competitiveness Council in 2007, have recommended renewed attention to STEM initiatives and gifted education as US policies in this area have been counterproductive to remaining globally competitive (Loveless, 2008; Jolly, 2009).

INFORMAL SCIENCE LEARNING ENVIRONMENTS FILL IN THE GAPS

Informal science learning environments have not been restricted by the ongoing changes in legislative priorities that have impacted gifted learners' science education opportunities and have served as a constant resource for these students. As noted above, there is tremendous potential to creatively meet the needs of gifted students within informal science learning environments. Crane et al. (1994), defined informal learning as

> ...those activities that occur outside the school setting, are not developed primarily for school use, are not developed to be part of an ongoing school curriculum, and are characterized by voluntary as opposed to mandatory participation as part of a credited school experience. (p. 3)

Among the goals of informal education is the intention to foster a more informed public who may be more likely to become interested in science careers and issues. Informal curriculum and learning opportunities might also target specific populations, and might not be readily available to all students in the traditional school environment. Museums, which may have had the reputation of being distant and unapproachable in the past, have been especially eager to be seen as more audience-centered (Ciotti, 2010) and have modified their formatting to include hands-on and participatory aspects in addition to the more traditional un-touchable, and seemingly, dusty displays. Bell et al. (2009), referenced a report by the Committee on Learning Science in Informal Environments that found that science learning can occur in everyday activities in a variety of social settings; can develop positive science-related attitudes, emotions, and identities; and can involve knowledge developed by cultural groups in settings that feature distinct cultural traits or environments of a given group. Likewise, due to the flexibility afforded to this educational avenue, informal science education practitioners and communities can foster informal science learning opportunities that create a powerful impact on less dominant, underrepresented groups in a society.

Tours of science centers, museums, zoos, aquariums, planetariums, botanical gardens, and national parks are all denoted as informal learning experiences and can occur either as part of school field trip or a family outing. In either case, such activities are generally seen as a fun way to participate in an authentic science experience that produces voluntary learning outside of the classroom environment. The informal structure provides an opportunity for a more creative approach to science education for gifted learners since the process is partially guided by the individual learner's interests, prior experience, and prior knowledge. The gifted learner is freed from waiting for the other students to catch up and from the limiting structure provided in the classroom which imposes an imaginary boundary to topics learned. The engagement of gifted students in science learning, in this environment, is mostly limited by students' self-motivation and imagination, as they are provided with many more choices and have greater control over their own learning (Rennie, Feher, Dierking, & Falk, 2003).

INTERSECTION OF FORMAL AND INFORMAL SCIENCE LEARNING

The National Science Teachers Association's position statement on informal science education states that informal science education complements, enhances, and supplements formal science studies and also aids in developing literacy in science (1998). The National Science Education Standards emphasized the incorporation of informal science learning activities into the formal science curriculum because these activities can bridge the "conceptual gap" between real world experiences and the classroom (National Research Council, 1996; Melber & Abraham, 2002).

Since so many important decisions require STEM knowledge, informal science institutions attempt to facilitate the production of a scientifically literate public. When the public is better informed, they are able to make meaningful and important decisions about the world (Melber & Abraham, 2002). Given the relative lack of creativity in classroom based STEM education, and the demonstrated importance of creativity for the advancement of STEM areas, it is critical that science educators take full advantage of the interconnections between formal and informal environments to maximize student learning and enhance science literacy among students. In formal environments, learning is most often achieved through a combination of student-centered and teacher-centered pedagogy, while learning in informal environments is more likely to follow the student-centered constructivist paradigm whereby participants make use of prior knowledge to actively build upon existing cognitive frameworks and actively construct new meanings (Osborne & Wittrock, 1989). In both environments, and in all environments for that matter, learning also occurs through interaction with the physical world, and cannot be entirely separated from the physical and social constructs present. However, in the formal learning environment, interactions are purposefully reduced, while in the informal learning environment, interactions are typically enhanced, thus leading to the possibility of more flexibility, richness, and creativity. Furthermore, learning is mediated through the socio-cultural interactions of peers, family, teachers, and others within the society, all of which can be filtered through the lens of prior knowledge (Falk & Dierking, 1992). The construction of new knowledge is dependent upon the integration of the learner's prior knowledge with new knowledge brought about by the physical and social contexts of the discrepant events presented in the learning environment.

Gifted students often supplement their science leaning with informal science opportunities; and research on visitor learning in informal free choice institutions indicates that the knowledge, experience, and interest of the learner greatly influence what is learned and how it is learned in these settings (Falk & Dierking, 2000). Therefore, science educators using informal free choice institutions need to be aware that there may be discrepancies between students' prior knowledge and content to be presented while also taking advantage of the potential to allow the learner to engage with the content creatively. Informal learning settings provide an important opportunity for students to connect concrete learning experiences with higher levels of cognitive learning and these opportunities benefit from guided instruction

(Ausubel, 1968; Folkomer, 1981; MacKenzie & White, 1982; Vinci, 1968). Guided instruction in these settings is generally presented either one-on-one, or in very small groups, which provides the opportunity for gifted learners to engage creatively with the STEM content at a relevant level and depth. Ausubel (1968) indicated in his learning theory the need for concrete experiences as a transition from primary to secondary concepts. This idea is similar to hands on experiences that Piaget (1970) identified as helpful in order to have transition from concrete to abstract levels of cognition. As suggested in the literature, the role of informal learning is to provide direct exposure to concrete materials and phenomena and to provide an avenue for the integration of creativity within STEM education.

STATUS OF PRESENT DAY GIFTED SCIENCE EDUCATION

Contemporary directions in gifted science education attempt to shift the focus from traditional approaches to science learning in which science principals, laws, and theories as outcomes of inquiry are emphasized. Research has demonstrated that gifted science learners are often able to quickly and easily understand; master and integrate content; solve problems creatively; challenge established assumptions; and be taught to effectively apply self-regulated learning strategies (Brandwein, 1955; Maker & Neilson, 1996; Tang & Neber, 2008; Yoon, 2009; Genc, 2013; Kahyaoglu, 2013). However, a number of recent studies have revealed several important needs of gifted science learners that include both academic as well as social factors, including studies by Tang and Neber (2008), which explored differences in motivation and self-regulated learning as related to ethnicity, gender, and grade level. Pride (2014) examined using learning stories to characterize gifted and "hard-worker" mindsets in gifted high school students. Coxon (2012) studied spacial and creative abilities in learners, while Anderson (2014) documented an inattention to visual-spacial ability in gifted education despite its importance in STEM education. These factors, among others, highlight some of the challenges with gifted science education and provide some insight into the importance of integrating creativity into gifted STEM education.

Furthermore, other researchers examined teacher concerns on challenging gifted students, teacher-student interactions, and student attitudes towards subjects such as chemistry and physics (Lang, Wong, & Fraser, 2005; Coates, 2006). Given what is known about gifted learner needs and the demand for future STEM professionals in the global landscape, current science education, especially for gifted science learners, needs to foster scientific literacy, interest in the nature of science, and appreciation for processes of inquiry where discovery, knowledge production, and active investigation occur. The skills for conducting original research, learning advanced content, and developing critical thinking abilities should be greatly emphasized (National Research Council, 1996; VanTassel-Baska, Bass, Ries, Polan, & Avery, 1998; Yoon, 2009). Researchers have also recognized that the success of authentic inquiry requires deliberate attempts on the part of teachers and facilitators to cultivate all learners' creative abilities, their desire to take risks, and even a willingness to fail (Erez, 2004; Hennessey, 2004; Neu, Baum, & Cooper, 2004; Seo, Lee, & Kim, 2005; Yuk & Cramond, 2006; Park, 2011).

RECOMMENDED FUTURE DIRECTIONS FOR GIFTED SCIENCE AND CREATIVITY EDUCATION

Research suggests that early experiences in science encourage development of problem-solving skills, which are critical as the lay public interacts with a technical environment (DeWitt & Osborne, 2010) and are critical for gifted STEM learners. However, university outreach efforts by scientists are often aimed at middle school students, and more commonly, high school students even though students have often made up their minds about their scientific abilities and affinities by middle school (Maltese & Tai, 2010; Thiry et al., 2008). Thus, it is important for the scientific community to provide opportunities for elementary students to interact with scientists. Laursen et al. (2007) reported that having scientists interact with students leads to increased student engagement and enhanced interest in science. Therefore, it is important for gifted students of all ages to interact with working scientists so that they gain insight into the creative process of science.

As mentioned in the introduction, Benjamin Bloom's Taxonomy of Student Learning, later revised by Anderson and Krathwohl (2001), attempts to classify the cognitive behaviors associated with learning into a hierarchy (Bloom et al., 1956). These levels are often illustrated as a pyramid suggesting that the lower levels of this taxonomy are foundational to the higher levels of learning much like Maslow's hierarchy of needs whereby an individual must have their basic physiological needs meet before they can address higher order needs such as self-actualization or esteem (Maslow, 1943). Although there have been many advancements in identifying the needs of gifted learners, many gifted programs have returned to the "factory floor" ideology of having students slog through repetitive material. Clearly, most students are capable of performing at several levels of cognitive functioning, and our gifted students should be encouraged to utilize all of these stages throughout their lifetime, including their early elementary and middle school years. The mastery of science process skills, especially the higher level process skills, is necessary for success in the sciences and many components of these process skills, for example, careful observation and measurement, can be mapped back onto Bloom's taxonomy of learning and are important for all students to learn and be able to use. However, as mentioned in the introduction, great scientists also have the ability to ask the right questions (Create), to discern the way to investigate these questions (Analyze), and to evaluate their findings (Evaluate). Thus, higher level thinking skills and higher level process skills of science should be emphasized with gifted learners rather than having the focus of TAG education being on content knowledge accumulation.

The importance of this focus is captured in the National Science Education Standards (1996) which notes that science as inquiry requires students to combine processes and scientific knowledge while using scientific reasoning and critical thinking to develop their understanding of science. Engaging students in inquiry enables students to develop a number of important scientific skills such as gaining an understanding of scientific concepts, an appreciation of "how we know" what we know in science, and the affective quality to be able to function scientifically.

Inquiry also facilitates an understanding of the nature of science and provides practice of the skills required to become independent investigators of the world.

In 2000, The National Academies followed up the National Science Education Standards with Inquiry and the National Science Education Standards: A Guide for Teaching and Learning. This document outlines the research supporting the importance of the use of inquiry for learning science. While it is beyond the scope of this chapter to thoroughly review the importance of inquiry, a summary of these issues are provided in their list of research findings:

- Research Finding 1: Understanding science is more than knowing facts.
- Research Finding 2: Students build new knowledge and understanding on what they already know and believe.
- Research Finding 3: Students formulate new knowledge by modifying and refining their current concepts and by adding new concepts to what they already know.
- Research Finding 5: Effective learning requires that students take control of their own learning.
- Research Finding 6: The ability to apply knowledge to novel situations, that is, transfer of learning, is affected by the degree to which students learn with understanding.

In the discussion of the importance of inquiry, the National Academies' publications make reference to Donovan and colleagues (1999) work on learning. In this, the premise that we have reached a point in history whereby it is not humanly possible for an individual to remember the set of accumulated knowledge that is available is well described. If we accept this premise, it then follows that the goal of education should not be so much about accumulating facts and content knowledge. Nobel laureate Herman Simon (1996) postulated in his presentation at Carnegie Mellon University that

> the goal of education is better conceived as helping students develop the intellectual tools and learning strategies needed to acquire the knowledge that allows people to think productively about history, science and technology, social phenomena, mathematics, and the arts. Fundamental understanding about subjects, including how to frame and ask meaningful questions about various subject areas, contributes to individuals' more basic understanding of principles of learning that can assist them in becoming self-sustaining, lifelong learners.

One of the important concepts discussed is the idea of the ability to transfer learning from one situation, environment, or subject to another. This is critical since the unmitigated volume of information precludes the possibility of learning facts exhaustively. Thus, it is important to understand how to promote learning that is transferable rather than static.

In order to provide the most flexible and transferable learning experiences, the focus should be on the process of science and on the habits of accomplished scientists rather than on specific content knowledge. The Polycyclic Inquiry Approach is used to transcend content knowledge and to facilitate students gaining a fundamental understanding of the process of science (Demetrikopoulos, Thompson, Morris, & Pecore, 2011b). This allows students to formulate questions that guide learning as well as provides a mechanism with necessary scaffolding to seek answers to their questions. This scaffolding is critical as the ability to process information supersedes the ability to remember information for both accomplished scientists and global citizens. The Polycyclic Inquiry Approach somewhat borrows from and builds upon the Coupled-Inquiry Cycle in that there is a movement between teacher-centered and student-centered aspects within the approach. However, a fundamental distinction between these models includes the focus on process knowledge rather than on content knowledge. This approach is particularly helpful when differentiating instruction for TAG students who have been described as being capable of self-management (Sternberg, 1997) since it encourages broader student ownership of the process through modified self-directed learning. Although gifted students generally have the intellectual capability to be successful scientists, they often lack sufficient knowledge of the scientific process. For example, they may have difficulty in developing a novel scientific question or developing an appropriate experimental design once they formulate their questions. Gifted students benefit from the Polycyclic Inquiry Approach which provides opportunities to design and refine their own inquires by conducting experiments in phases including: collection and exploration of preliminary data, re-examination of experimental design, modification of methods and questions, collection and exploration of experimental data, and finally drawing conclusions. This approach reduces frustration of gifted learners as students come to understand the iterative nature of the scientific process and learn how knowledge builds upon prior incremental discoveries. It is important to allow students to both work in groups, which allows students to collaborate as part of a research team, and to conduct some studies individually to ensure each student understands the entire experimental design process. Additionally, having students conduct their experiments in phases allows gifted students to explore their preliminary data and then consider if their experimental design allowed them to answer their proposed question or how it may need to be modified in order to answer their question (Demetrikopoulos, Pecore, Morris, & Thompson, 2011a). This allows gifted students to reflect creatively on how the methods and questions can be further refined and what additional factors or questions can be explored within their experimental models.

REFERENCES

Anderson, L. (2014). Visual-spatial ability: Important in STEM, ignored in gifted education. *Roeper Review, 36*(2), 114–121.

Anderson, L. W., & Krathwohl, D. R. (Eds.). (2001). *A taxonomy for learning, teaching and assessing: A revision of Bloom's taxonomy of educational objectives: Complete edition.* New York, NY: Longman.

Anderson, R. H. (1961). Arousing and sustaining the interest of gifted children in the study of science. *Gifted Child Quarterly, 5*, 35–41.

Anonymous. (1998). Informal science education. *Journal of College Science Teaching, 28*(1), 17.

Ausubel, D. P. (1968). *Educational psychology: A cognitive view.* New York, NY: Holt, Reinhardt & Winston.

Bell, P., Lewenstein, B., Shouse, A. W., & Feder, M. A. (Eds.). (2009). *Learning science in informal environments: People, places, and pursuits.* Washington, DC: National Academies Press.

Bloom, B. S., Engelhart, M. D., Furst, E. J., Hill, W. H., & Krathwohl, D. R. (1956). *Taxonomy of educational objectives: The classification of educational goals; Handbook I: Cognitive domain.* New York, NY: Longmans, Green.

Brandwein, P. F. (1955/1981). *The gifted student as future scientist: The high school student and his commitment to science.* New York, NY: Harcourt Brace. (Reprinted in 1981, retitled The gifted student as future scientist and with a new preface, as vol 3 of A perspective through a retrospective, by the National/State Leadership Training Institute on the Gifted and the Talented, Los Angeles, New York.)

Butz, W. P., Kelly, T. K., Adamson, D. M., Bloom, G. A., Fossum, D., & Gross, M. E. (2004). *Will the scientific and technology workforce meet the requirements of the federal government?* Pittsburgh, PA: RAND.

Ciotti, J. E. (2010). Museums and planetariums: Bridging the gap between Hawaiian and astronomy through informal education—A case study. *Forum on Public Policy Online, 2010*(2), 1–14.

Coates, D. (2006). Science is not my thing: Primary teachers' concerns about challenging gifted pupils. *Education 3–13, 34*(1), 49–64.

Coxon, S. V. (2012). Innovative allies: Spatial and creative abilities. *Gifted Child Today, 35*(4), 277–284.

Crane, V., Nicholson, H., Chen, M., & Bitgood, S. (1994). *Informal science learning: What the research says about television, science museums, and community-based projects.* Dedham, MA: Research Communications Ltd.

Demetrikopoulos, M. K., Pecore, J. L., Morris, L. G., & Thompson, W. D. (2011a, May). *Modification of marine science education to meet the needs of gifted learners.* Jacksonville, FL: Florida Marine Science Educators Association.

Demetrikopoulos, M. K., Thompson, W. D., Morris, L. G., & Pecore, J. L. (2011b, November). *Scientific process and learning the habits of scientists essential in the teaching of neuroscience to gifted students.* Washington, DC: Society for Neuroscience.

DeWitt, J., & Osborne, J. (2010). Recollections of exhibits: Stimulated-recall interviews with primary school children about science centre visits. *International Journal of Science Education, 32*(10), 1365–1388.

Donovan, M. S., Bransford, J. D., & Pellegrino, J. W. (Eds.). (1999). *How people learn: Bridging research and practice.* Washington, DC: National Academy Press.

Dow, P. (1997). *The sputnik-inspired reforms of the 60's.* Retrieved from http://www.nationalacademies.org/sputnik/dow2.htm

Erez, R. (2004). Freedom and creativity: An approach to science education for excellent students and its realization in the Israel arts and science academy's curriculum. *Journal of Secondary Gifted Education, 15*(4), 133–140.

Falk, J. H., & Dierking, L. D. (1992). *Learning from museums: Visitor experiences and the making of meaning.* Walnut Creek, CA: Alta Mira Press.

Falk, J. H., & Dierking, L. D. (2000). *Learning from museums: Visitor experiences and the making of meaning.* Walnut Creek, CA: Alta Mira Press.

Flattau, P. E., Bracken, J., Van Atta, R., Bandeh-Ahmadi, A., de la Cruz, R., & Sullivan, K. (2006). *The national defense education act of 1958: Selected outcomes.* Washington, DC: Science and Technology Policy Institute.

Folkomer, T. H. (1981). Comparison of three methods of teaching geology in junior high school. *Journal of Geological Education, 29*, 74–75.

Genc, M. (2013). Science diary of a gifted student. *Educational Research and Reviews, 8*(8), 390–395.

Havinghurst, R., Stivers, E., & DeHaan, R. F. (1955). *A survey of the education of gifted children*. Chicago, IL: University of Chicago Press.

Hennessey, B. A. (2004). *Developing creativity in gifted children: The central importance of motivation and classroom climate (RM04202)*. Storrs, CT: The National Research Center on the Gifted and Talented, University of Connecticut.

Jolly, J. L. (2009). The national defense education act, current STEM initiative, and the gifted. *Gifted Child Today, 32*(2), 50–53.

Kahyaoglu, M. (2013). A comparison between gifted students and non-gifted students' learning styles and their motivation styles towards science learning. *Educational Research and Reviews, 8*(12), 890–896.

Lang, Q. C., Wong, A. F. L., & Fraser, B. J. (2005). Teacher-student interaction and gifted students' attitudes toward chemistry in laboratory classrooms in Singapore. *Journal of Classroom Interaction, 40*(1), 18–28.

Laursen, S., Liston, C., Thiry, H., & Graf, J. (2007). What good is a scientist in the classroom? Participant outcomes and program design features for a short-duration science outreach intervention in K-12 classrooms. *CBE-Life Sciences Online, 6*, 49–64. Retrieved from http://www.lifescied.org/cgi/content/full/6/1/49 doi: 10.1187/cbe.06-05-0165

Loveless, T. (2008). *High-achieving students in the era of NCLB*. Washington, DC: Thomas B. Fordham Institute.

Maker, C. J., & Nielson, A. B. (1996). *Curriculum development and teaching strategies for gifted learners* (2nd ed.). Austin, TX: PRO-ED.

Marland, S. P., Jr. (1972). *Education of the gifted and talented: Report to the congress of the United States by the U. S. commissioner of education* (Vol. 1). Washington, DC: U. S. Government Printing Office.

MacKenzie, A., & White, R. (1982). Fieldwork in geography and long term memory structure. *American Educational Research Journal, 19*, 623–632.

Maltese, A. V., & Tai, R. H. (2010). Eyeballs in the fridge: Sources of early interest in science. *International Journal of Science Education, 32*(5), 669–685.

Maslow, A. H. (1943). A theory of human motivation. *Psychological Review, 50*(4), 370–396.

Melber, L. M., & Abraham, L. M. (2002). Science education in U.S. natural history museums: A historical perspective. *Science and Education, 11*(1), 45–54.

National Research Council. (1996). *National science education standards*. Washington, DC: National Academies Press.

National Research Council. (2000). *Inquiry and the national science education standards: A guide for teaching and learning*. Washington, DC: National Academies Press.

Neu, T. W., Baum, S. M., & Cooper, C. R. (2004). Talent development in science: A unique tale of one student's journey. *Journal of Secondary Gifted Education, 16*(1), 30–36.

Osborne, R., & Wittrock, M. C. (1989). The generative learning model and its implications for science education studies. *Science Education, 12*, 57–87.

Park, J. (2011). Scientific creativity in science education. *Journal of Baltic Science Education, 10*(3), 144–145.

Piaget, J. (1970). *Structuralism*. New York, NY: Columbia University Press.

Pride, L. D. (2014). Using learning stories to capture "Gifted" and "Hard Worker" mindsets within a NYC specialized high school for the sciences. *Theory into Practice, 53*(1), 41–47.

Rennie, L. J., Feher, E., Dierking, L. D., & Falk, J. H. (2003). Toward an agenda advancing research on science learning in out-of-school settings. *Journal of Research in Science Teaching, 40*(2), 112–120.

Seo, H., Lee, E. A., & Kim, K. H. (2005). Korean science teachers' understanding of creativity in gifted education. *Journal of Secondary Gifted Education, 16*(2–3), 98–105.

Simon, H. A. (1996). *Observations on the sciences of science learning*. Paper prepared for the committee on developments in the science of learning for the sciences of science learning: An interdisciplinary discussion. Department of Psychology, Carnegie Mellon University, Pittsburgh, PA.

Stenberg, R. J. (1997). A triarchic view of giftedness: Theory and practice. In N. Colangelo & G. Davis (Eds.), *Handbook on gifted education* (2nd ed.). Needham Heights, MA: Allyn and Bacon.

Tang, M., & Neber, H. (2008). Motivation and self-regulated science learning in high-achieving students: Differences related to nation, gender, and grade-level. *High Ability Studies, 19*(2), 103–116.

Thiry, H., Laursen, S. L., & Hunter, A. (2008). *Professional development needs and outcomes for education-engaged scientists: A research-based framework.* Retrieved from http://www.redorbit.com/news/science/1529666/professional_development_needs_and_outcomes_for_educationengaged_scientists_a_researchbased/

VanTassel-Baska, J., Bass, G., Reis, R., Polan, D., & Avery, L. D. (1998). A national study of science curriculum effectiveness with high ability students. *Gifted Child Quarterly, 42*, 200–211.

Vinci, T. G. (1968). *Educational field trips for disadvantaged pupils in nonpublic schools: Evaluation of ESEA Title I projects in New York City 1968–1969* (No. ERC-E010e). New York, NY: Center for Urban Education, Educational Research Committee.

Wiszowaty, K. W. (1961). A special science program for gifted elementary school children. *Gifted Child Quarterly, 5*, 121–126.

Yoon, C. H. (2009). Self-regulated learning and instructional factors in the scientific inquiry of scientifically gifted Korean middle school students. *Gifted Child Quarterly, 53*(3), 203–216.

Yuk, K. C., & Cramond, B. (2006). Program for enlightened and productive creativity illustrated with a Moire patterns lesson. *Journal of Secondary Gifted Education, 17*(4), 272–283.

Lynne M. Bailey
Pinellas County Schools
USA

Lee G. Morris
Division of Arts and Sciences
Institute for Biomedical Philosophy
USA
and
Department of Biology
Piedmont Technical College
USA

Wesley D. Thompson
Division of Arts and Sciences
Institute for Biomedical Philosophy
USA

Stephen B. Feldman
Division of Arts and Sciences
Institute for Biomedical Philosophy
USA

Melissa K. Demetrikopoulos
Division of Program Development and Assessment
Institute for Biomedical Philosophy
USA

MIHYEON KIM

2. IMPORTANCE OF CREATIVE THINKING FOR PARADIGM SHIFTS THAT FOSTER SCIENTIFIC ADVANCES

There is a strong attraction among scientists, educators, and psychologists to creativity, and this attraction is supported by theories of creativity, theories of intelligence, or education because of current, rapid societal changes, global economic reformation, and technology development that require creative solutions. Because advances in science and technology are likely to affect leadership in industrial productivity, the appeal of creativity for scientists has grown among international competitors.

Scientific advances or scientific discoveries are often mentioned within paradigms. A paradigm represents "a model, template, or matrix for making or evaluating something" (Nickles, 1999, p. 335). A paradigm is a belief system in a field that holds true for a certain time period. However, a paradigm does not remain forever, and paradigms change over time based on conceptual changes in the field. For example, Darwin was born at a time when the creation paradigm was dominant. People, including scientists, believed that God had created all species and the world, so that the universe could not be changed. However, Darwin observed marine fossils thousands of feet above sea level and reasoned that the land had been raised up by earth movements, and he found that the fossil mammals he discovered in South America looked like living mammals from the same area. If each species was created in particular, he questioned why this should be; and he wondered why there were so many species in an island group that looked very similar but with slight differences from island to island. The elegant simplicity of Darwin's observation, reasoning, and questioning guided him to create his evolution theory, which changed the paradigm in science from creation to evolution (Berra, 2008). Scientists call this type of conceptual change a paradigm shift. A paradigm shift means that a new set of rules or standards replace existing rules or beliefs among professionals or within a community. A breakthrough discovery or invention often changes the conceptual framework of a field, and a new accepted conceptual framework leads to a scientific conceptual leap such as the change from creation theory to Darwin's evolution theory. Scientific advances are made from innovative ideas that replace previous understandings; the creativity of scientists plays a critical role for paradigm shifts in science (Nickles, 1999).

How, then, do scientists come up with a new rule, and how do paradigm shifts happen? Originality or novelty is often discussed to explain the cognitive capacities

M. K. Demetrikopoulos & J. L. Pecore (Eds.), Interplay of Creativity and Giftedness in Science, 15–28.

that may contribute to a paradigm shift. So what helps scientists to generate original ideas? Gardner (1983) explained originality within the context of a domain-specific skill that shapes an unfamiliar but worthy product or performance. Originality and novelty are the characteristics associated with creativity, and scientists who want to change a paradigm need to produce new ideas or answers to the questions or discover new things that impact the current conceptual framework. However, scientists cannot produce original thinking like a sudden rain from a blue sky. Clouds should be formed in advance, in order to have rain. Likewise, novelty cannot be accomplished without prior reasoning and thinking in the field of study. Knowledge in the field of study is considered to be a critical element to generating novel ideas, meaning that creative scientists need to acquire a mastery of the knowledge in their research areas (Gardner, 1983). After acquiring knowledge in the field, scientists may formulate adequate problems for scientific discovery, and those discoveries will have an impact on existing understanding.

The possibility of creative solutions in science comes from the investigation of previous sets of beliefs. Scientists need to know how to make inquiries and to look at the existing agreed upon beliefs from different angles in order to answer questions and make scientific advances (Kuhn, 1999). Not every scientist finds appropriate questions from different perspectives, from within the predominant beliefs. Then how are creative scientists capable of asking the right questions and seeing the same things in different ways? The sources of original ideas or the ability to think from different perspectives than the established beliefs are not clearly identified. There have been efforts to discover attributes or strategies for making significant scientific discoveries. With the increased importance of advances in science, it seems clear that having creativity that generates new ideas and seeing things differently is an essential ability in science. The "Aha" moment of accidental discovery has been mentioned throughout scientific history, demonstrating the same importance of creativity in the field of science as exists for poets and artists.

For a better understanding of creativity, many definitions and various approaches have been explored. According to some definitions, creative thinking allows scientists to view things differently from the common understanding of observed phenomena and contribute to scientific advances (Dunbar, 1999). One of the most common ways to investigate scientific creative work is to analyze creative scientists themselves. The characteristics or cognitive styles of creative scientists include elements such as imagination, intuition, insights, and inspirations, which are often examined to learn about the creative strategies the scientists applied (Piirto, 2011). Researchers interested in special scientific ability and creativity have examined creative scientists' thinking processes and personalities to try to understand the exceptional performances that allowed them to make scientific conceptual leaps (Dunbar, 1999).

When scientific discovery is discussed, note that there are two different kinds: factual discovery and conceptual discovery (Noe, 2001). Kuhn (1999) distinguished these two different discoveries as normal science and scientific revolution. Normal science involves researching within a paradigm and finding pieces for solving a puzzle, but scientific revolution results in a conceptual change in normal scientific research. Revolutionary scientific discovery creates a paradigm shift, producing problem-solving efforts and representations of belief systems affecting common knowledge during certain time periods (Gruber, 1993). Belief systems do not, however, always represent the shared understandings of all community members. While anyone whose experience corresponds to claims made by common knowledge may not have problems with shared generalizations, others whose imagination or understanding of current scientific agreement is not commensurate with prevalent knowledge may be frustrated by these shared beliefs. The frustrated person may attempt to produce a justification or interpretation of a different type of observation or understanding than that of the existing paradigm. Through the process of a new interpretation and inspection of a new assumption, a conventional paradigm can be replaced by a new set of paradigms, and a paradigm shift happens.

In examining how considerable scientific advances have been made, scientists raised questions and investigated problems with different points of view from those of traditional knowledge. Scientists are remarkable for their rational process, but the rational process is not enough to explain the capability of asking original questions to produce breakthrough discoveries. Original questions end up providing new conceptual understandings of problems being worked on. Therefore, as many researchers have explored ways to identify a process of scientific discovery that would create changes in conception to frameworks in a field, creativity has been examined as a factor in scientists' capabilities that go beyond the rational.

To see how closely creative thinking processes are intertwined in scientific advances, creative scientists are often analyzed to examine how they differ and what characteristics allow them to produce extraordinary performances, as well as to find ways of nurturing those characteristics and the role of creativity in scientific discoveries. Common to all creative scientists is that spirit of wondering and questioning that constitutes the creative scientific attitude. For example, Einstein loved asking probing questions with a curious mind. In his autobiography, he stated that:

> For me it is not dubious that our thinking goes on for the most part without use of signs (words) and beyond that to a considerable degree unconsciously. For how, otherwise, should it happen that sometimes we wonder quite spontaneously about some experience? This wondering seems to occur when an experience comes into conflict with a world of concepts which is already sufficiently fixed in us. Whenever such a conflict is experienced hard and intensively it reacts back upon our thought world in a decisive way. The development of this thought world is in a certain sense a continuous flight from wonder. (Robinson, 2005, p. 29)

As shown in his statement, Einstein drove himself into the world of wondering and questioning. He loved intense discussions with his students and colleagues. He also recognized that the Newtonian Laws could not be applied to light. Einstein questioned the idea that time was absolute. If time was absolute, one hour on Earth would be the same as one hour on Mars; however, Einstein imagined himself chasing a beam of light (Banerji, 2006). From his imagination, he understood the nature of light and time, which was only later proven true through mathematics. This was not because Einstein did not have a deep understanding of quantum physics and mathematics, but because his imagination, insight, and intuition guided him to produce innovative thinking. Einstein stressed the importance of imagination and said that "Imagination is more important than knowledge. For knowledge is limited to all we now know and understand, while imagination embraces the entire world, and all there ever will be to know and understand" (Smith, 2003). The traditional procedure in physical science is to make observations first. Then scientists create hypotheses and collect systematic data to explain what they observed. Based on the results of the analysis of the data, scientists develop principles and theories therefrom. However, Einstein started with a higher level of abstraction through imagination first, and then drew empirical inferences and linked to other laws. The thinking process of Einstein was not aligned with the traditional scientific research process. Beyond rational thinking capability, and knowledge, he applied creative thinking strategies to his innovative discovery. Imagination, intuition, insight, enjoyment, and inspiration played essential roles in his innovative thinking. This chapter explores imagination, insight, and intuition, the core fundamental characteristics for creative analogical reasoning and dynamic connections among different cultures and ideas, in order to enable creative leaps in science.

NATURE AND ROLE OF IMAGINATION: HOW IMAGINATION IMPACTS SCIENTIFIC ADVANCES

The Merriam-Webster (1999) dictionary defines imagination as "the act or power of forming a mental image of something not present to the senses or never before wholly perceived in reality." From this definition, imagination is considered as a way of exploration and potential experience beyond what we can perceive. Imagination has been meaningful to all creative people. Psychologists have studied imagination to identify its origin, role, and meaning for their psychological therapies. However, in the field of creativity, imagination is considered a strategy for allowing people to open their minds and get over barriers between reality and thoughts, concepts, ideas, images or whatever we experience. With imagination, creative people may express their thinking more freely in unique ways so that they can produce significant outcomes compared to other people. It is exciting to watch science fiction movies or read science-fiction stories because their imaginings make us think of capabilities beyond reality, and their descriptions are vivid enough to make us think, "What if that story is true?" Many inventions were born out of the imagination. Robert Goddard was captivated by the novel, *War of the Worlds*, and the book evoked his imagination to invent the first liquid-fueled rocket (Strauss, 2012). A science-fiction novel, *Twenty Thousand Leagues Under the Sea*, stimulated Simon Lake, who is the father of the modern submarine, to imagine undersea travel (Strauss, 2012). John Horgan (1996) mentioned in his book, *The End of Science*, that "No matter how far empirical science goes, our imaginations can always go further" (p. 30).

Imagination is sometimes discussed and more highly valued in the fields of art and literature than in science. In the field of business, people have dedicated themselves to improving imagination in order to solve problems in creative ways. However, in more recent times, imagination has also been appreciated in the area of science. Polanyi (1958) addressed the similarity between novel writing and questions of mathematics that can also be applied to science. A theory and hypothesis grow depending on imaginative thinking, empirical findings, and systematic verifications that take a long period of time; in the same way, a novel grows with the author's imagination, research on the settings, and the story synopsis while the writing is occurring. Visualization helps people in both areas, the scientists creating original questions and writers creating original themes and stories that take them beyond what they are seeing in their reality.

Numerous references to the imagination can be found in history. Cocking (1991) identified that Quintilian referred to imagination related to *phantasiai.* A person who is very sensitive to impressions will have the greatest power over emotions, and this power of vivid imagination is presented to the sensitive person in the most realistic manner. Imagination encompasses an elaborate wish for a future arrangement of events, fantasy stories, or daydreaming (Singer, 1999). Affective awakening induces people to imagine what they long for. Barker (1938) recognized that Spinoza identified the imagination as the lowest level of knowledge or cognition; however, imagination provides higher levels of knowledge when it is furnished with distinct and adequate ideas. Creative performances are accomplished through imagination in various forms depending on the discipline. While the process of the imaginative thinking process in the fields of art or literature may be involved with unique and brilliant works, imagination in the field of science inspires creative scientists to wonder and even provides esthetic aspects to their theories. Streminger (1980) examined Hume's theory of imagination and identified that imagination is in the area of subjective abstraction and plays a role in the context of discovery and in the context of justification. A famous example of Kekule's discovery of the molecular structure of benzene may be examined to represent how imagination empowers scientists to build a new concept or an idea. Kekule experienced a daydream in which he imagined the atoms forming into snakelike chains, and then he saw one of the snakes biting its own tail, which made him realize that benzene was a ring-shaped molecule (Weisberg, 2006).

There are many other examples showing the role of imagination in scientific discovery. Isaac Beeckman (1588–1637), who introduced the concept of the compressibility of the lower layers of the atmosphere and the weight and elasticity of the whole mass of air, imagined the air as a large sponge surrounding the earth. Imagination is the power of creating new intellectual conceptions (Webster, 1965).

However, imagination without knowledge in the field of science and without the scientist's intention or will to discover cannot produce a significant output that will contribute to a paradigm shift.

Imagination is seen as a single magical event; however, as Kuhn (1996) pointed out, scientific thinking advances over time. There is a kind of chain feedback of observations, which lead to theories, which in turn lead to a world view. New scientific theories and products start from the imagination and build their foundations step-by-step, over a long period of time, for scientific verification. No single experiment can make a generalization of a new idea. Imagination makes individuals ask "what if" questions and guides them to investigate their questions through observations. The unconscious process is geared up by scientists' intention of what they want to achieve and to believe (Lieberman, 2003; Modell, 2003).

Imagination represents an understanding and insight into the world that is not readily acceptable to others as an important and meaningful experience. Murray (1986) examined Sartre's thought about imagination, and addressed the idea that we always project ourselves toward expected future potentials, and imagination leads the projection. Imagination is the essential part of our life to reach better understanding of knowledge, including in the realm of science. Imagination allows people to envision other courses of action in the real world as well as diverse ranges of responses toward problems; this helps creative people to predict the immediate and long-term future. This is the important role of imagination in scientific thinking (Person, 1995).

Mental capabilities, furnished with desire and knowledge, shape an idea with beauty and meaning. Imagination often allows scientists to come to the limits and exceed the boundaries of the self, of the communicable, and of the sequential (Maguire, 2006) and to go beyond rational understanding. Einstein had special talent for envisioning problems, and he enjoyed fantasizing: "When I examine myself and my methods of thought I come to the conclusion that the gift of fantasy has meant more to me than my talent for absorbing positive knowledge" (Gardner, 1993, p. 105). He enjoyed visualization and imagination to build understanding and inquiries rather than following mental models created by other scientists. For him, imagination was the important thinking strategy, and his enjoyment of exploring worlds within his own mind was a source of his productivity. However, imagination is not the only thing that enables creative scientists to produce significant performance. Even for Einstein, his knowledge, his capacity to integrate empirical phenomena, his rational thinking strategies, and his other creativity characteristics including intuition and insight enabled him to think innovatively.

INTUITION

Everybody has intuition. Intuition is used in daily life to make decisions such as choosing which direction to drive. However, many don't trust their intuition for making important decisions because it cannot be explained using reason or logic.

Intuition is considered knowledge that is beyond logical explanation, like other creative characteristics. Intuition is an immediate understanding of something vague or some kind of natural knowing independent of rational thought. Intuitive thinking is frequently interchanged with the word "hunch," because intuition does not provide proof or analysis for the result of thinking. Usually, rational thinking takes formalized steps to achieve new knowledge, and this knowledge is the extension of what we already know. However, intuition happens in a sudden moment, where one reaches a conclusion or develops insights into a problem (Weisberg, 1993).

Many researchers in the field of creativity and science have taken up the challenge of defining intuition and identifying the intuitive thinking process so that they can understand the role of intuitive thinking in scientific discoveries. Policastro (1999) defined intuition as:

> A tacit form of knowledge that orients decision making in a promising direction. In the context of problem solving, a promising direction is one that leads to potentially effective outcomes. In the context of innovation, a promising direction is one that leads to potentially creative result. (p. 89)

This definition suggests that intuition is a form of knowledge that guides decision making toward creative outputs or performances. As Policastro describes it, intuition enables people to select appropriate information or make proper connections of information from numerous alternatives for decision making. Intuition guides people to explore the most valuable connections of information, and intuition sets boundaries for investigation. In this context, Policastro identifies intuition as a tacit form of knowledge that largely confines the creative search by setting its major scope. However, Policastro's definition does not provide an explanation of how intuition happens.

Some scientists explain the intuitive thinking process using the concept of an intelligent memory. The neuroscientists Gordon and Berger (2003) explained the process of engaging in intuitive thinking with the concept of intelligent memory as:

> Intelligent Memory…is like connecting dots to form a picture. The dots are pieces or ideas, the lines between them are your connections or associations. The lines can coalesce into larger fragments, and these fragments can merge to form a whole thought. This whole thought may be a visual image, a piece of knowledge, and idea, or even a solution to a problem. Individual pieces, the connections, and the mental processing that orchestrates them generally work together so they appear to be a single cognitive event. That's what happens when ideas or concepts "pop" into your mind. (pp. 8–9)

Bowers and his colleagues (1990) examined intuitive judgment and stated that intuitive judgment can be a result of previous logical thinking processes, meaning that intuition is based on intelligence collected from experiences and from previous knowledge. In this regard, intuition may stem from prior knowledge and the logical thinking process rather than being a natural form of knowing. In any case, intuition does not involve conscious determination but comes in the relaxed moment after struggling with a problem. Intuition happens unexpectedly through continuous mental activity and a connection between unconscious awareness and the logical thinking process (Isenman, 1997). Although intuition suggests that a sudden idea is true, the result of intuitive thinking cannot be verified. Intuitive thinking output cannot be accepted as a new scientific discovery or solution to the problem without any verification. Scientists need to change their intuitive assumption to a testable hypothesis in order for experimental results to be accepted among scientists and to contribute to shaping a new paradigm in their field (Eysenck, 1995).

NATURE AND THE ROLE OF INSIGHT

Understanding insight is important for identifying the creative thinking process, so a consistent line of research has tried to understand the process of problem solving including insight. However, insight is difficult to understand, and there is no more clear agreement about how insight occurs than there is with many other creative thinking processes. Several different theories have been proposed to illustrate insight for problem solving. The first view on insight is to consider insight as a mystic power. This approach describes insight as something that can be experienced when unnecessary constraints are relaxed and a new portion of the solution space becomes available for investigation. Sometimes, restructuring of a problem is necessary to solve that problem; however, when a person attempts to solve a problem, that person is fixed to the current experience so that he or she cannot think of alternative ways to see the problem (Jansson & Smith, 1991; Weisberg, 2006). In this viewpoint, insight is the mystic ability of restructuring problems so that creative scientists having insight can accomplish significant problem solving and contribute to scientific advances.

The second approach is to consider insight as the capability to associate different ideas or knowledge. If we assume knowledge as a node in a knowledge graph (Pols, 2002; Schilling, 2005), finding a solution to problems is to discover the correct associations of knowledge nodes. Insight is experienced when an unlikely association of knowledge for solving the problem is retrieved, and this is a result of analytic reasoning. Appropriate study design and analytic reasoning enable people to come up with new paths for solving problems. A capability of strong association of knowledge is the matter of insight (Weisberg, 2006). In this approach, insight into solutions happens with a gradual accumulation of knowledge and experience. Insight is not a mystical power but an analytic reasoning process. Insight is like a growing organism in our minds. Ideas and information undertake recombination as a person experiences and acquires more knowledge. A creative person selects only useful information or ideas for further cognitive processes.

A third approach is to understand insight as a special information process. Insight is a stretched, unconscious leap in thinking, a critically accelerated mental process, or a short-circuiting of the normal reasoning process (Perkins, 1981). A leap in thinking means that someone tries to understand phenomena, but there is a gap between acquired information and the understanding of observed phenomena, and insight fills this gap through an accelerated mental process. In the 1700s, Isaac Newton built a significant intellectual structure with the discovery of the laws of gravity. When he was under the tree watching an apple fall from a tree near him, it suddenly struck him to make a connection between the moon and the apple and to understand that the earth's gravitation must be curving the moon's path in a way resembling the apple's path in falling toward the earth (Gleick, 2003). Insight enabled him to fill in the gap between the apple's falling and moon's path, so that he could understand the inner nature of gravity clearly.

The fourth theory about insight recognizes it as three information processes: selective encoding, selective combination, and selective comparison (Sternberg & Davidson, 1999). The selective encoding process constitutes a separation of relevant information from irrelevant information. A creative person with insight can identify relevant information from a vast store of information in order to solve problems. The selective combination process comprises competency in combining relevant information for problem solving. This competency is an important ability of connecting important information to come up with new pathways to the solution. Currently, the importance of interdisciplinary study has been suggested, and selective combination may be a critical thinking process for creative performance. Finally, selective comparison involves connecting old information to new information in order to understand observed phenomena better and to solve problems.

All of the proposed theories related to insight have strengths and weaknesses, and no theories thoroughly explain how insight happens and what insight is. However, they share several characteristics of insight, demonstrated through science history:

- Insight happens suddenly. In many scientists' experience, a problem is solved in the flash of a moment. Insight takes place suddenly, finding a new direction to the problem or finding connections between separate pieces of knowledge.
- Insight happens when there is no progress after intensive work. However, preparatory work is essential in order to experience insight. Insight does not occur without previous efforts.
- Insight happens after an incubation period. This is the stage in which unusual connections are made. When we try to solve problems consciously, the thinking is done in a logical fashion. However, in the moment of insight, all the pieces of information which seemed to be separate are connected. Sometimes, this is called an "Aha" moment (Weisberg, 2006).
- Insight enables people to see a new approach to the problem (Weisberg, 2006). When a scientist has a question based on his or her observation, the person works on the problem in order to answer the question with in-depth knowledge and reasoning. However, the scientist arrives at a point of not having made any progress in solving the problem. Then, insight occurs allowing a scientist to look at the problem in a different way and from unusual perspectives. Insight helps people to reconceptualize problems to come up with an "Aha" moment.

- Insight helps people make connections between old and new information.
- Insight should be verified in order to create a theory or meaningful knowledge.

Whether insight is an experience beyond acquired knowledge or is a result of previous knowledge and an elaborate study design, insight plays an important role in coming up with significant solutions for creative scientists. Many scientific advances have been made through insight, enabling scientists to achieve a new pathway of thinking to the solutions. Without insight, no matter how new, scientists would slip into confusion.

Although insight guides a scientist to solutions, the solutions or ideas derived from insight cannot be accepted within the professional community without verification. A verification process, after insight occurs, needs to be followed to find out if the connections made through insight make sense. Scientists may go through calculations or experiments to verify whether their possible solutions from insight would work or not. Cskszentmihalyi (1996) suggested four main conditions for this stage. The person who gets an idea to solve a problem through insight should continue to be flexible and open to new ideas. Also, one must keep in mind one's goals and feelings to make sure that the work is processed as intended. The third condition is to acquire updated knowledge in the field in order to make use of new techniques, information, and theories. A deliberate effort to answer a question enables individuals to understand the inner nature of thinking clearly and come up with restructured inquiries. The last condition is to communicate with others involved in similar problems, so that one can focus on the idea and have it accepted by other people in the field. This stage is critical for making a paradigm shift, because it involves the process of changing people's predominant understanding so the idea can be recognized as a new paradigm.

INSPIRATION

Inspiration is the energy that makes creators keep working on the problem and inquiring. Inspiration provides the motivation for scientists to make intentional efforts to keep their minds open. Although it is hard to prove and define the best conditions for creative work, creative people seem to be inspired by their environments. There are many things suggested as inspiring elements for creators. Piirto (2011) listed possible elements that can inspire creators such as love, nature, transcendent experiences like a mysterious force, substances, dreams, travel, the dark side of emotions such as death or illness, other creative works or creators, being thwarted, and a sense of injustice. Some of these inspirations are geared toward artistic creators, but many of them are also related to scientific work.

Scientific advances and the success of scientists do not stem exclusively from the internal operations of the individual mind, but often come from interacting with other minds and cultures that support the creative working process. In the field of creativity, chance has been considered a possible element of creative work.

One example of chance is to be involved in highly professional group work in a field. Meeting mentors, having an opportunity to work in a specific lab, or having connections to a professional network are examples of things that allow the chance for creative work. Certain environments have more intense interactions among professionals and provide more excitement so that people have more opportunities to produce more ideas and get new ideas from others. For example, John Bardeen was the first winner of two Nobel Prizes in the same field. He won his first Nobel Prize in 1956 with two colleagues while he was working at Bell Labs on the invention of the transistor. Then he moved to the University of Illinois, where he became captivated by superconductivity and won another Nobel Prize with two researchers in other universities for a fundamental theory of conventional superconductivity. As Csikszentmihalyi (1996) suggested, if scientists are involved with major research laboratories, journals, departments, institutes, and other professional networks, they tend to have more access to new voices that are heard and appreciated, so that they are inspired to keep working intensively on problems. As another example, Niels Bohr, who contributed to understanding atomic structure and quantum mechanics and received the Nobel Prize in Physics in 1922, liked to develop his ideas through interacting with others and discussing various subjects. His debate and friendship with Einstein became one of the famous stories in physics history, and their interaction inspired developments in science as well as inspiring each other (Segre, 2012).

However, even being involved in professional groups does not always guarantee the production of creative breakthroughs. Current academic systems require many publications and results produced with recognized methodological skills. Researchers must work for grants or publications each year, and it is difficult to get government funding without a specific methodology or specific timelines. As Loehle (1990) pointed out, output pressures and organizational requirements can be barriers to creative production. What if Einstein wrote a proposal for a government funding in order to explore relativity theory? If he specified his method as abstract mathematical thinking in a comfortable chair and asked for funding for 30 years of research, could he get funding? The structure of the academic world may hinder scientists' breakthrough performances, even though professional networking and communication inspire scientists to come up with new ideas.

Because of the significant role of interactions in the field of science, the scientific institutional environment should be designed to build the optimum atmosphere for creative performance. Hollingsworth (2012) recognized the significant role of diversity within scientific institutions on creative performance. Diversity allows a scientist to think in flexible ways and be open to new ideas and perspectives. Communication and interaction among scientists with diverse interests allows them more easily to adapt new perspectives and to facilitate creativity in making improvements. Hollingsworth analyzed the organizational context in which major discoveries occurred or did not occur throughout the twentieth century in Britain, France, Germany, and the United States, and he listed the characteristics of organizational contexts that facilitated the making of major discoveries. The first organizational characteristic supporting major discoveries is to have reasonably high diversity. The second characteristic is to have the capacity to integrate diversity. The third characteristic is to support communication and social integration of scientists from different fields through frequent and intense interaction. The fourth characteristic is to have the capacity to recruit scientists who adopt diversity. The fifth characteristic is to have flexibility and independence in the institutional environment. All of these suggest that the main organizational characteristic supporting major discoveries are diversity and flexibility. Diversity and flexibility are the major factors for breakthrough work in scientific organizations. New ways of thinking emerge when individual scientists have intense interactions with other scientists from different backgrounds. Intense and frequent interactions among scientists with diverse backgrounds allow them to look at new aspects and pathways to solutions.

CONCLUSION

Creativity has been a critical element for revolutionary scientific discoveries that result in paradigm shifts. In the history of science, scientific advances were made through the creative thinking process as well as through rational reasoning. Creativity enables scientists to raise questions about what most people believe about a phenomenon. Imagination enables scientists to think about alternative ways of understanding. Intuition and insight help scientists to make valuable connections between knowledge and ideas from various sources and reach toward new solutions to a problem. Inspiration motivates scientists to sustain energy for intensive work. In this chapter, many aspects of creativity, including imagination, intuition, insight, and inspiration, were examined to identify the roles of creativity in scientific paradigm shifts. Although there are many unanswered questions about the process of creative thinking, it is clear that the creative thinking process helps scientists to fulfill their duty to make sound progress in developing theories and contributing to revolutionary scientific discoveries for new scientific conceptions.

REFERENCES

Banerji, S. (2006, February). How Einstein discovered the special theory of relativity. *Resonance*, 27–42.

Barker, H. (1938). Notes on the second part of Spinoza's ethics (III), *Mind, 47*. Reprinted in S. Paul Kashap (Ed.), *Studies in Spinoza: Critical and interpretive essays*. Berkeley, CA: University of California Press.

Berra, T. (2008). Charles Darwin's paradigm shift. *The Beagle, Records of the Museums and Art Galleries of the Northern Territory, 24*, 1–5.

Bowers, K. S., Regehr, G., Balthasard, C., & Parker, N. (1990). Intuition in the context of discovery. *Cognitive Psychology, 22*, 71–110.

Cocking, J. (1991). *Imagination: Study in the history of ideas*. New York, NY: Routledge.

Csikszentmihalyi, M. (1996). *Creativity*. New York, NY: Harper Collins.

Dunbar, K. (1999). Science. In M. A. Runco & S. R. Pritzker (Eds.), *Encyclopedia of creativity* (pp. 525–531). Boston, MA: Academic Press.
Eysenck, H. (1995). *Genius: The natural history of creativity*. New York, NY: Cambridge University Press.
Gardner, H. (1983). *Frames of mind*. Cambridge, MA: BasicBooks.
Gardner, H. (1993). *Creating minds*. Cambridge, MA: BasicBooks.
Gleick, J. (2003). *Isaac Newton*. New York, NY: Vintage Books.
Gordon, B., & Berger, L. (2003). *Intelligent memory*. New York, NY: Viking Press.
Gruber, H. E. (1993). Aspects of scientific discovery: Aesthetics and cognition. In J. Brockman (Ed.), *Creativity* (pp. 48–74). New York, NY: Simon & Schuster.
Hollingsworth, R. (2012). Factors associated with scientific creativity. *Euresis Journals*, *2*(Winter), 77–112.
Horgan, J. (1996). *The end of science: Facing the limits of knowledge in the twilight of scientific age*. New York, NY: Broadway Books.
Isenman, L. D. (1997). Toward an understanding of intuition and its importance in scientific endeavor. *Perspectives in Biology and Medicine*, *40*(3), 395–403.
Jansson, D. G., & Smith, S. M. (1991). Design fixation. *Design Studies*, *12*, 3–11.
Kuhn, T. S. (1999). *The structure of scientific revolutions* (3rd ed.). Chicago, IL: The University of Chicago Press.
Lieberman, M. D. (2003). Reflective and reflexive judgment processes: A social cognitive neuroscience approach. In J. P. Forgas, K. R.Williams, & W. von Hippel (Eds.), *Social judgments: Explicit and implicit processes* (pp. 44–67). New York, NY: Cambridge University Press.
Loehle, C. (1990). A guide to increased creativity in research: Inspiration or perspiration? *BioScience*, *40*(2), 123–129.
Maguire, M. W. (2006). *The conversion of imagination: From Pascal through Rousseau to Tocqueville*. Cambridge, MA: Harvard University Press.
Modell, A. H. (2003). *Imagination and the meaningful brain*. Cambridge, MA: MIT Press.
Murray, E. L. (1986). *Imaginative thinking and human existence*. Pittsburgh, PA: Duquesne University Press.
Nickles, T. (1999). Paradigm shifts. In M. A. Runco & S. R. Pritzker (Eds.), *Encyclopedia of creativity* (pp. 335–346). Boston, MA: Academic Press.
Noe, K. (2001). The structure of scientific discovery: From a philosophical point of view. In S. Arikawa & A. Shinohara (Eds.), *Progress in discovery science* (Vol. 2281, pp. 31–39). Berlin, Germany: Springer.
Perkins, D. N. (1981). *The mind's best work*. Cambridge, MA: Harvard University Press.
Person, E. S. (1995). *By force of fantasy*. New York, NY: Penguin Books.
Polanyi, M. (1958). *Personal knowledge*. Chicago, IL: University of Chicago Press.
Policastro. (1999). Intuition. In M. A. Runco & S. R. Pritzker (Eds.), *Encyclopedia of creativity* (pp. 89–94). Boston, MA: Academic Press.
Pols, A. J. K. (2002). *Insight in problem solving* (doctoral dissertation). Department of Psychology, University of Utrecht, Utrecht, The Netherlands.
Robinson, A. (2005). *Einstein: A hundred years of relativity*. New York, NY: Harry N. Abrams, Inc.
Schilling, M. A. (2005). A small-world network model of cognitive insight. *Creativity Research Journal*, *17*(2–3), 131–154.
Segre, G. (2012). Niels Bohr and his physics institute: An example of creativity and creative inspiration in science. *Euresis Journals*, *2*(Winter), 21–36.
Singer, J. L. (1999). Imagination. In M. A. Runco & S. R. Pritzker (Eds.), *Encyclopedia of creativity* (pp. 13–26). Boston, MA: Academic Press.
Smith, P. D. (2003). *Einstein*. London, England: Haus Publishing Limited.
Strauss, M. (2012). Ten inventions inspired by science fiction. *Smithsonian*. Retrieved from http://www.smithsonianmag.com/science-nature/Ten-Inventions-Inspired-by-Science-Fiction.html
Streminger, G. (1980). Hume's theory of imagination. *Hume Studies*, *6*(2), 91–118.
The Merriam-Webster. (1999). *The Merriam-Webster dictionary*. Springfield, MA: Merriam-Webster, Inc.

Webster, C. (1965). The discovery of Boyle's law, and the concept of elasticity of air in the seventeenth century. *Archive for History of Exact Sciences, 2*, 441–502.
Weisberg, R. W. (1993). *Creativity: Beyond the myth of genius*. New York, NY: W. H. Freeman and Company.
Weisberg, R. W. (2006). *Creativity: Understanding innovation in problem solving, science, invention, and the arts*. Hoboken, NJ: John Wiley & Sons, Inc.

Mihyeon Kim
Center for Gifted Education
William and Mary
USA

TRUDI GAINES, JENNIFER MESA AND JOHN L. PECORE

3. TWENTIETH CENTURY SCIENTISTS WHO EXEMPLIFY THE INTERPLAY OF CREATIVITY AND GIFTEDNESS

INTRODUCTION

In the 20th century, innovative scientists made many contributions to our society and well-being, as well as expanded our basic understanding of natural phenomena. Children today study the structure and function of cells, DNA, and the atom using computer simulations and videos at home and at school. In this chapter, we focus on the life histories and accomplishments of three 20th century Nobel Prize-winning scientists to gain an understanding of the factors that may have influenced their interest in science and unlocked their creative potential. In particular, we examine the lives and creative achievements of Luis Walter Alvarez, Barbara McClintock, and Peter Mitchell. Although these eminent scientists were not formally identified as gifted, they can clearly be thought of as highly intelligent by any standard in light of their considerable accomplishments.

Intelligence, Giftedness, and Creativity

Counter to the 20th century phenomenon of emphasizing intelligence in the analytic domain alone, Sternberg (1985) conceptualized intelligence as residing in three domains – the creative, the practical, and the analytic. Intelligence quotient (IQ) tests such as the Stanford-Binet and the Wechsler Intelligence Scales have been used to identify children with above average intellect for advanced and accelerated coursework, especially in mathematics and science. A common critique of such an approach is that many children with high potential in other domains are not identified. Sternberg (2003) noted that individuals with strong creative abilities may not be the ones with the highest IQ scores. Indeed, one scientist, Walter Alvarez, as will be discussed, failed to meet the minimum required IQ score to be a participant in a well-known study on intelligence (Trower, 2009).

Besides superior analytical ability, various definitions of giftedness also include creativity, imagination, inventiveness, and problem-solving (Gagne, 1985; Renzulli, 1978). In school settings, early identification and educational programming are designed to ensure that gifted children develop their analytical and creative abilities and become successful adults. This is especially important as giftedness has also

M. K. Demetrikopoulos & J. L. Pecore (Eds.), Interplay of Creativity and Giftedness in Science, 29–43.

been identified as a non-static characteristic; that it may ebb and, sadly, wane over time if not cultivated. Giftedness rarely metamorphoses into genius (Simonton, 2003). Experience, or context, plays an important role in sustaining giftedness and is also considered to be a key element in developing creative abilities.

According to Sternberg and Lubart (1998), conceptions of creativity focus on the ability of the individual to make associations from among existing knowledge to arrive at new questions, ideas, interpretations, or conclusions about what is already known. Creativity is the product of the interactions among multiple components, including both domain-specific (e.g., knowledge and skills) and domain-general (e.g., personality traits) components (Sternberg & Lubart, 1998). With respect to knowledge, an individual's knowledge is central to creativity with a broader knowledge base enabling an individual to make more novel associations. Exploration of diverse interests and engagement with interesting people from other disciplines are among the recommendations to expand the knowledge base of gifted children and thus support the development of their creative tendencies (Epstein, 1996). At the same time, a certain level of structure and predictability is needed in order to not have creativity be diluted by too many stimuli or be stifled by activities that are too narrowly focused. Epstein also recommended that the optimal learning environment for fostering creativity be one that incorporates formal liberal arts instruction balanced by informal opportunities to explore and cultivate individual interests.

Davis (2003) noted that certain personality characteristics are common among highly creative individuals. These characteristics include being original, artistic, independent, motivated, curious, open-minded, and intuitive. Being able to recognize creativity, a sense of humor, an attraction to complexity, and a willingness to take risks are also among personality traits associated with creativity. Furthermore, creative individuals often persist at exploring complex and challenging problems that interest them. They may enjoy spending time alone in order to think about and work on these problems.

Creativity is often automatically associated with the arts and not necessarily with the sciences, especially the so-called "hard" sciences such as biology, physics and chemistry. The commonly held idea about scientific pursuits is that of highly analytical, methodical investigation, omitting the creative element which is so integral to the process. Scientists use creativity when they generate questions and hypotheses as well as design investigations and technology to study these questions and hypotheses (Deboer, 1991). Bickmore (2010), for instance, held that creativity in scientific research has to do with the process by which the researcher is able to more narrowly define a large question or problem by formulating a list of questions that address the parts of the whole, then deciding which of those questions might be more answerable than others. Furthermore, scientists use creativity when they are generating possible explanations for their results. The construction of theories involves much creativity as they are broad explanations that re-frame current thinking about natural phenomena and offer new insights on existing evidence (Lederman, Abd-El-Khalick, Bell, & Schwartz, 2002).

The three featured scientists, Alvarez, McClintock, and Mitchell, were selected to represent the various scientific disciplines. Alvarez received the Nobel Prize in Physics in 1968, McClintock received the Nobel Prize in Physiology or Medicine in 1983, and Mitchell received the Nobel Prize in Chemistry in 1978. In addition, we sought to include a diverse sample of scientists. Alvarez was a Hispanic-American man, McClintock was an European-American woman, and Mitchell was a British man. By examining the stories of these scientists, we wished to gain insight into how their creative abilities were developed and manifested in the sciences.

BIOGRAPHIES

Luis Walter Alvarez (1911–1988)

An American experimental physicist, inventor, and professor, "Luis Alvarez was one of the most brilliant and productive experimental physicists of the twentieth century" (Whol, 2007, p. 968). He is perhaps most well known for developing the Alvarez hypothesis to propose an asteroid impact as the cause of the dinosaur extinction event. Other lifetime achievements include his work on the Manhattan Project to develop detonators now standard in the explosives industry and being a recipient of the Collier Trophy for inventing the radar system used to assist in blind landing of airplanes. Among his many accomplishments, Luis Alvarez was awarded the Nobel Prize in 1968 for his contributions to elementary particle physics which included working with hydrogen bubble chambers to photograph particle interactions, developing computerized detectors to measure and analyze the interactions, and discovering new particles and resonance states (Martínez, 2011).

Early years. Born in San Francisco in 1911 to well educated parents, his physician father, Walter C. Alvarez, had much influence in his life. He was homeschooled by his mother through the second grade and skipped the third grade. In his autobiography, Alvarez (1987) credits his father taking him to the San Francisco Pan-American Exposition in 1955 and his fascination with the Machinery Hall exhibits as the beginnings of his lifelong interest in hardware. He spent Saturdays with his father who conducted physiological research at the Hooper Foundation. While young ten-year-old Alvarez did not find his father's work on the exposed stomach and intestines of anesthetized dogs interesting, the electrical equipment in an adjoining room fascinated him.

At the age of eleven, Alvarez's father gave him a *Literary Digest* article describing how to make a crystal radio using a cylindrical ice cream carton, shellacked copper wire, a galena crystal, and a pair of earphones, which they built together. Due to his keen interest in all things mechanical, Alvarez was sent to Polytechnic High School, a vocational training school for students not preparing for college where Alvarez found himself misplaced as one of the few students enrolled in an academic program.

Interestingly, Alvarez was interviewed by Stanford psychologist Lewis Terman for his famous study of the gifted, but was not selected, nor did he qualify for Mensa membership (Trower, 2009). Of note, none of Terman's 1,528 gifted participants received a Nobel Prize. When Alvarez's father was offered a full-time research position at the Mayo Clinic, the family, which included his father, mother Harriet, older sister Gladys, and younger siblings Bob and Bernice, moved to Rochester, Minnesota. Life at Rochester High School was more social and Alvarez began to come out of his shell. He skated every afternoon, played mixed-doubles tennis, and attended dances. While his high-school science courses were adequately taught, Alvarez did not find them particularly interesting.

Alvarez's father hired one of the machinists at the Mayo Clinic to provide private weekend lessons for him and during the summers the young Alvarez worked in the clinic instrument shop. In Rochester, Alvarez and a friend would sneak past security guards to climb towers and buildings and explore the power house. According to Alvarez, "a controlled disrespect for authority is essential to a scientist" because all good experimental scientists have had "an intense curiosity that no Keep Out sign could mute" (Alvarez, 1987, p. 14). He credits his youth for developing a judicious skepticism about authority and regulations.

College years. In 1928, Alvarez entered the University of Chicago where he lettered in gymnastics by practicing two hours a day every day for four years, and pledged Phi Gamma Delta, which became the center of his social life. During his undergraduate years, he lived in the fraternity house. By his junior year, Alvarez found physics, a relatively new science discipline, writing "the physics library was so engrossing that I had to force myself to leave it for food and friends" (Alvarez, 1987, p. 22). According to Alvarez (1987) and supported by Trower (2009), his ability to retain material published in physics journals was excellent. He was readily able to reproduce equations or text from memory, to recall author's names and recall locations of important graphs in an article.

For his first undergraduate research project, Alvarez constructed a Geiger counter with limited known details or the aid of specifications (Alvarez, 1987). The task tested the limits of his skills as he spent countless hours in contented solitude on the project. As an undergraduate, Alvarez learned persistence, found inventing enjoyable, and discovered a passion for optics (Trower, 2009). In 1932, Alvarez enrolled in graduate school at the University of Chicago and moved into the graduate students scientific house, which contained a piano. A fellow housemate was an accomplished musician and taught Alvarez the basics of harmony. As with mathematics, Alvarez discovered that with sustained effort, he eventually could play by ear any music he ever heard (Alvarez, 1987). While in graduate school, Alvraez constructed a cosmic ray telescope using his Geiger counter tubes. At the request of his academic advisor Arthur Compton, Alvarez traveled with his telescope to Mexico City where he spent a month measuring the East-West effect of cosmic rays and making the significant basic physics discovery that primary cosmic rays were positively charged (positrons). This work resulted in a widely referenced paper with Compton putting Alvarez as first author (Alvarez, 1987).

Creative achievements. After completing his Ph.D. in 1936, Alvarez married Geraldine Smithwick, with whom he would have two children, Walter and Jean, and traveled to Radiation Laboratory in California to work with Ernest Lawrence. Upon his arrival, Alvarez learned to operate and repair the cyclotron and was soon challenged by his mentor, Lawrence, with designing the magnet for a new cyclotron. With respect to what helped him become a professional nuclear physicist at Radiation Laboratory, Alvarez (1987) writes of his intense curiosity about how everything works and Lawrence's journal club, a weekly gathering of physicists to discuss the nuclear-physics literature. On the advice of his father, every few months Alvarez would spend an evening with his eyes closed as he tried to think of new problems to solve. His first wonderings were of the problem of slow-neutron capture (i.e., resonance), which led to his invention of time-of-flight techniques to make the first measurement of the magnetic moment of the neutron. Another accomplishment at Radiation Laboratory included include devising a set of experiments to observe K-electron capture in radioactive nuclei as predicted but not observed by beta decay theory, discovering the radioactivity of tritium and measuring its lifetime (Nobelprize.org).

With America's imminent involvement in World War II, Alvarez was dispatched to the Massachusetts Institute of Technology, and a summer in England, to help develop war-fighting technologies (Trower, 2009). There he developed three important radar systems: the microwave early warning system, the Eagle high altitude bombing system, and a blind landing system known as Ground-Controlled Approach (Nobelprize.org). In recognition of this work, Alvarez received the National Aeronautic Association's Collier Trophy. Upon his return from England, Alvarez went to work at Los Alamos on the Manhattan project. There he devised an intelligence gathering system carried on an airplane for monitoring fission products by detecting radioactive gases. When Alvarez arrived at Los Alamos, he became involved with the design of "Fat Man" (a plutonium bomb) since work on "Little Boy" (a uranium bomb) was well developed. His tasks involved finding a way to simultaneously and symmetrically explode the tiles that surround the plutonium pit required to initiate a nuclear explosion (which led to the development of detonators now standard in the explosives industry) and a way to measure the energy of the nuclear bombs. Alvarez flew in the observation plane and deployed the pressure sensor gauges used to measure the bombs energy on both the Trinity, New Mexico test flight and the Hiroshima raid (Alvarez, 1987).

Subsequent to returning to Berkeley in 1946, Alvarez was elected to the National Academy of Sciences on the nomination of his mentor, Ernest Lawrence. In addition to providing technical advice to the U. S. government as an active member of JASON[1], most of Alvarez's post war work involved hydrogen bubble chambers to photograph particle interactions, for which he received his Nobel Prize in 1968.

During this time, Alvarez advised, as an outside director, the newly public Hewlett-Packard Corporation and invented a stroboscopic golf swing analyzer. He also formed Schwem Instruments to commercialize his inventions in stabilized optics and Humphrey Instruments for inventions in virtual optics including a device to automatically determine a person's eyeglass prescription (Trower, 2009, p. 12).

Toward the end of his career, Alvarez applied his talents to solving problems that interested him. He showed that sufficient evidence existed for Oswald to be the single shooter in the J. F. Kennedy assassination (Trower, 2009; Whol, 2007). Perhaps his most joyful achievement was working alongside his son, Walter, to explain "the extraterrestrial boloid explanation of the extinction of the dinosaurs" (Trower, 2009, p. 17) known as the K-T extinction hypothesis.

Barbara McClintock (1902–1992)

A review of Barbara McClintock's biography showcases not only her contributions to the field of cytology but also those personal characteristics and experiences common among creative individuals. At the same time as she grew intellectually through her studies, she developed other interests such as sports, outdoor activities, and music – all with the encouragement and support of her parents. McClintock valued her time alone, enjoyed thinking about alternative solutions to problems, and was able to retain a sense of humility about her own achievements with a willingness to pass credit for them to others.

Early years. Barbara McClintock was born on June 16, 1902 to Dr. Thomas Henry McClintock and Sara Handy McClintock in Hartford, Connecticut. Thomas and Sara initially named their third daughter, Eleanor, a delicate and feminine sounding name (Keller, 1983, p. 20). However, they changed the baby's name to the more masculine Barbara at the age of four months after observing her temperament to be quite stoic. The new baby did not cry for anything and was content to be left alone.

In 1908, the family moved to Brooklyn, New York where McClintock grew into an independent yet active child. She often preferred to simply sit alone and think, read, or take long solitary walks. In accordance with their approach to parenting, Sara and Thomas supported and encouraged their daughter's differences in personality and interests. McClintock was allowed to play as she wished, and was not made to play with girls' toys which held little interest for her. When she asked for tools at age five, her father gave her a set of toy tools. Furthermore, Sara and Thomas gave much credence to their daughter's preferences regarding her activities. They provided her with the proper clothes for exploring the outdoors and playing outdoor sports of all kinds. When interviewed about her childhood, McClintock recalled, "I could do anything I wanted. I could play baseball, I could play football, I could climb trees, I could just have a completely free time that my brother and the people on the block had" (Keller, 1983, p. 24). Her parents even allowed her to stay home from school for days or weeks at a time to do the things she enjoyed most such as ice skating. They saw school as just one part of their children's lives, and believed that it should not minimize other opportunities for exploration and learning.

Like her older sisters, McClintock was an exceptional student at Erasmus Hall High School. Unlike her sisters, she loved learning and became absorbed in finding novel approaches for solving difficult problems in her science and mathematics classes. McClintock was thus committed to the idea of attending Cornell University to study science when she graduated a semester early at the beginning of 1918. Her mother, however, believed that a college education was not appropriate for women, and refused to support her desire to continue her education. She feared that McClintock might become a female professor, would not marry, and would have no place in society (Keller, 1983, p. 27). In addition, the family was struggling financially and there was no money for college tuition. At the time, McClintock's father was serving overseas as a military doctor in World War I. When her father returned home in the summer of 1918, he was able to convince her mother to allow her to attend college.

College years. In the fall of 1918, McClintock entered the College of Agriculture at Cornell University where tuition was fortunately free. She threw herself into her studies, taking an overload of courses many semesters. In her first two years, she studied a wide array of sciences including botany, zoology, and geology. She also studied music and showed a flare for composition that surprised her harmony professor. She played in a jazz band in her free time. In her junior year, McClintock enrolled in courses in genetics and cytology. At the end of the genetics course, the professor, C.B. Hutchinson, invited her to take the graduate course in genetics which set her on the path towards becoming a geneticist. In her autobiographical statement for the Nobel Prize, McClintock stated,

> By the time of graduation, I had no doubts about the direction I wished to follow for an advanced degree. It would involve chromosomes and their genetic content and expressions, in short, cytogenetics. This field had just begun to reveal its potentials. I have pursued it ever since and with as much pleasure over the years as I had experienced in my undergraduate days. (Nobel Media AB, 2014)

In 1923, McClintock graduated from Cornell University with a Bachelor of Science in Agriculture. She immediately registered as a graduate student, declaring a major of cytology and a minor in genetics. Her thesis advisor was Lester Sharp, a cytologist. He provided her with additional training in cytological techniques, and allowed her to determine the focus of her own research. In 1924, L.F. Randolph, a recent student of Sharp's, hired McClintock who was only in the second year of her graduate studies to assist him with a study of maize chromosomes (Kass, 2003). McClintock was able to refine a technique for effectively examining individual maize chromosomes in a matter of days, accomplishing what Randolph could not in years of work (Keller,

1983). Together they used this refined technique to examine the chromosomes of a unique maize plant that McClintock had located in the Cornell corn fields. They discovered that the plant was triploid; it possessed three sets of chromosomes instead of two (Randolph & McClintock, 1926). McClintock continued the study of the triploid maize plant's chromosomes in her dissertation entitled "A Cytological and Genetical Study of Triploid Maize" (McClintock, 1927). Shortly thereafter, she refined the technique further and was able to clearly distinguish among the ten chromosomes of the maize plant for the first time (McClintock, 1929).

Creative achievements. In addition to the innovations she developed to establish the cytology of maize, McClintock similarly developed techniques for identifying the seven chromosomes of the bread mold *Neurospora* in 1944. At the time, she was a researcher at the Department of Genetics of the Carnegie Institution of Washington at Cold Spring Harbor in New York. Her colleague and friend, George Beadle, invited Barbara to Stanford University in California to solve a problem that was holding back his own research, the behavior of the chromosomes of *Neurospora* during meiosis. The chromosomes were so small that no one had even been able to determine their number, let alone how they underwent meiosis.

At first, McClintock had her own doubts that she would be able to solve the challenge. Indeed, five days into her studies, she was so frustrated that she felt the need to go outside and sit under some eucalyptus trees to cry and do "very intense, subconscious thinking" (Keller, 1983, p. 115). A short half hour later, she had the solution and was able to modify her techniques developed with maize to prepare slides that clearly showed the full complement of *Neurospora* chromosomes (Perkins, 1992). Over the next week, she was able to distinguish among the chromosomes and examine their actions during meiosis. Furthermore, in an interview she recalled being able to imagine herself inside of the nucleus with the chromosomes,

> ...when I was really working with them, I wasn't outside, I was down there. I was part of the system. I was right down there with them, and everything got big. I even was able to see the internal parts of the chromosomes. (Keller, 1983, p. 117)

Beadle later wrote, "Barbara, in two months in Stanford, did more to clean up the cytology of *Neurospora* than all other cytological geneticists had done in all previous time on all forms of mold" (as cited in Keller, 1983). Besides her ability to develop new cytological techniques, this example points to McClintock's supreme ability to integrate her past experiences and observations of meiotic chromosomes in maize towards analyzing the behavior of unfamiliar chromosomes.

McClintock's 1983 Nobel Prize in Medicine was awarded for her discovery of transposition, or "jumping genes," in maize in the mid-1940s. Plausible explanations are offered by Keller (1983) and Comfort (2008) among others as to why this discovery was so slow to be recognized by the scientific community. In an interview, McClintock acknowledged, "Transposition was absolutely nonsensical to biologists then" (Keller, 1983). Biologists of this era were convinced that the genes were fixed in their position on chromosomes like beads on a string. In contrast, McClintock asserted the revolutionary claim that genes were able to detach from and reinsert themselves into chromosomes, which regulated the function of other genes. She inferred this from single-handedly performing, analyzing, and synthesizing observations and cytological studies of a unique variegated maize plant over six years. When asked about how she was able to persist at this endeavor for so long, McClintock stated in an interview,

> It never occurred to me that there was going to be any stumbling block. Not that I had the answer, but [I had] the joy of going at it. When you have that joy, you do the right experiments. You let the material tell you where to go, and it tells you at every step what the next has to be because you're integrating with an overall brand new pattern in mind. You're not following an old one; you are convinced of a new one. And you let everything you do focus on that. You can't help it, because it all integrates. (Keller, 1983, p. 125)

McClintock was taken aback when she realized her contemporaries did not understand her reasoning and even doubted her sanity at symposium presentations in 1951 and 1956 (Keller, 1983). After all, she had been elected to the National Academy of Sciences in 1944, and served as the president of the Genetics Society of America in 1945. She eventually decided to largely withdraw from the scientific community that had rebuffed her, but to continue her research on transposition and gene regulation at Cold Spring Harbor. After publishing a 1961 paper drawing comparisons between her work and the work of Jacques Monod and Francois Jacob on bacterial gene control (McClintock, 1961), she only reported her findings in the Cold Spring Harbor annual reports and attended few professional meetings in her discipline prior to receiving the Nobel Prize (Keller, 1983; McClintock, 1987). McClintock's early realization that transposition was a common phenomenon in all sorts of organisms was not widely recognized by biologists until the late 1970s.

Peter Dennis Mitchell (1920–1992)

Arguably the highest honor awarded to Peter Mitchell was the Nobel Prize for chemistry in 1978. His chemiosmotic theory about energy conversion in mitochondria, which was initially rebuffed and criticized by the scientific establishment, came after years of development and modification based on the very criticisms of those who rejected it.

Early years. Peter Mitchell was born in 1920 to a middle class family in England. According to his biographers Prebble and Weber (2003), his upbringing was largely without any noteworthy traumatic events, apart from the continuously deteriorating relationship between his parents. Mitchell benefited from the cultural influence of his mother who ensured that he was exposed to music and the arts. From his father, who was mathematically inclined and university educated, Mitchell seems to have acquired a similar inclination. His childhood interest in cobbling together bits and pieces found outside and around his home into mechanical devices would continue throughout his life. Mitchell engaged in many hours of creating simple experiments at home which was encouraged by his mother with whom he was considerably closer than with his father. His paternal grandparents were modestly wealthy and ran a relatively formal home, making family visits there less enjoyable for young Peter than visits to his maternal grandparents whose middle class home environment was more relaxed and comfortable albeit less luxurious.

With the exception of studying mathematics, Mitchell preferred being at his workshop at home to attending school. His parents' marriage steadily deteriorated over the years and Mitchell and his brother were sent to boarding school in an effort to remove them from the tension and stress of the family home.

He was sent to Queens College, Taunton, in 1931 to study engineering or science. In addition to education, it was intended that the boys learn upper class manners and speech/language patterns. Headmaster Wiseman became a father figure to replace Mitchell's own absentee father and became very influential in Mitchell's life by fostering an ongoing love for music as well as mathematics. Wiseman arranged for him to have a workshop on campus so that Mitchell could spend free time applying the math and science he was learning to his gadgets and creations. Mitchell also became an activist against the hazing that had been traditional at Queens Collegeand succeeded in having it abolished. His social conscience and activism contribute to his broad range of interests beyond the workshop and the laboratory. Mitchell was not especially interested in competitive, team sports, having a propensity for more solitary activities; however, he was encouraged by a teacher and athletic coach to pursue rugby in order to not show fear to his classmates. Mitchell went on to be captain of the team.

While he excelled in math and science, his academic record in the humanities was not stellar. His complaint about history, for example, was that it only focused on wars and battles and that Newton was nowhere to be found. Mitchell did, however, have an interest in certain areas of literature, especially poetry and Shakespeare, having played the part of Macbeth in a school production.

In general, Mitchell reported that his greatest interest lay in arriving at solutions to problems or questions by beginning with first principles and avoiding textbooks. His experience with creating physical objects – his "devices"— transferred to his experience with learning in general. Interestingly, the subject which interested him least among the sciences was chemistry which he described as being taught in isolation from other subjects and to be a string of facts and experiments that were not well related to one another or to anything else.

College years. Although Mitchell was initially rejected due to his performance on his scholarship admission examination, he began his university studies at Cambridge in 1939 upon the strong recommendation of the ever supportive Wiseman. Mitchell pursued his studies in the sciences at the same time that he continued to be very involved with the arts, especially with music. In addition to already being able to play the violin, Mitchell taught himself to play the piano. He preferred the company of artists during these years to that of his fellow scientists and had a reputation for flamboyant dress and non-traditional appearance including the unusual length of his hair for the times. In spite of his colorful and extravagant appearance, Mitchell was a diligent and committed student of science, having been especially inspired by instructors who were able to conceptualize and communicate their subjects as operating within a greater context. Mitchell became a member of the very selective Cambridge natural sciences club and one of his presentations there was on the topic of "meaning," exploring the relationship between principles of biological science and philosophical concerns.

What Mitchell found most stimulating about the department of biochemistry at Cambridge was that he objected to many of the views expressed there, spurring him on to pursue his own investigations and to formulate the beginnings of his own views and theories. He did not distinguish himself during his years at Cambridge, which extended beyond his undergraduate experience through to his doctoral studies. Although his first doctoral thesis was rejected, Mitchell went on to complete the degree and obtained a teaching and research position. Mitchell subsequently moved to the University of Edinburgh and his research steadily proceeded toward the ultimate formulation of his chemiosmotic theory, moving through various barriers and deflections necessitated by external demands for research funding.

Creative achievements. In 1961, he and his wife purchased a manorly "fixer-upper" – Glynn House in Cornwall – which not only served as their residence but also as his new laboratory independent of an university and the attendant demands. Establishing a private research institute was made possible using Mitchell's own funds as well as those donated by his brother. It was here that Mitchell conducted the research which was to result in the formulation of his theory for which he was ultimately awarded the Nobel Prize for chemistry.

In his later years, Mitchell's focus turned to broader concerns about how science and research could or would best serve the greater good and how perspectives of the scientific community were shaped. He became interested in behavioral research and in the well-being of the research community within higher education institutions – the very environment which he had abandoned many years previously. Mitchell especially did not support what he perceived to be a centralization of the direction of scientific research, believing that too much planning would result in stultification of creativity among researchers. In Mitchell's own words:

> We don't do science because we are scientists, because of science—we do it because we are human beings. It is a most wonderful romantic, cultural activity, just as much as being a sculptor. It's problem solving.

Mitchell died in 1992. His passion for an array of interests and disciplines fostered the creativity and commitment that made his accomplishments possible. Mitchell's life work was an on ongoing synthesis of his deep interest in philosophy, social issues, the arts, and science. Although he did not excel academically in subjects outside of mathematics and the sciences, he pursued what educators today promote as "lifelong learning" in the humanities. He was equally a participant and a connoisseur of music; he also learned glass-blowing so that he could create laboratory equipment and build models to support his theories. He often espoused the importance of imagination in scientific endeavors and believed strongly that centralized planning of research would be detrimental.

In addition, Mitchell's dedication to the private research facility he founded and administered stands as tribute to his entrepreneurial skills. At the same time that he engaged in research that would result in the Nobel Prize, he created a financial foundation which supported the Glynn research laboratory for many years.

DISCUSSION / FINAL THOUGHTS

An analysis of the life histories and creative achievements of these three renowned scientists provides insights into how creativity both ignited and sustained their interest not only in science, but also in other disciplines and activities. Alvarez, McClintock, and Mitchell were all attracted to tinkering as children, to building devices, and to problem-solving in their early years. As posited by Davis (2003) and Simonton (2003), context and experience are critical elements in promoting creativity. A consideration of the family context of the three highlighted scientists points to an experience of being encouraged to be an independent thinker and to undertake activities that nurtured their imagination and curiosity. Each of the families was at least of middle class socioeconomic status, affording their children opportunities for education and advancement that would not necessarily be available otherwise. However, the salient feature of these parental influences is that of shaping attitude and encouraging a wide range of interests in their children.

Indeed, Simonton (2003) asserted that "all of the diverse components of exceptional achievement–intellect, motivation, personality, developmental experiences, education, etc.—are multiplied together rather than merely added" (p. 361). The family experiences of Alvarez, McClintock, and Mitchell as well as their educative experiences (Dewey, 1938) thus served to shape their creative potential. These scientists' mentors, teachers, or parents afforded flexibility in their educational and training opportunities providing challenges and allowing them to develop expertise in areas of interest that were later essential for their creative achievements. In the case of McClintock, her graduate advisor provided her with additional training to hone her skills in cytology. Mitchell's headmaster arranged a workshop for Mitchell to use to construct mechanical and electrical devices, allowing him to discover foundational principles in the physical sciences. Alvarez's father similarly arranged mechanics lessons for him and found him summer work in the clinic instrument shop, both of which allowed him to gain skill and confidence in building complex machines and instruments.

Alvarez, McClintock, and Mitchell showed evidence of diverse talents and creative thinking outside of science. Their stories all include participation in sports and an affinity for time spent alone. Alvarez was a gymnast, McClintock ice skated and played sports of all kinds, and Mitchell was the captain of his rugby team. In addition, they all shared a passion for music and played various instruments with great musicality. This musical interest and ability can be seen as an ideal counterpart to scientific endeavors affording each scientist the opportunity to activate a different aspect of their imagination and affective experience, playing perfectly into the advice given by Epstein (1996) to engage in activities that stimulate new kinds of thinking.

As university students, Alvarez, McClintock, and Mitchell discovered science as offering many creative challenges. They pursued challenges in their respective fields unswervingly, and as a result, advanced the understanding of fundamental scientific processes in their respective fields. These individuals described becoming absorbed in their research, and experiencing joy when they immersed themselves in solving a question or problem. Furthermore, McClintock's story includes a specific report of her ability to understand a problem once she had stepped away from it, putting some physical distance between herself and the matter at hand. Reports of creative moments often include this aspect of inspiration when not immediately involved with the problem. Even when faced with extreme resistance to their ideas, Alvarez, McClintock, and Mitchell were undeterred and continued to pursue their research. Their persistence in finding answers to their questions in spite of external discouragements is another feature of creativity which their accomplishments epitomize.

What is absent from these stories may be as important as what is found in them. There is a decided lack of rigidity in their attitude, of needing their environment to be neatly described and organized. There is an absence of a need for control even of their own experimental endeavors in the sense that they were highly collaborative individuals who even welcomed their critics and used the criticisms to improve their work. In fact, Alvarez was reluctant to be listed as coauthor on publications when collaborators contributed a greater portion of the effort, a practice that could have cost him the Nobel prize. Such a lack of personal ego may likely be the result of the broad experience of their lives, of their ability, for example, to become good at playing a musical instrument which requires much practice after many mistakes. That feature of creativity which is openmindedness requires a level of humility which by its very nature does not allow for an overdeveloped ego.

The richness of experience seen in the stories presented here reaffirms what researchers in the field of giftedness, intelligence, and creativity suspect – the role of the environment must be emphasized as educators seek to foster creativity in their students. The "budding scientist" may well fall by the wayside absent someone – family, teachers, and community leaders – deliberately providing context and experience upon which she or he may develop creatively.

NOTE

[1] The JASON society brought together the most prominent physicists to consult for the United States government on scientific questions.

REFERENCES

Alvarez, L. W. (1987). *Alvarez: Adventures of a physicist*. New York, NY: Basic Books, Inc.

Bickmore, B. R. (2010). *Creativity in science: How scientists decide what to study*. Retrieved from http://www.visionlearning.com/library/module_viewer.php?mid=182

Comfort, N. (2008). Rebellion and iconoclasm in the life and science of Barbara McClintock. In O. Harman & M. Dietrich (Eds.), *Rebels, mavericks, and heretics in biology* (pp. 137–153). New Haven, CT: Yale University Press.

Davis, G. A. (2003). Identifying creative students, teaching for creative growth. In N. Colangelo & G. A. Davis (Eds.), *Handbook of gifted education* (pp. 311–324). Boston, MA: Pearson.

Deboer, G. E. (1991). *A history of ideas in science education: Implications for practice*. New York, NY: Teachers College Press.

Dewey, J. (1938). *Education and experience*. New York, NY: Collier Books.

Epstein, R. (1996). Capturing creativity. *Psychology Today, 29*(4), 41–43.

Gagné, F. (1985). Giftedness and talent: Reexamining a reexamination of the definitions. *Gifted Child Quarterly, 29*, 103 –112.

Kass, L. (2003). Records and recollections: A new look at Barbara McClintock, nobel-prize winning geneticist. *Genetics, 164*, 1251 –1260.

Keller, E. (1983). *A feeling for the organism: The life and work of Barbara McClintock*. New York, NY: W.H. Freeman and Company.

Lederman, N., Abd-El-Khalick, F., Bell, R. L., & Schwartz, R. S. (2002). View of nature of science questionnaire: Toward valid and meaningful assessment of learners' conceptions of nature of science. *Journal of Research in Science Teaching, 39*(6), 497–521.

Martínez, J. R. (2011). *The adventures of Luis Alvarez: Identity politics in the making of an American science*. Retrieved from http://repositories.lib.utexas.edu/bitstream/handle/2152/ETD-UT-2011-12-4485/MARTINEZ-DISSERTATION.pdf?sequence=1

McClintock, B. (1927). *A cytological and genetical study of triploid maize* (Unpublished doctoral dissertation). Cornell University, Ithaca, NY.

McClintock, B. (1961). Some parallels between gene control systems in maize and bacteria. *American Naturalist, 95*, 265–277.

McClintock, B. (1987). *The discovery and characterization of transposable elements*. New York, NY: Garland Publishing.

Nobel Media AB. (2014). Barbara McClintock-Biographical. *Nobel Prize.org*. Retrieved from http://www.nobelprize.org/nobel_prizes/medicine/laureates/1983/mcclintock-bio.html

Perkins, D. (1992). Neurospora chromosomes. In N. Federoff & D. Botstein (Eds.), *The dynamic genome: Barbara McClintock's ideas in the century of genetics* (pp. 33–44). Cold Spring Harbor, NY: Cold Spring Harbor Laboratory Press.

Prebble, J., & Weber, D. (2003). *Wandering in the gardens of the mind: Peter Mitchell and the making of Glynn*. New York, NY: Oxford University Press.

Randolph, L., & McClintock, B. (1926). Polyploidy in Zea mays. *American Naturalist, 60*, 99–102.

Renzulli, J. S. (1978). What makes giftedness? Re-examining a definition. *Phi Delta Kappa, 60*, 180–181.

Simonton, D. (2003). When does giftedness become genius? An when not? In N. Colangelo & G. A. Davis (Eds.), *Handbook of gifted education* (3rd ed., pp. 358–370). Boston, MA: Allyn & Bacon.

Sternberg, R. J. (1985). *Beyond IQ: A triarchic theory of intelligence*. Cambridge, UK: Cambridge University Press.

Sternberg, R. J. (2003). Giftedness according to the theory of successful intelligence. In N. Colangelo & G. A. Davis (Eds.), *Handbook of gifted education* (pp. 311–324). Boston, MA: Pearson.

Sternberg, R. J., & Lubart, T. I. (1998). The concept of creativity: Prospects and paradigms. In R. J. Sternberg (Ed.), *Handbook of creativity* (pp. 3–15). Cambridge, England: Cambridge University Press.

Trower, P. W. (2009). *Luis Walter Alvarez: 1911–1988*. Washington, DC: National Academy of Sciences.

Whol, C. G. (2007). Scientists as detective: Luis Alvarez and the pyramid burial chambers, the JFK assassination, and the end of the dinosaurs. *American Journal of Physics*, *75*(11), 968–977.

Trudi Gaines
Department of Teacher Education and Educational Leadership
University of West Florida
USA

Jennifer Mesa
Department of Teacher Education and Educational Leadership
University of West Florida
USA

John L. Pecore
Department of Teacher Education and Educational Leadership
University of West Florida
USA
and
Division of Program Development and Assessment
Institute for Biomedical Philosophy
USA

SECTION 2

ENCOURAGING SCIENTIFIC CREATIVITY IN GIFTED LEARNERS

ERIN E. PETERS-BURTON AND LISA M. MARTIN-HANSEN

4. IMPLICATIONS OF GIFTED STUDENT SELECTION TECHNIQUES FOR SCIENTIFIC CREATIVITY

INTRODUCTION

When schools attempt to identify gifted children, quantitative measures are part of that identification process. While exploring the nexus of gifted children and science, it is worthwhile to ask, "How well are scientific ways of thinking represented on a commonly used metric to identify a gifted child?" This question has become even more important as the U.S. expands its attention on STEM and STEM education.

Creativity is valued on a personal level because of its empowering nature (Kind & Kind, 2007; Newton, 2000) and on a societal level due to its role in forwarding a successful economy, especially in Western cultures (Pink, 2005). Student creativity can be enhanced by school environments (Carey & Shavelson, 1988; Penick, 1996), and is encouraged in science education reform documents globally and at all levels of study (Fleming, 2008; Walker & Gleaves, 2008). For example in the United States, the *National Science Education Standards* (1996) state, "Creativity, imagination, and a good knowledge base are all required in the work of science and engineering" (p. 192) and the *Benchmarks for Scientific Literacy* (1993) state, "Mathematics, creativity, logic, and originality are all needed to improve technology" (p. 47). England's educational document, *Excellence and Enjoyment* (2003), encourages teachers to exercise young children's creativity and problem solving skills. The recently revised South Korean National Science Curriculum stated one of the major educational goals as fostering creative scientific problem-solving capacity (Ministry of Education & Human Resources Development [MOE & HRD], 2007).

There is also interest in creativity due to research regarding brain disorders affecting the frontalstriatal system output including Turret's syndrome, Autism or Asperger's Syndrome, Attention Deficit Hyperactive Disorder, Obsessive-Compulsive Disorder (Bradshaw & Shepherd, 2000), which have been found to contribute in many cases to an individual's creative output. These individuals think and notice different things. For example, patients with Turret's syndrome that transfer brain activity to other regions to prevent ticks have reported increases in imagination (Sacks, 1992). Furthermore increased lateral thinking has been reported in Asperger's Syndrome and other Autistic cases, and increased quick web-like thinking in has been reported with ADD patients (Sacks, 1992). Interestingly, there is a statistical trend for autism to be more prevalent in families with parents in STEM career fields of physics, engineering, and mathematics (Wheelwright & Baron-Cohen, 2001; Baron-Cohen,

M. K. Demetrikopoulos & J. L. Pecore (Eds.), Interplay of Creativity and Giftedness in Science, 47–69.

Wheelwright, Stott, Bolton, & Goodyer, 1997). Bradshaw and Shepherd (2000) state that individuals with autism and Asperger's disorders "may acquire a natural ability to select out detail, as they seem to lack the normal preference for focusing on the overall gestalt or configuration" (Rinehart, Bradshaw, Moss, Brereton, & Tonge, 2001). Creativity is a phenomenon that has captured the attention and resources of the scientific community on many different levels.

Defining Creativity

The characterization of human creativity can be traced back to a variety of sources, and descriptions of creativity vary depending on context (European University Association, 2007). Although some researchers consider creativity as an indefinable concept (Bohm, 1998; Craft, 2003), it is necessary to provide an explicit definition of creativity in order to measure the existence of the entity. In the field of psychology, descriptions of creativity are generated from two lines of thinking: creativity as a divergent process, and creativity as the ability to draw unique solutions from logic. Each of these lines of research has a propensity to rely on different measurement techniques.

The group of researchers that consider creativity as a process grounded in intellectual activities such as remote associations, rich imagery, and divergent thinking (Guildford, 1967; Mednick, 1962; Rothenberg, 1986; Simonton, 1997; Suler, 1980; Torrance, 1990) tend to investigate human creativity using correlational studies (Cattell, 1963; Eiduson, 1962; Eysenck, 1995; Roe, 1953). More recent investigations have confirmed such processes such as mental imagery to the depiction of creativity (Finke, 1990; Newcombe & Learmonth, 2005). Highly creative individuals are described as having a flat hierarchy of associations as compared with a less creative individual who would have steep hierarchy of associations. Individuals that have flat hierarchies of associations among ideas have more ideas to retrieve and to attempt to connect together. Individuals that have steep hierarchies do not consider ideas that are out of range, resulting in fewer associations among ideas (Eysenck, 1995; Simonton, 1999). Highly creative individuals also tend to have an openness to experience (King, McKee Walker, & Broyles, 1996; McCrae, 1987), a preference of complexity and novelty, and a tolerance of ambiguity (Barron, 1963; Davis, 1975; Gough, 1979). All of these qualities are related to a variety of experiences, from which an individual can develop rich imagery and have many instances to connect ideas.

Although remote associations and rich imagery are bounded concepts that are well-defined enough to be measurable, divergent thinking tends to be more complex. Divergent thinking has been described as having four dimensions: (a) fluency, which is the number of ideas one can present, (b) flexibility, which is the number of different methods to solve a problem, (c) originality, which is the uniqueness of a solution to a problem, and (d) elaboration, which is how detailed a solution is presented. The characteristics of a creative person in this line of research can be described as one who can make connections that are not usually made between factors in a situation (remote associations), one who can develop detailed mental models (rich imagery), one who can develop a large number of ideas (fluency), one who can put forward many different methods to solve a problem (flexibility), one who can describe unique solutions to a problem (originality), and one who can explain the detail in a situation (elaboration). The measurement of these characteristics lends itself to correlations among these factors in a context to best describe if a factor is present and to what extent it is present.

Another line of research views creativity as a straightforward form of problem solving that relies on logic (Klahr & Simon, 1999; Newell & Simon, 1972) and these researchers tend to conduct experiments in a lab setting (Hayes, 1989; Klahr, 2000; Newel & Simon, 1972; Weisberg, 1988). For example, a person having creativity as defined by these types of characteristics may suggest a unique experimental technique based on current scientific knowledge that takes a different perspective, but still remains logical. Although definitions of creativity are numerous, a central idea of creativity is "the ability to offer new perspectives, generate novel and meaningful ideas, raise new questions, and come up with solutions to ill-defined problems" (Beghetto, 2007, p. 1). Literature reviews have revealed that over 100 instruments have been authored to measure aspects of creativity. However, characteristics specific to scientific creativity are not considered in these measures. These two different lines of research demonstrate the complexity that is undertaken when trying to capture scientific creativity.

Many researchers have evidence that creativity is a domain-specific activity (Alexander, 1992; Amabile, 1987; Findlay & Lumsden, 1988; Albert, 1983; Gardner, 1983; Feldman, 1986). In order to be creative, one must be creative *about* a content area, thus the need for the knowledge base of the creative person to be well-organized for efficient retrieval that is relevant to the problem or situation (Mumford et al., 1991). In order to solve problems in science, one needs to be able to examine the catalog of background knowledge relevant to a problem and imagine a variety of routes to a solution by combining ideas in a modular way. Creative scientists must first acquire the facts, concepts, techniques, and theories that make up the body of knowledge known as science (Amabile, 1983; Hayes, 1989). A University of Toronto professor, Jordan Peterson (Carson, Peterson, & Higgins, 2003; Science Daily, 2003) had found that creative persons, "remain in contact with the extra information constantly streaming in from the environment." Whereas an average or "normal" person will name the object and may cognitively tag the object or idea and then move on. In contrast, the creative person, says Peterson, is "always open to new possibilities." Additionally, specific regions of the brain have been reported to be important for creativity such that turning off an inhibitory neurotransmitter results in large increases of creativity (Limb, 2010; Johns Hopkins Medicine, 2008).

Patterns in when and how bursts of creativity occur can be seen when studying scientists and inventors – strategies that can be learned through content understanding. However, the ability to measure these patterns to develop generalizations about creative persons is less sophisticated. In a comparison of measures of creativity, Sternberg (1996) found that there is only a correlation of 0.37 across tests, demonstrating the domain-based nature of these measures. Due to the difficulty in measuring general creativity, we must turn to domain-based scientific creativity in order to better articulate how people are scientifically creative.

Scientific Creativity

The development of scientific knowledge relies on scientists being able to think beyond current knowledge and techniques to progress current understandings. Creativity is a key element in building scientific knowledge (Csikszentmihalyi, 1996; Innamorato, 1998; Popper, 1959), and the subject of science has been hailed to foster creativity because of the wide range of activities that a person must be proficient in to produce scientific knowledge (Torrence, 1992).

Robert and Michele Root-Bernstein have researched creativity in relation to the enterprise of science in the book *Sparks of Genius: The Thirteen Thinking Tools of the World's Most Creative People* (1999) that includes a large number of scientists. A summary of the thirteen thinking tools can be found in Table 1.

Table 1. The 13 tools for thinking from sparks of genius (adapted from Robert & Michele Root-Bernstein, 1999)

Observing	Honing ALL the senses to perceive acutely and accurately.
Imaging	Creating mental images using any or all of the senses.
Dimensional Thinking	Translating between 2, 3 or n dimensions; shrinking or expanding within a dimension (e.g. size or duration).
Abstracting	Discovering simplicity in complexity by eliminating all but one essential characteristic.
Recognizing Patterns	Perceiving similarities in structure or property among different things.
Forming Patterns	Creating or discovering new ways to organize or arrange things.
Analogizing	Discovering functional similarities between structurally different things.
Modeling	Creating a simple analog of a complex thing in order to test, modify or play with its properties.
Body Thinking	Using motor memory, physical feelings and emotional states to recognize and address problems.
Empathizing	Becoming the thing you study, be it animate or inanimate.
Playing	Goal-less activity performed for fun that incidentally develops skill, intuition, and knowledge.
Transforming	Using some set of the previous tools to think and make in a serial, integrated manner.
Synthesizing	Knowing things in multiple ways simultaneously, subjectively as well as objectively, intuitively as well as intellectually; i.e. the result of fully using your imaginative tool box.

Observation is a large part of what scientists and other creative persons do. However, observation and imagery can also yield new results, discoveries and interpretations. At a workshop, Root-Bernstein (2012) described how Louis Pasteur who, as a student, observed crystalline structure much differently than previous students as well as professors. He was the first to note that tartaric crystals formed in both a left-handed and right-handed orientation (Debré, 1998) which was quite astounding as the crystals had been studied by several well-known scientists at that time. Additionally, as observation does not only happen with the eyes, it can be part of listening or with touch, individuals have reported that they have been able to develop their senses over time to compensate for the loss of another sense. For instance, the loss of sight typically means that the individual compensates with increased sensitivity and awareness of other senses (Voss, Collignon, Lassonde, & Lepore, 2010) as was the case with Geerat Vermeij (Cole & Vermij, 1998). Vermeij, a biologist, studied shells using other senses, namely touch to describe their characteristics. He was able to find new relationships between ancient shells to today's modern versions of shells and was able to research the development of the shell over time and advantages or disadvantages in a snail shell's design.

Imaging before creating a model, is something a scientist may do to create images in her head regarding the scientific phenomenon. Max Planck (1932) wrote of the importance of imagination and science, "[the] researcher has an entirely free hand. He may give rein in to his own spirit of initiative and allow the constructive powers of the imagination to come into play without let or hindrance. This naturally means that he has a significant measure of freedom ..." (p. 86). And later on p. 215, "Again and again the imaginary plan on which one attempts to build up order breaks down and then we must try another. This imaginative vision and faith in the ultimate success are indispensable. The pure rationalist has no place here." Imagination is not only helpful, but also necessary for the scientist to begin to think of her ideas before planning an investigation or creating a new model of an idea.

Considering *dimensional* thinking, it has been documented that training in art has positively influenced scientific thinking as the mind is trained to see things differently. At Michigan Technological University, struggling engineering students were encouraged to enroll in a course designed to develop students' spatial skills (Hill, Corbett, & St. Rose, 2010; Sorby, 2009; Sorby & Baartmans, 2000). Historically, women had had more difficulty with the spatial skills aspects of the introductory engineering graphics course. When researchers examined the predictors of what allowed someone to be successful on the Purdue Spatial Visualization Test: Rotations (PSVT:R) (Hill, Corbett, & St. Rose, 2010; Sorby, 2009; Guay, 1977), they found that previous experiences in "design-related courses such as drafting, mechanical drawing, and art as well as play as children with construction toys such as Legos, Lincoln Logs, and Erector sets" all predicted good performance on the spatial skills assessment. The other predictor was being male. Women were three times more likely to fail the test (39% of women failed compared with 12% of the men). Therefore, the spatial-skills boot camp course was created. The graduates of that course had statistically significant increases in their ability to see and think spatially (ten times better than those who had no intervention) and the numbers of students who remained in engineering majors also increased (Hill, Corbett, & St. Rose, 2010; Sorby, 2009). Interestingly, this has also decreased the gender gap in the sciences where spatial skills are required. Since this time, Virginia Tech and Purdue have also begun offering this course. Creativity is not an innate construct, and such case studies support the understanding that spatial skills can be nurtured and learned.

Root-Bernstein & Root-Bernstein (1991) have documented several ways that scientists have used *abstracting* in their creative thinking. Biologists have sometimes cut objects of their study into different forms to examine their fundamental structures. Botanists will at times characterize flowers in abstract sections, by cutting them piece by piece and then laying them out flat. After creating these types of abstractions, other ways of knowing can be applied such as mathematics as the Fibonacci sequence (generated by beginning with 0 and 1 and adding the last two numbers to generate the next number: 0, 1, 1, 2, 3, 5, 8, 13 and so forth) is found in many of nature's common patterns (Root-Bernstein & Root-Bernstein, 1999).

Other aspects of creativity in science are the abilities to *recognize and form patterns*. The Root-Bertsteins (1999) researched the importance of recognizing patterns and formation and found that different cultures foster different types of experiences with patterns. Additionally, knowing what you don't know – understanding the pattern of one's ignorance – can be valuable. "It is [the unknown] that spurs scientific progress," said Nobel laureate Thomas Weller. Babies see facial patterns. Humans create constellations out of star patterns. Patterns were noted in the shapes of continents leading to the ideas of continental drift. There are patterns in inheritance, and patterns in molecular structure. Playing with molecular models and experimenting with patterns were important when Watson and Crick explored the possible structure of DNA, although the X-ray image provided by Rosalind Franklin proved to be a central ingredient as well. Many scientists who are noted for creativity were also accomplished musicians or artists (consider Albert Einstein with his violin, Leonardo da Vinci, physicist and renowned artist, Richard Feynman who was an accomplished bongo drummer) all requiring pattern formation and recognition.

Analogizing in scientific thinking is closely tied to pattern recognition. If one examines a phenomenon and attempts to describe it, it can be helpful if a parallel association can be found in order to explain, consider, and communicate the phenomena in terms of a relationship. Nancy Leys Stephen (1986) wrote about the connections of culture, gender, analogy and science. She describes the importance of the metaphor.

> … because a metaphor or analogy does not directly present a preexisting nature but instead helps "construct" that nature, the metaphor generates data that conform to it, and accommodates data that are in apparent contradiction to it, so that nature is seen via the metaphor and the metaphor becomes part of the logic of science itself.

Creating scientific *models* and model interpretation are more typical ideas that people may think of when considering creativity in science. Watson and Crick used molecular models to build the possible structures for DNA. They continuously looked for a model that was simple and elegant or "pretty" (Watson, 1968). Models are useful and necessary tools in science for communication and explanation. It is entirely appropriate for several models to exist for one scientific concept so that different models emphasize different details of the phenomena. Such is the case with atomic models where one model focuses on possible orbits, another model resembles a cloud, and another model is constructed with sticks and balls. While the use of models may sometimes cause a misconception, such as when students think that bonds between atoms are actually sticks, models are a useful conception in science (Serban & Strugariu, 2011). As mentioned earlier, the actual experiences of building models have been documented to enhance individuals' spatial skills abilities (Hill, Corbett, & St. Rose, 2010; Sorby, 2009; Sorby & Baartmans, 2000).

An aspect of creativity not typically given much attention, especially in scientific creativity, is *body thinking*. With body thinking, one combines thought with an extension of perception in a way that allows one to consider the possible feeling of being in a certain situation. For instance, influences of body thinking have been documented (Root-Bernstein & Root-Bernstein, 1999) when surgeons perform an operation using the Telepresence Surgery System (TeSS), a virtual-reality machine that enables doctors to perform a surgery miles away by controlling a machine that responds to the doctor's hand and arm movements. Doctors reported that even though they were not in the room with the patient, by looking at the screen and later encountering resistance when the machine touches the patient's tissue, the doctor begins to feel as if she or he is actually touching the patient. This is similar to how a car can feel like an extension of the driver – it becomes part of you. Root-Berstein continues to explain that physicists working with atomic-force microscopes that can manipulate matter almost sense and feel the molecules and bonds as they work.

The best clinicians are reportedly the ones who *empathize* with their patients to such a degree that they feel and truly understand the patient's experiences. They are the ones who can be so sensitive and understanding that the patient is able to reveal embarrassing details and to listen to announcements about their prognoses that they would rather not hear as a base of trust and openness is established (Quince, Parker, Wood, & Benson, 2011). Strange as it may seem, empathy may also be extended to vegetation or inanimate objects. French philosopher Henri Bergson posited (in Root-Bernstein & Root-Bernstein, 1999) that important insights arose, "We here call intuition the *sympathy* by which one transports oneself to the interior of an object in order to coincide with its unique and therefore ineffable quality." In social work, clients who experience empathy from their practitioners are found to have improved outcomes (Gerdes & Segal, 2011). Additionally, Karl Popper, a philosopher well known for his study of the nature of science wrote (Root-Bernstein & Root-Bernstein, 1999, p. 187; Popper in Krebs & Shelly, 1975), "I think the most helpful suggestion that can be made … as to how one may get new ides in general [is] … 'sympathetic intuition' or 'empathy.' .... You should enter into your problem situation in such a way that you almost become part of it."

Interestingly, *play* is something that a lot of people do not associate with their ideas of what scientists do. However, the media may be assisting to rectify this issue. The ultra-popular Discovery Channel show *MythBusters* takes common ideas or myths that exist in popular culture and tests these myths – scientifically. While the myths are tested, it is clear that the scientists often play in their work. They make jokes, they mess around with materials and equipment just to "see what it does". Spatial reasoning tests have found that individuals who played with model-building type of toys as children, scored higher in spatial thinking (Hill, Corbett, & St. Rose, 2010; Sorby, 2009). Play is very much a part of what scientists do (Lazlo, 2004) as a lot of science is puzzle-solving. Joseph Lambert wrote to Lazlo (2004),

> When I grew up, every kid put in some serious sandbox time, and it often involved building (what seemed like) complex sand structures around which fantasies were composed and competitions took place with neighborhood kids. The organic chemistry labs (at Yale during the junior year) were fun in the same way. We constructed molecules and competed with each other in the class on speed and yield. We mixed things up, and chemical transformations took place. We separated, we isolated, we analyzed. The odors were pleasant, and the physical process of working with our hands, as with sand, was satisfying. The biweekly organic labs became the high points of my week. By the end of the year, I knew that I wanted to be an organic chemist, as I realized one could play in the sandbox for a living.

In many aspects of creativity in science, individuals need to *transform* their ideas from one type to another. When fossilized footprints are found, a scientist must interpret those two dimensional fossil prints transforming the ideas in her head about what made those particular footprints in that particular way. A three dimensional creature must then be considered that could match those footprints in the way they currently appear. How tall must the creature be? How did it walk? Or crawl? In the same way, a forensic scientist may see evidence of a violent act on the bones of a crime victim. How did the person die? What instrument was inflicted upon the person? Creativity often lies in the creation of a new tool in order to see differently.

In transformational thinking one must be both original and have ideas that have value (Amabile, 1998; Lubart & Mouchiroud, 2003). When new knowledge is created through transformative thought processes, those processes are by definition creative.

Root-Bernstein's framework of creativity in science lastly features the role of *synthesis*. Synthesis is found where scientists take ideas from multiple places creating new innovations. The innovators at Google, Apple, and 3M purposefully modified their practices at their companies to prompt innovation. At Apple, workers must go to a central location to eat, therefore allowing for impromptu hallway conversations to take place. At 3M, teams working on different projects are often sent to another team in order for ideas to cross-pollinate. Additionally, both 3M and Google spend time on speculative projects (15%–20%) as long as they share their ideas with their colleagues (Fries, 2009). Google hosts a conference called Crazy Search Ideas (Finch, 2012) where hundreds of posters from multiple fields of study are presented about widely different concepts where teams working on glue can end up talking to teams studying nanotechnology. The result is conceptual blending where new insights can occur. Interestingly, adults with attention-deficit disorder are also well-suited for examining many ideas in a short period of time. This meandering among concepts and ideas can lead to scientific insight (White & Shah, 2011).

Solving problems in science requires one to ask relevant questions, be aware of the repertoire of tools from which to measure phenomena, anticipate appropriate measurements that will provide information for a conclusion, make assumptions regarding the analysis of the measures, and translate trends in the analysis to make meaning in a conclusion, among other mental activities. Asking relevant questions requires identification of knowledge that is not yet known, which means imagining what concepts could fill in the gaps of current knowledge. Being aware of the tools and analyses available requires one to think flexibly from process to product and the possible implications of the choices. Translating trends from data into meaningful conclusions requires one to visualize the trends that are not often readily apparent in the data, and consider all possible options of a conclusion given the data patterns.

SCIENTIFIC CREATIVITY

The activities of scientists are not exactly the same as the activities of students conducting scientific investigations (National Research Council, 1996). Scientists have more expertise in a particular area of research and as experts, have the ability to generate more relevant options than their naïve student counterparts (Kind & Kind, 2007; NAE, 2005; Simonton, 1999; Taylor, Smith, & Gheselin, 1975). That is, in generating different creative scenarios and manipulating them for unique problem solving opportunities, scientists have more material to draw from and are more adept at ignoring unproductive paths of reasoning. In considering these differences between mature scientists and students, the same basic framework to describe creativity exists (Hu & Adey, 2002). Boden (2004) situates this comparison between the creativity scientists have and the creativity children have by distinguishing between something novel to the larger community (scientists) and something novel to the person (children). In school science, children are often asked to be creative by re-discovering what is already known about the natural word, resulting in original ideas about phenomena to them. It is reasonable to ask children to be creative scientifically in this context, which includes developing habits of mind of scientists (Newton & Newton, 2008).

The re-discovery of phenomena is not to be confused with reproductive thought, which is the recalling of information, following algorithms, or compiling information without adding any new insight (Moseley et al., 2005). An example of reproductive thought is to define the form and function of cell organelles. Creative processes in science can include speculation that is supported by background knowledge, such as tentative explanations, hypotheses, proposing alternative situations, or constructing empirical ways of evaluating ideas (Barron, 1988; Beghetto, 2007; Givens, 1962; Lubart & Mouchiroud, 2003; Metcalfe, 1983; Spearman, 1931). For example, students could offer evidence to support or to refute the classification of Tyrannosaurus Rex as a predator or a scavenger. Reproductive thought is distinguished here from scientific creativity by the creation of a novel idea to the person, in this case to the student, which is backed up by empirical evidence. Scientific creativity has two combined conceptions, the production of a novel idea and the support of empirical evidence (Costello & Keane, 1992).

However, stating that scientific thinking has two combined conceptions that are inherent is not measurable. Measuring creativity in science is more than documenting a student's understanding of a body of facts and a set of procedures. Despite the large variety of perceptions of creativity in the literature (Taylor, 1988; Welsh, 1981), there are several themes that emerge from the field of measurement of creativity that can then be organized in a way to measure scientific creativity. These themes, the characteristics of creative persons, creative products, and creative processes, will be explored in the following sections

Creative Persons

Creativity in science identifies three dimensions of personal traits which a person can possess and these traits come from both cognitive and affective sources (Simonton, 2003). These personality traits include fluency, flexibility and original thinking. Fluency is the ability to produce a large number of ideas. The idea of fluency is different than divergent thinking, which is being able to think in ways that may not lead to a single answer. Divergent thinking is considered a component of creative potential (Runco, 1991). In addition to fluency, a person who possesses creative traits has flexibility of thinking, which is the ability to change orientation and perspective. Lastly, original thinking is considered a dimension of creative personal traits. Original thinking is interpreted statistically in measures, and an answer that is rare statistically is considered original (Torrence, 1990). Hudson (1966) in his famous "how many ways can you use a brick?" questionnaire would give higher scores on his measurement to infrequent answers than to common answers. An example question from Hu and Adey's (2002) science-specific creativity instrument that addresses all three dimensions of creative persons is "Please write down as many possible scientific uses as you can for a piece of glass. For example, make a test tube" (p. 394). This question examines fluency, flexibility and original thinking but by using a common piece of scientific materials and eliciting a scientific purpose for the new ideas generated.

Creative Products

It is well known that if science as an enterprise is going to progress, then scientists must generate new and creative products, which include methodologies of research, tools of research, empirical evidence, models for understanding evidence, and theories to synthesize the knowledge production. This type of scientific production is a result of not using a recipe and using unique ways to solve problems (Lubart, 1994), and also of being able to not only generate ideas, but put these ideas into context of the body of knowledge we know as science. Creativity has been shown to be enhanced during scientific production when the individual is exposed to incongruous stimuli (Finke, Ward, & Smith, 1992; Proctor, 1993; Rothenberg, 1987; Sobel & Rothenberg, 1980). The creation of technical products includes new variables stemming from newly discovered phenomena, novel tools for measurement often through the use of technology to extend our senses, advances in scientific knowledge by uniquely combining previously known knowledge or adding innovative ideas, understanding of scientific phenomena (Cattell, 1971; Oschse, 1990; Einstein & Infield, 1938) through new perspectives, and scientific problem solving by being adaptive in manipulating what is currently known to produce original approaches to the problem. An example question from Hu and Adey's (2002) science-specific creativity instrument that addresses scientific products is "Please design an apple picking machine. Draw a picture, point out the name and function of each part." In this assessment, the participant must be scientifically-oriented to know the forces and motion that are involved and be able to communicate this knowledge in scientifically-appropriate ways. The focus is on being able to use what is already known in science and applying it to a technical product. An alternative open-ended question that can be asked about a different kind of scientific product, a methodology can be asked, "You have two magnets that have different magnetic strengths. How can you use empirical methods to support this claim and to determine the relative strength of each? Please write down as many possible methods as you can and the instruments, principles and simple procedure." This question, although based in eliciting creativity, directly addresses the inherent guidelines that govern scientific thinking.

Creative Processes

Scientific processes include imagination (Gardner, 1983; Johnson-Laird, 1987) and creative thinking (Einstein, 1952). Imagination refers to the ways that one can integrate new thoughts into a conceptual framework or the ways that the conceptual framework can be recombined for a different result. Creative thinking refers to imagery that is combined in new ways. A well-known example of a description of imagination and creative thinking comes from the French mathematician and physicist Henri Poincaré (1921) when he explained his thinking processes, "Ideas rose in crowds; I felt them collide until pairs interlocked, so to speak, making a stable combination" (p. 387).

Creative processes are the least developed area of measurement. For example, Hu and Adey (2002) were able, through their 3-dimensional Scientific Structure Creativity Model (SSCM), to identify questions that would measure secondary students' outcomes on products by traits by processes. Since the products category had four indicators, the traits category had three indicators, and the process category had two indicators, this model was a 4 by 3 by 2 model and had overall 24 cells each identifying a different indicator. However, the only cells in the three dimensional model that they could not address via asking scenario-based questions was science knowledge by imagination cells.

If creativity is a domain-specific phenomena, then domain-specific measures must be used to identify creative individuals so that they may be fostered appropriately in educational settings. In light of the enormous efforts being placed on developing the Science, Technology, Engineering and Mathematics (STEM) pipeline, which encourages students in the K-12 setting to pursue STEM college majors and ultimately careers, early identification of scientifically creative students is advantageous to identifying and supporting students. Measures that help to predict students who have the capacity for creative scientific thought are needed in the current education environment.

IDENTIFYING CREATIVITY/GIFTEDNESS IN CHILDREN

It is productive to identify student creative ability early in school careers, as there is empirical evidence that creative students become successful as adults. Milgram (1993) studied students in a longitudinal study that spread across 18 years, and found that creativity was associated with accomplishment in adult life, and that the contribution of creativity was more important in this association than intelligence ratings or school grades. Cicirelli (1965) investigated 641 primary school students and found that both intelligence quotient and creativity summed in a linear way to result in the effect of academic achievement. Creativity, along with information processing speed, intelligence, and school performance all contributed to academic success later in life (Rindermann & Neubauer, 2004). In a review of studies on creativity, Stephens and Karnes (2000) found a strong correlation among studies between creativity and giftedness. More recently, Palaniappan (2007) found a positive correlation between intelligence and creativity.

Although there are no specific universal identification systems for creativity, assumptions of creativity are made in identification of gifted students (Lam, Yeung, Lam, & McNaught, 2010). Perhaps the best known test of general creativity is the Torrance Test of Creative Thinking (Torrance, 1990) which is a paper-and-pencil test measuring fluency, flexibility, and original thinking. Methods used by Hocevar and Bachelor (1989) consisted of eight categories to identify creativity: tests of divergent thinking; attitude and interest inventories; personality inventories; biographical inventories; rating by teachers, peers, and supervisors; judgments of products; eminence; and self-reported creative activities and achievements. It is notable that these methods use self-report, rating scales and attitude inventories, which tap into the affective and social aspects of creativity. Additionally, activity-based tests have been used in assessing creativity (Kitto, Lok, & Rudowicz, 1994). In terms of domain-based scientific creativity tests, a test establishing divergent thinking responding to a situation about animal and plants relevant to planning experiments, hypothesis construction, data collection, and problem solving was developed by Friedlander (1983). Sinha and Singh (1987) developed an assessment for scientific creativity involving flexibility, novelty, observation abilities, imagination, analysis capabilities, and transformation abilities. More recently, Hu and Adey (2002) designed a scientific creativity test identifying the components of product, personal trait, and process. However, all of the mentioned assessments are designed for secondary students, and identification of scientific creativity and of giftedness should occur earlier in a school career in order to be fostered throughout the grades.

IDENTIFICATION OF GIFTEDNESS IN EARLY CHILDHOOD

A common instrument to identify gifted students in public schools systems in the United States is the *Cognitive Abilities Test* (CogAT) (Lohman & Hagen, 2001). The CogAT measures general and specific cognitive developed abilities, and the results of the CogAT is then used by school districts to place high ability students into different levels of educational service. To date there have been seven forms of the CogAT (CogAT7, available fall 2012) and the assessment's psychometric properties are well-known, as they were studied using large-scale pilot studies resulting in scores of peer-reviewed research publications (Lohman & Gambrell, 2012). The revisions that are incorporated into the CogAT7 take English language learners into account and make attempts to accommodate language issues, which demonstrate how the CogAT has kept up with the changing needs of society. Administration of the CogAT usually takes place in the second grade with students aged 8–9. Although intact examples of the whole test are not available because it is a protected test, there is a great deal of information about the psychometrics of the test that can be found in peer-reviewed literature.

The CogAT has three batteries of tests that are designed to measure verbal, quantitative, and nonverbal ability. The verbal reasoning section focuses on verbal classification, sentence completion, and verbal analogies. The quantitative reasoning section focuses on quantitative relations, number series, and equation building. The nonverbal reasoning section focuses on figure classification, figure analogies, and figure analysis. Each reasoning ability section (verbal, quantitative, and nonverbal) is measured in multiple formats with multiple items so that there is a high reliability in the measure (Lohman, 2006). The reported score is a Standard Age Score (SAS), which compare an individual to other children based on their age.

The verbal reasoning section measures verbal classification, sentence completion, and verbal analogies. The verbal classification questions ask students to demonstrate their knowledge of how words relate to each other and their hierarchical classification. The sentence completion questions assess students' ability to recognize a relevant word to be placed in a missing space given the context of the sentence and the word function. The verbal analogies questions test students' ability to match a word to another using a parallel relationship as two given words. The entire verbal reasoning section is heavily reliant on students' experiences with the written English language.

The quantitative reasoning section focuses on quantitative relations, number series, and equation building. The quantitative relations questions assess students' knowledge about number analogies and students' abilities to determine numerical organization. Number series questions assess students' knowledge regarding continuation of numerical patterns. Equation building questions assess students understanding of meaning of mathematical statements of equality using number puzzles. The quantitative reasoning section relies on knowledge about number values, patterns, and relationships.

The nonverbal reasoning section focuses on figure classification, figure analogies, and figure analysis. Figure classification provides information on students' ability to recognize similar characteristics of a group of figures and classify another figure from the options that has similar characteristics, but is not exactly the same. Figure analogies are similar to verbal analogies and quantitative analogies, but the analogies are communicated with figures. In figure analogies, a student shows their ability in distinguishing a relationship between two given figures and replicating that relationship with another given figure and the resulting choice. Figure analysis is expressed on the CogAT through diagrams of paper folding, cutting into the paper and unfolding the paper. The student must choose the correct resulting paper with cutouts after the described diagrams of cutting and folding. The nonverbal reasoning section assesses knowledge about geometric figure characteristics, relationships, patterns and mental manipulation of dimensional objects.

Although there are some commonalities, there is an overwhelming misalignment of aspects of scientific creativity and cognitive abilities measured with the CogAT as seen in Table 2. The point is not to denigrate the CogAT as a measure, but rather to point out that although the CogAT is used successfully to identify gifted students, the reliance on a few measures to accelerate or enrich students ignores the characteristics of scientific creativity. By highlighting the gap between what has been recognized in the research literature as scientific creativity and the intended measures on the CogAT, we emphasize the lack of identification of scientifically creative and by default scientifically gifted students. As a result, early identification of students who are scientifically-minded is not accomplished and as a result students who are scientifically creative are not being supported in a systematic way.

Scientific creativity can be considered to be domain-specific and the only specific portion of the test involves mathematics. Table 2 provides a matrix of scientific creativity characteristics by CogAT substests and displays the intersection between the two. Note that the blank cells in the matrix indicate no overlap between the characteristic and subtest. Also note the limited areas of overlap between scientific creativity and the CogAT subtests. Due to the limitations of forced answer format questions such as multiple choice given on the

CogAT, the characteristics of fluency, flexibility and original thinking are not addressed as they are defined in the field of scientific creativity. Individuals have few opportunities to generate a large number of ideas or to think of these ideas in different perspectives when only four options are available. One opportunity does arise in the figure analysis portion of the CogAT because in order to choose the correct option, one must manipulate the diagram in their mind. Mental manipulation of the diagram requires flexibility to generate many different versions of the solution to see which options fit the possibilities. Creation of technical products, advances in scientific knowledge, and understanding of scientific phenomena may be too subject-specific to be tested on the CogAT. The CogAT does not claim to test subject-specific matter, but the implication is that there remains a gap between measures used to identify gifted students and characteristics of scientific creativity. Scientific problem solving, which requires quantitative reasoning, has some overlap with the CogAT if interpreted loosely. Quantitative reasoning is necessary for some types of scientific problem solving (such as in biology, probability of phenotypes; in physics, kinematics; in chemistry, stoichiometry) because having a sense of numbers and how quantities are manipulated is a skill that facilitates the logical choices in decision making. In the same way, equation building is necessary in some topics of science because mathematical modeling creates efficiency and elegance in problem solving. In a small way, imagination is needed for figure analysis because there are situations on the CogAT where combining ideas about figures is unique to the individual. Imagination as defined by the field of creativity in science means to put a conceptual framework together in a unique way. Having a forced answer response again limits the ability to determine imagination using the CogAT. The CogAT does not claim to measure imagination, but if schools are choosing this measure to identify gifted students, it should be noted that students who may have high ability in scientific creativity and who do not have high verbal or qualitative ability would not be selected for accelerated or enriched programs. Therefore, there is a high possibility that we are ignoring our scientifically creative students for future support because there is no measure in place to identify these students. Lastly, thinking about scientific processes is not measured on the CogAT because neither metacognition nor scientific process skills are part of this assessment.

If we are to increase the STEM pipeline in the United States, it is strategic to begin with elementary children. However, if there is very little alignment between scientific creativity and measures used to identify gifted students, we have no systematic features in place to identify and foster students who may show scientific

Table 2. Common aspects of CogAT and scientific creativity

		CogAT								
		Verbal classification	*Sentence completion*	*Verbal analogies*	*Quantitative relations*	*Number series*	*Equation building*	*Figure classification*	*Figure analogies*	*Figure analysis*
Scientific creativity	Fluency									
	Flexibility									1
	Original thinking									
	Creation of technical products									
	Advances in scientific knowledge									
	Understanding of scientific phenomena									
	Scientific problem solving				2		3			
	Imagination									4
	Thinking about scientific processes									

Note: Empty cells indicate that there is no direct connection.
1. Ability to manipulate diagrams to fit a situation
2. Logical reasoning used in comparisons
3. Logical reasoning to develop mathematical modeling
4. Drawing from many different situations

mindedness. All children should be encouraged to practice creative thinking. A strategy that scientists have used historically is to study something very deeply, step back from it for a while, allow the brain to consolidate the information to allow for divergent thinking to occur and thus be open to more creative thought. Explicitly teaching strategies to encourage creative thinking may be productive in helping more students have positive experiences with STEM content.

IMPLICATIONS

As scientific creativity is central to the habits of mind of scientists, it is necessary for more measures with adequate psychometric properties be developed in the areas of creativity. As creativity is a major component of thinking scientifically, typical identifiers for gifted students may not always identify gifted children who excel in scientific thinking. Several of the types of thinking that scientists consider central to their work are not currently included in the COGAT as shown in Table 2. Additionally, many types of scientific thinking, as highlighted in Root-Bernstein & Root-Bernstein's work (1999) is not yet measurable in psychometric means. We currently lack the ability to measure: careful observation, abstraction, modeling, and transforming on a large scale. As this is currently the case, it is important for us to consider the implications of an educational system that selects gifted students without consideration of scientific creativity. With much emphasis in other domains, it is possible that we are not identifying students who are gifted in scientific creative aspects.

With STEM-focused schools whether charter, magnet, or simply schools that emphasize STEM, one must ask, "How do we support students who are creative or gifted in science when we don't know who they are?" Additionally, students who are highly creative may not be industrious in the type of school activities that we typically would ask students to engage in (textbook reading, detailed writing in areas of the teachers' choice rather than student's choice). Students who are twice-exceptional often are motivated by specific teachers and specific subjects making it difficult to recognize their talents when in a mismatch of either (VanTassel-Baska, Feng, Swanson, Quek, & Chandler, 2009). Poverty adds additional challenges as students often seemingly lack in motivation, and organizational skills (VanTassel-Baska, 2010) have difficulty in abstracting, and may answer questions creatively by choosing an answer that the teacher did not anticipate, yet was correct from their viewpoint (Slocumb & Payne, 2000). Lack of progress in school subjects sometimes, perhaps often, influences teachers if there must be a teacher recommendation involved in identifying a child as gifted. Additionally, measures of giftedness such as the CogAT admittedly test students' experiences rather than abilities, and often students who come from low socio-economic statuses have fewer academic experiences than students from high socio-economic statuses.

Lastly, as many traits of scientific creativity can be learned, it is important to consider that female students and students from different cultures may value different creative tasks during child development. A measure could display what looks like a "deficit" if it focused on those aspects of scientific creativity that simply were not yet developed. If model building was not part of a child's play in upbringing, it is likely that the child would struggle with figure analysis as measured by the CoGAT. To complicate matters further, as recess is cut due to more emphasis being placed on language arts (Barth, 2008), there are fewer chances to develop body thinking. Consider the experiences of how one feels on a swing (pendulum) or a Merry-Go-Round (centrifugal force). Play helps to forge the body thinking aspects of science. Another area of education important to scientific creativity is the arts. However, arts programs are usually the first cut due to budget constraints (AJE, 2012). When this happens, students have fewer opportunities to further develop their understandings of careful observation, perspective, and model-building. This therefore may affect the potential identification of children who could be gifted in these areas, but have simply not gained the experience necessary to fully develop these skills. Research has shown that those who were described as creative thinkers as adults often had an experience that led them to practice and value a specific type of creativity (Root-Bernstein & Root-Bernstein, 1999) and thereby developed those skills. Practitioners should consider that all children need to be fostered in creative thought, thus developing the gifts of each child. Often the schools that are affected by these choices are low-income and high minority settings (AJE, 2012). It is very possible that students at these schools are under-identified as gifted in the sciences as they continue to miss out on the experiences that can develop these areas.

We as a nation are missing an opportunity to build up the STEM pipeline when the measures we use to channel gifted students into supportive programs are not specifically designed to foster students in STEM. Few resources are directed toward the development of measures with more specific attention to creativity in science and the aim to identify persons who possess different levels of creative thinking. Further we should be cautious in our interpretation or use of current metrics to identify scientifically creative children, because there is little overlap in the qualities of scientifically creative people and the skills measured on tests used to identify gifted students. Children grow up to become scientists in spite of the educational system, rather than having the educational system identify, nurture, and support scientific creativity. Systematically developing tests of scientific creativity to recognize students and then fostering students' ways of thinking will create an environment that will allow children to fully access their connections to STEM and could increase the numbers of students interested in STEM as a career.

REFERENCES

American Journal of Education Forum. (2012, February 14). *Budget cuts and the future of early education: A conversation with Maryellen Schaub*. Retrieved from http://www.ajeforum.com/archives/30

Albert, R. S. (1983). *Genius eminence*. New York, NY: Pergamon.

Alexander, P. A. (1992). Domain knowledge: Evolving themes and emerging concerns. *Educational Psychology*, *27*(1), 33–51.

Amabile, T. M. (1987). The motivation to be creative. In S. G. Isaken (Ed.), *Frontiers of creativity research: Beyond the basics* (pp. 223–254). Buffalo, NY: Bearly.
American Association for the Advancement of Science. (1989). *Science for all Americans*. New York, NY: Oxford University Press. Retrieved from http://www.project2061.org/publications/sfaa/online
American Association for the Advancement of Science. (1993). *Benchmarks for scientific literacy*. New York, NY: Oxford University Press.
Hill, C., Corbett, C., & St. Rose, A. (2012). *Why so few: Women in science, technology, engineering, and mathematics?* Washington, DC: AAUW.
Barron, F. X. (1963). The needs for order and for disorder as motives in creative activity. In C. W. Taylor & F. X. Barron (Eds.), *Scientific creativity: Its recognition and development* (pp. 153–160). New York, NY: Wiley.
Barron, F. (1988). Putting creativity to work. In R. J. Sternberg (Ed.), *The nature of creativity* (pp. 76–98). Cambridge, England: Cambridge University Press.
Baron-Cohen, S., Wheelwright, S., Stott, C., Bolton, P., & Goodyer, I. (1997). Is there a link between engineering and autism? *Autism, 1*(1), 101–109.
Barth, P. (2008). Time out: Is recess in danger? *Center for public education*. Retrieved from http://www.centerforpubliceducation.org/Main-Menu/Organizing-a-school/Time-out-Is-recess-in-danger
Beghetto, R. A. (2007). Does creativity have a place in classroom discussions? Perspective teachers' response preferences. *Thinking Skills and Creativity, 2*(1), 1–9.
Boden, M. A. (2004). *The creative mind: Myths and mechanisms*. London, England: Routledge.
Bohm, D. (1998). *On creativity*. London, England: Routledge.
Bradshaw, J. L., & Shepherd, D. M. (2000). The neurodevelopmental frontostriatal disorders: Evolutionary adaptiveness and anomalous lateralization. *Brain Language, 73*, 297–320.
Carey, N., & Shavelson, R. (1988). Outcomes, achievement, participation, and attitudes. In R. J. Shavelson, L. M. McDonnell, & J. Oakes (Eds.), *Indicators for monitoring mathematics and science education* (pp. 147–191). Los Angeles, CA: Rand Corporation.
Carson, S. H., Peterson, J. B., & Higgins, D. M. (2003). Decreased latent inhibition is associated with increased creative achievement in high-functioning individuals. *Journal of Personality and Social Psychology, 85*(3), 499–506.
Cattell, R. B. (1963). The personality and motivation of the researcher from measurements of contemporaries and from biography. In C. W. Taylor & F. Barron (Eds.), *Scientific creativity: Its recognition and development* (pp. 119–131). New York, NY: Wiley.
Cattell, R. B. (1971). *Abilities: Their structure, growth and action*. New York, NY: Houghton Mifflin.
Cicirelli, V. G. (1965). Form of the relationship between creativity, IQ, and academic achievement. *Journal of Educational Psychology, 56*(6), 303–308.
Cole, R., & Vermij, G. J. (1998). *Privileged hands: A scientific life*. New York, NY: W. H. Freeman and Company.
Costello, F., & Keane, M. T. (1992). Concept combination: A theoretical review. *Irish Journal of Psychology, 13*, 125–140.
Craft, A. (2003). The limitation to creativity in education: Dilemmas for the educators. *British Journal of Educational Studies, 51*, 113–127.
Csikszentmihalyi, M. (1996). *Creativity: Flow and the psychology of discovery and invention*. New York, NY: HarperCollins.
Davis, G. A. (1975). In frumious pursuit of the creative person. *Journal of Creative Behavior, 9*, 75–87.
Debre, P. (1998). *Louis Pasteur*. Baltimore, MD: Johns Hopkins University Press.
DfES (Department for Education and Skills). (2003). *Excellence and enjoyment*. London, England: DfES.
Eiduson, B. T. (1962). *Scientists: Their psychological world*. New York, NY: Basic Books.
Einstein, A. (1952). Letter to Jacque Hadamard. In B. Ghiselin (Ed.), *The creative process* (pp. 43–44). Berkeley, CA: University of California Press.
Einstein, A., & Infield, L. (1938). *The evolution of physics*. New York, NY: Simon & Schuster.
European University Association. (2007). *Creativity in higher education: Report on the EUA creativity project 2006–2007*. Retrieved from http://tinyurl.com/25qv8mz

Eysenck, H. J. (1995). *Genius: The natural history of creativity*. Cambridge, England: Cambridge University Press.

Feldman, D. H. (1986). *Nature's gambit*. New York, NY: Basic Books.

Finch, A. (2012). *Why Stellar SEO is so important*. Retrieved from http://www.titan-seo.com/NewsArticles/googlesdominance.html

Findlay, C. S., & Lumsden, C. J. (1988). The creative mind: Toward an evolutionary theory of discovery and innovation. *Journal of Social and Biological Structures, 11*, 3–55.

Finke, R. A. (1989). *Principles of mental imagery*. Cambridge, MA: MIT Press

Finke, R. A., Ward, T. B., & Smith, S. M. (1992). *Creative cognition: Theory, research, applications*. Cambridge, MA: MIT Press.

Fleming, M. (2008). *Arts in education and creativity: A review of the literature*. London, England: Creative Partnerships/Arts Council England.

Friedlander, M. (1983). A natural science creativity as a prediction of creative thinking in science. *Creative Child and Adult Quarterly, 8*, 211–215.

Fries, A. (2009). *Daydreams at work: Wake up your creative powers*. Herndon, VA: Capital Books, Inc.

Gardner, H. (1983). *Frames of mind: The theory of multiple intelligence*. New York, NY: Basic Books.

Gerdes, K. E., & Segal, E. A. (2011). The importance of empathy for social work practice: Integrating new science. *Social Work, 56*(2), 141–148.

Givens, P. R. (1962). Identifying and encouraging creative processes. *Journal of Higher Education, 33*(6), 295–301.

Gough, H. G. (1979). A creative personality scale for the adjective check list. *Journal of Personality and Social Psychology, 37*(8), 1398–1405.

Guay, R. (1997). *Purdue spatial visualization test: Rotation*. West Layfeyette, IN: Purdue Research Foundation.

Guilford, J. P. (1967). Some theoretical views of creativity. In H. Helson & W. Beva (Eds.), *Contemporary approaches to psychology* (pp. 419–459). Princeton, NJ: Van Nostrand.

Hayes, J. R. (1989). Cognitive processes in creativity. In J. A. Glover, R. R. Ronning, & C. R. Reynolds (Eds.), *Handbook of creativity* (pp. 135–145). New York, NY: Plenum Press.

Hocevar, D., & Bachelor, P. (1989). A taxonomy and critique of measurements used in the study of creativity. In J. A. Glover, R. R. Ronning, & C. R. Reynolds (Eds.), *Handbook of creativity* (pp. 53–75). New York, NY: Plenum.

Hu, W., & Adey, P. (2002). A scientific creativity test for secondary school students. *International Journal of Science Education, 24*(4), 389–403.

Hudson, L. (1966). *Contrary imaginations*. London, England: Methuen.

Innamorato, G. (1998). Creativity in the development of scientific giftedness: Educational implications. *Roeper Review, 21*(1), 54–59.

Johns Hopkins Medicine. (2008). *This is your brain on jazz: Researchers use MRI to study spontaneity, creativity*. Retrieved from http://www.hopkinsmedicine.org/news/media/releases/this_is_your_brain_on_jazz_researchers_use_mri_to_study_spontaneity_creativity

Johnson-Laird, P. N. (1987). Reasoning, imagining, and creating. *Bulletin of the British Psychological Society, 40*, 121–129.

Kind, P. M., & Kind, V. (2007). Creativity in science education: Perspectives and challenges for developing school science. *Studies in Science Education, 43*, 1–37.

King, L. A., McKee Walker, L., & Broyles, S. J. (1996). Creativity and the five-factor model. *Journal of Research in Personality, 30*, 189–203.

Kitto, J., Lok, D., & Rudowicz, E. (1994). Measuring creative thinking: An activity-based approach. *Creativity Research Journal, 7*, 59–69.

Klahr, D. (2000). *Exploring science: The cognition and development of discovery processes*. Cambridge, MA: MIT Press.

Klahr, D., & Simon, H. A. (1999). Studies of scientific creativity: Complementary approaches and convergent findings. *Psychological Bulletin, 125*, 524–543.

Krebs, H. A., & Shelley, J. H. (1975). *The creative process in science and medicine*. Amsterdam, The Netherlands: Excerpta Medica / American Elsevier.

Lam, S. L., Yeung, A., Lam, P., & McNaught, C. (2010). Creativity and science learning in a science enrichment programme in Hong Kong. *The International Journal of Learning, 17*(2), 429–437.
Lazlo, P. (2004). Science as play. *The American Scientist*. Retrieved from http://www.americanscientist.org/issues/pub/science-as-play
Leys Stepan, N. (1986). Race and gender: The role of analogy in science. *Isis, 77*(2), 261–277.
Limb, C. J. (2010). Your brain on improv. *TED: Ideas worth spreading: Ted Talks*. Retrieved from http://www.ted.com/talks/charles_limb_your_brain_on_improv.html
Limont, W. (2003). Creative imagination in science and science education. In P. Csermely & L. Lederman (Eds.), *Science education: Science and talent recruitment and public understanding* (Vol. 38, pp. 53–59). Amsterdam, The Netherlands: IOS Press.
Lohman, D. F. (2006). Beliefs about differences between ability and accomplishment: From folk theories to cognitive science. *Roeper Review, 29*(1), 32 40.
Lohman, D. F., & Gambrell, J. (2012). Use of nonverbal measures in gifted identification. *Journal of Psychoeducational Assessment, 30*(1), 25–44.
Lohman, D. F., & Hagen, E. P., (2001). *Cognitive abilities test (Form 6): Research handbook*. Itasca, IL: Riverside.
Lubart, T. I. (1994). Creativity. In R. J. Sternberg (Ed.), *Thinking and problem solving* (pp. 289–332). London, England: Academic Press.
Lubart, T. I., & Mouchiroud, C. (2003). Creativity: A source of difficulty in problem solving. In J. E. Davidson & R. J. Sternberg (Eds.), *The psychology of problem solving* (pp. 127–148). Cambridge, England: Cambridge University Press.
McCrae, R. R. (1987). Creativity, divergent thinking, and openness to experience. *Journal of Personality and Social Psychology, 52*, 1258–1265.
Mednick, S. A. (1962). The associative basis of the creative process. *Psychological Review, 69*(3), 220–232.
Metcalfe, R. J. A. (1983). *Assessment of creativity*. Oxford, England: TRC-Rediguides.
Milgram, R. M. (1993). Creative thinking and creative performance in adolescents as predictors of creative attainments in adults: A follow-up study after 18 years. *Roeper Review, 15*(3), 135–139.
Ministry of Education & Human Resources Development of Korea. (2007). *National science curriculum*. Retrieved from http://cutis.moe.go.kr/edu/edu-pro/
Moseley, D., Baumfield, V., Elliott, J., Gregson, M., Higgins, S., Miller, J., ... Gregson, M. (2005). *Frameworks for thinking*. Cambridge, England: Cambridge University Press.
Mumford, M. D., Mobley, M. I., Uhlman, C. E., Reiter-Palmon, R., & Doares, L. M. (1991). Process analytic of creative capacities. *Creativity Research Journal, 4*(2), 91–122.
National Research Council. (1996). *National science education standards*. Washington, DC: National Academy Press.
Newcombe, N. S., & Learmonth, A. E. (2005). The development of spatial competence. In P. Shah & A. Miyake (Eds.), *Handbook of visuospatial thinking* (pp. 213–256). Cambridge, England: Cambridge University Press.
Newell, A., & Simon, H. A. (1972). *Human problem solving*. Englewood Cliffs, NJ: Prentice-Hall.
Newell, A., Shaw, J. C., & Simon, H. A. (1958). Elements of a theory of human problem solving. *Psychological Review, 65*(3), 151–166.
Newton, D. P. (2000). *Teaching for understanding*. London, England: Routledge-Falmer.
Newton, L. D., & Newton, D. P. (2008, July). Conceptions of creativity in elementary school science. *Excellence in education, 2008: Future minds and creativity*. Proceedings of the conference at the International Centre for Innovation in Education, Paris.
Ochse, R. (1990). *Before the gates of excellence: The determinants of creative genius*. Cambridge, England: Cambridge University Press.
Palaniappan, A. K. (2007). Academic achievement of groups formed based on creativity and intelligence. In Taxén Lars (Ed.), *The 13th international conference on thinking*. Norrköping, Sweden. Retrieved from http://www.ep.liu.se/ecp/021/vol1/020/exp2107020.pdf
Penick, J. E. (1996). Creativity and the value of questions in STS. In R. E. Yager (Ed.), *Science/technology/society: As reform in science education* (pp. 84–94). Albany, NY: State University of New York Press.

Pink, D. H. (2005). *A whole new mind.* New York, NY: Penguin.

Planck, M. (1932). *Where is science going?* (J. Murphy, Trans.). New York, NY: Norton & Co.

Poincaré, H. (1921). *The foundations of science: Science and hypothesis, the value of science, and science and method* (G. B. Halstead, Trans.). New York, NY: Science Press.

Popper, K. (1959). *The logic of discovery.* New York, NY: Basic Books.

Proctor, R. A. (1993). Computer simulated associations. *Creativity Research Journal, 6*, 391–400.

Quince, T., Parker, R., Wood, D., & Benson, J. (2011). Stability of empathy among undergraduate medical students: A longitudinal study at one UK medical school. *BMC Medical Education, 11*, 90–102.

Rindermann, H., & Neubauer, A. C. (2004). Processing speed, intelligence, creativity, and school performance: Testing of causal hypothesis using structural equation models. *Intelligence, 32*(6), 573–589.

Rinehart, N. J., Bradshaw, J. L., Moss, S. A., Brereton, A.V., & Tonge, B. J. (2001). A deficit in shifting attention present in high-functioning autism but not Asperger's disorder. *Autism, 5*(1), 67–80.

Roe, A. (1953). *The making of a scientist.* New York, NY: Dodd, Mead.

Root-Bernstein, R. (2012, September 15). *13 thinking tools.* Workshop held at Georgia State University, College of Education, Atlanta, GA.

Root-Bernstein, R., & Root-Bernstein, M. (1999). *Sparks of genius: The 13 thinking tools of the world's most creative people.* New York, NY: Mariner Books.

Rothenberg, A. (1986). Artistic creation as stimulated by superimposed versus combined-composite visual images. *Journal of Personality and Social Psychology, 50*, 370–381.

Rothenberg, A. (1987). Einstein, Bohr, and creative thinking in science. *History of Science, 25*, 147–166.

Runco, M. A. (1991). *Divergent thinking.* Norwood, NJ: Ablex.

Sacks, H. (1992). In G. Jefferson (Ed.), *Lectures on conversation* (Vols. 2) [with an introduction by E. A. Schegloff]. Oxford, England: Basil Blackwell.

Sacks, O. (1992). Tourette's syndrome and creativity. *British Medical Journal, 305*(12), 1515–1516.

Science Daily. (2003, October 1). Biological basis for creativity linked to mental illness. *Science Daily.* Retrieved from http://www.sciencedaily.com/releases/2003/10/031001061055.htm

Serban, S., & Strugariu, L. (2011). Educational software "ChimUniv" for chemical bonds. *Annals of Faculty Engineering Hunedoara – International Journal of Engineering, 9*(3), 485–488.

Simon, H. A. (1973). Does scientific discovery have a logic? *Philosophy of Science, 40*(4), 471–480.

Simonton, D. K. (1997). Creative productivity: A predictive and explanatory model of career trajectories and landmarks. *Psychological Review, 104*, 66–89.

Simonton, D. K. (1999). *Origins of genius: Darwinian perspectives on creativity.* Oxford, England: Oxford University Press.

Simonton, D. K. (2003). Scientific creativity as constrained stochastic behavior: The integration of product, person, and process perspectives. *Psychological Bulletin, 129*, 475–494.

Sinha, A. K., & Singh, C. (1987). Measurement of scientific creativity. *Indian Journal of Psychometry and Education, 18*, 1–13.

Slocumb, P., & Payne, R. (2000). *Removing the mask: Giftedness in poverty.* Highlands, TX: aha! Process, Inc.

Sobel, R. S., & Rothenberg, A. (1980). Artistic creation as stimulated by superimposed versus separated visual images. *Journal of Personality and Social Psychology, 39*, 953–961.

Sorby, S. A. (2009). Educational research in developing 3-D spatial skills for engineering students. *International Journal of Science Education, 31*(3), 459–480.

Sorby, S. A., & Baartmans, B. J. (2000). The development and assessment of a course for enhancing the 3-D spatial visualization skills of first year engineering students. *Journal of Engineering Education, 89*(3), 301–307.

Spearman, C. (1931). *The creative mind.* London, England: Nisbet.

Stephens, K. R., & Karnes, F. A. (2000). State definitions for the gifted and talented revisited. *Exceptional Children, 66*(2), 219–238.

Sternberg, R. (1996). *Successful intelligence: How practical and creative intelligence determine success in life.* New York, NY: Simon & Schuster.

Sternberg, R. J. (2006). Introduction. In J. C. Kaufman & R. J. Sternberg (Eds.), *The international handbook of creativity* (pp. 1–9). Cambridge, England: Cambridge University Press.
Suler, J. R. (1980). Primary process thinking and creativity. *Psychological Bulletin, 88*, 144–165.
Taylor, C. W. (1988). Various approaches to and definitions of creativity. In R. J. Sternberg (Ed.), *The nature of creativity* (pp. 99–121). Cambridge, England: Cambridge University Press.
Taylor, C. W., Smith, W. R., & Gheselin, B. (1975). The creative and other contributions of one sample of research scientists. In C. W. Taylor & F. Barron (Eds.), *Scientific creativity: Its recognition and development* (pp. 53–76). Huntington, NY: R. E. Krieger.
Torrance, E. P. (1990). *Torrance tests of creative thinking*. Beaconville, IL: Scholastic Testing Services.
Torrance, E. P. (1992). A national climate for creativity and invention. *Gifted Child Today, 15*(1), 10–14.
VanTassel-Baska, J., & Stambaugh, T. (Eds.). (2007). *Overlooked gems: A national perspective on low-income promising learners*. Washington, DC: National Association for Gifted Children.
VanTassel-Baska, J., Feng, A. X., Swanson, J. D., Quek, C., & Chandler, K. (2009). Academic and affective profiles of low-income, minority, and twice-exceptional gifted learners: The role of gifted program membership in enhancing self. *Journal of Advanced Academics, 20*(4), 702–739.
Voss, P., Collignon, O., Lassonde, M., & Lepore, F. (2010). Adaptation to sensory loss. *Wiley Interdisciplinary Reviews: Cognitive Science, 1*(3), 308–328.
Walker, C., & Gleaves, A. (2008). An exploration of students' perceptions and understandings of creativity as an assessment criterion in undergraduate-level studies within higher education. *Irish Educational Studies, 27*(1), 41–54.
Watson, J. D. (1968). *The double helix: A personal account of the discovery of the structure of DNA*. New York, NY: Touchstone.
Weisberg, R. W. (1988). Problem solving and creativity. In R. J. Sternberg (Ed.), *The nature of creativity* (pp. 148–176). Cambridge, England: Cambridge University Press.
Wheelwright, S., & Baron-Cohen, S. (2001). The link between autism and skills such as engineering, maths, physics, and computing. *Autism, 5*(2), 223–227.
Welsch, P. K. (1981). The nurturance of creative behavior in educational environments: A comprehensive curriculum approach. *Dissertation Abstracts International, 41*(9), 3870A.

Erin E. Peters-Burton
College of Education and Human Development
George Mason University
USA

Lisa M. Martin-Hansen
College of Education
University of California Long Beach
USA

ANTHONY M. WASHINGTON AND LORI ANDERSEN

5. EFFICACY OF CREATIVE TRAINING FOR GIFTED SCIENCE STUDENTS

FOSTERING STUDENT CREATIVITY IN THE SCIENCE CLASSROOM FOR GIFTED STUDENTS

Gifted students are often creative, however scientific experimentation is not typically viewed as a creative endeavor (Alexakos, 2010; Bell, 2008). The view of science as a linear set of prescribed steps often leads to inaccurate perceptions of the domain and undeveloped creativity skills within the sciences among gifted learners (Adams, 2003; Bell, 2008). When creative skills go undeveloped, gifted students are less likely to innovate within a scientific domain. Students need a science curriculum that uses creativity to develop scientific habits of mind. Development of students' creative ability in science is facilitated through instruction that fosters scientific habits of mind, engagement in solving real-world problems, and higher-level thinking.

In this chapter, methods for developing the creativity of gifted students in the science classroom will be explored. The use of learner-centered pedagogies such as problem-based learning and student-initiated science projects can improve the creative abilities of gifted students within the science classroom. Using complex, real-world problems to challenge gifted students and provide opportunities for divergent thinking scientific domains will increase teachers' ability to maximize their students' potential.

Problem-based activities nurture traits associated with scientific creativity such as risk-taking, persistence, originality, playfulness, and curiosity (Adams, 2003). However, to prepare exceptional students for the challenges of real-world problems and originative inquiry, teachers must scaffold the development of creative thinking skills by using classroom practices that support creative thinking and expression (Beghetto, 2006). Activities that allow students to build the types of creative abilities that are needed for creative production in sciences, such as the ability to create explanatory analogies, should be employed on a regular basis. This chapter will highlight effective creativity training methods that can be employed to develop these skill sets within the gifted population.

Inquiry-Based Science Teaching as Creativity Training

Creativity training focuses on specific cognitive processes or techniques that are used in creative production (Scott, Leritz, & Mumford, 2004). A meta analysis of 71

M. K. Demetrikopoulos & J. L. Pecore (Eds.), Interplay of Creativity and Giftedness in Science, 71–85.

creativity training programs led Scott et al. to categorize creativity into eight core processes: (1) problem finding, (2) information gathering, (3) concept selection, (4) conceptual combination, (5) idea generation, (6) idea evaluation, (7) implementation planning, and (8) solution monitoring (2004). Each of these eight core processes that comprise creativity are also an integral part of doing scientific investigations. Scientists must employ creative skills to find problems, bring different concepts together, generate ideas as to how to investigate the question, evaluate ideas by comparing them to extant data, plan experiments, predict potential outcomes, and monitor solutions to ensure consistency with other research data and theories. These processes illustrate several similarities between general creativity and scientific creativity; however, the scientific creative process does have some unique elements.

Mansfield and Busse (1981) modeled the creative process in science as five processes: (1) selection of the problem, (2) extended effort to solve the problem, (3) setting constraints, (4) changing constraints, and (5) verification and elaboration. These processes were identified through qualitative analyses of accounts of the work of four of the most creative scientists in the history of the field. This model of the creative process added techniques that are domain specific to scientific creativity, such as persistent effort and the setting and changing of constraints to the list of processes associated with creativity in general. Science curricula that are designed to strategically develop these process skills are actually creativity training programs that are domain specific to science.

Building Skills for Scientific Creativity

Gifted student's science experiences from early childhood to adolescence form the first stages of a talent development trajectory that lead to scientific, creative productivity in adulthood (Subtonik, Olszewski-Kubilius, & Worrell, 2011). Early science experiences must inspire student interest and build content knowledge while scaffolding the development of autonomous learning skills. Science process skills such as finding questions or problems, forming hypotheses, designing experiments, and performing procedures should also be included in these experiences (Adams, 2003).

Appropriate science curriculum for gifted students should include advanced content that deepens in complexity while facilitating increased amounts of independent investigation. For example, in the primary grades students should be given opportunities to make significant decisions in regard to pre-determined science projects, as these students may not yet be ready to design their own investigations from beginning to end. In Project Clarion, a series of inquiry-based science units were designed for gifted students to help these students gain confidence in their science process skills while transitioning toward self-directed learning and increasing creativity (Bland, Coxon, Chandler, & VanTassel-Baska, 2010; Kim, Bland, & Chandler, 2009). Students who participated in Project Clarion for three years showed significant increases in critical thinking compared to those students in traditional science curricula (Kim, VanTassel-Baska, Braken, Feng, Stambaugh,

& Bland, 2011). In later grades, structure should be lessened and autonomy expanded in increasingly open-ended, inquiry-based curriculum models such as problem-based learning, project-based science, and individual projects created for fairs or competitions. Teaching the creative methods of science requires teaching students how scientists are creative (Starko, 2005).

Students must be taught early that science is a creative endeavor, not a set of linear prescribed steps as it is presented in many science classrooms (Adams, 2003). According to Brandwein (1995), originative inquiry in science rarely follows the scientific method that is taught regularly in traditional science classes, and scientist's methods are as varied as the scientists themselves. Thus, the goal of science curricula should be to lead students to creative productivity in science, or said in another way, the ability to conduct independent and originative investigations. This ability will only develop when students are given opportunities to learn and practice these skills over time.

SCIENTIFIC CREATIVITY TRAINING

Creativity in Science Careers

The goal of public schooling is to prepare students for the workforce that will exist when they graduate from our schools. In the book *Creating Innovators: The Making of Young People Who Will Change the World*, Wagner (2012) discusses the importance of creativity as described by business leaders and product managers of some of the most innovative companies in the world. Ranging from multi-billion dollar conglomerate to Internet start-up gone viral, the CEO's and executives interviewed agreed that creativity and the ability to apply knowledge were more important than technical knowledge alone (Wagner, 2012). Wagner emphasizes that the critical qualities needed in successful innovators are skills and habits of mind that can be taught and developed, but that this rarely happens in traditional education programs (2012).

Unfortunately, in the U.S., more innovators have been produced accidentally than intentionally. A startling portrait in this text was the description of how some of the most innovative individuals of our lifetime (e.g. Bill Gates, Steve Jobs, and Mark Zuckerberg) chose to drop out of college to pursue their creative endeavors. This paints even our most distinguished educational institutions as woefully inept at nurturing innovative ideas and developing the talents of those students that create them. Institutions that seek to nurture creativity have adopted a teaching and learning style that mirrors the real-world problem-solving scenario that scientists encounter on a daily basis (Wagner, 2012).

Valuing Creativity

In order to embrace the creative nature of the 21st Century workforce, educators must change the way they approach delivering the content in core areas. If we are to prepare gifted students for the careers that will exist once they leave the ranks of education and enter the workforce, we must begin to develop their creative abilities in addition to their knowledge of the content. Scientific work has transitioned from working in an isolated laboratory setting to a team approach to problem solving that requires the ability to work in groups, problem solve, and define problems that need to be solved.

The next section of this chapter provides background on various approaches to teaching science that allow gifted students to develop problem defining and solving skills in addition to practice working as a part of a team toward a common goal.

INSTRUCTIONAL STRATEGIES

Problem-Based Learning

Problem-based learning (PBL) is an instructional method that grew out of the medical profession where it was shown to have positive effects on student engagement and learning (Allen, Donham, & Bernhardt, 2011; Larmer & Mergendoller, 2012). PBL was created to align the teaching and learning process in medical schools with the actual practices of the medical profession. The underlying belief in the creation of PBL was that students' would have a deeper learning experience when dealing with complex multifaceted problems based in a real-world setting (Allen et al., 2011).

As PBL was modified to K-12 educational settings it retained much of the same structure that was used in the medical schools. Students are organized into small groups and presented with an ill-defined problem related to the content area being studied. Students will define the problem and discuss what prior knowledge they have regarding the topic that has been covered by the teacher as well as what they bring from their personal experiences (Allen et al., 2011). Once they understand the problem, students will work in small groups to identify questions that must be answered to solve the problem. These questions will be the driving force behind the research they will conduct throughout the project.

Each student in the group will be accountable for specific tasks. These groups will be collectively held accountable for both their individual performance and the performance of the entire group. The results from these research projects are presented in written or oral form to demonstrate their learning and the results of their research. Assessments within the PBL structure are open-ended and performance based to allow room for growth across multiple skill levels (Bland et al., 2010). Open-ended questions can be structured to avoid a ceiling effect for high achieving students while them to use creative skills to investigate science related topics (Bland et al., 2010).

PBL allows gifted students to determine the level of complexity in which they engage with the content through intentionally differentiated activities. This allows for individual differentiation within the content, process, and product of the PBL lesson that can be utilized to meet the needs of gifted learners. As gifted students are able to steer themselves toward challenging material, their engagement will likely increase resulting in a more meaningful learning experience. PBL also allow teachers to create a structure that provides students the freedom to function at levels specific to their strengths and interests (Bland et al., 2010).

There are some cautions to be considered when implementing a project-based learning assignment. In PBL classrooms, the teacher must be willing to move from the role of delivering content to facilitating the problem solving process (Allen et al., 2011; King, 1993; Pecore, 2013). As a facilitator, the teacher is charged with ensuring that the students' exploration has relevance to the content, is reaching appropriate levels of depth of learning and analysis, and providing relevant information when necessary. This will require the teacher to be well prepared, although flexible, to take full advantage of the learning experiences that arise from the investigation (Pecore et al., 2012). Instructors also must exert considerable effort to ensure the problems presented to students are based on clear learning goals to ensure students gain mastery of key concepts.

In a PBL biology course, Palaez (2002) found that students had better acquisition of conceptual understandings than those who received lecture-based instruction. There have also been reports of increases in critical thinking (Tiwari, Lai, So, & Yurn, 2006) and greater gains in metacognitive skills (Downing et al., 2009) in classes using PBL. In addition, students' research skills, and collaborative effort have been shown to improve as a result of using PBL (Allen et al., 2011).

Science-Technology-Society Problems

Meyer (2012) provides suggestions as to how existing lessons and experiments can be modified to encourage creative production in the science classroom. First and foremost, the problem to be solved must have more than one solution. Merely providing a laboratory experiment in which students devise their own procedure does not qualify as teaching for creativity with gifted students. The activity must allow for individual and unique approaches to scientific experimentation that will allow gifted children to be creative within science related content in ways that can impact the outcome of the experiment.

Science-technology-society (STS), defined as the learning of science within the context of human experience (Lee & Erdogan, 2007), is a method that provides problems for students to solve that have a multitude of solutions. For example, students could investigate the potential consequences of the use of a new technology to decide if the potential benefits outweigh these consequences. The human context for the proposed use of a technology creates myriad complexities and different points of view for students to consider as they propose alternatives or justify implementation.

Lee and Erdogan (2007) found significantly larger gains for creativity scores in science students taught with an STS approach compared to a traditional stand and deliver instructional approach. In a summer enrichment program for gifted low-income students, a nanotechnology course taught using an STS approach led to large and significant gains in creativity scores (Andersen, Schmidt, & Tieso, 2012; Cross, 2014). Thus, the use of a STS approach in science teaching helps students build creativity skills and yields greater gains in creativity than traditional science teaching methods for gifted students regardless of SES.

Project-Based Science

Project-Based Science (PBS) is a learner centered instructional strategy that uses driving questions to guide students toward the creation of a tangible artifact (Colley, 2008). In PBS, students create and answer science content-based questions that are related to their own lives as well as their surrounding communities (Colley, 2008). Students are provided with several technological resources and guided through the learning process by a teacher who provides support and ensures the learning of content within the project.

PBS is based on the work of Piaget and Vygotsky that seeks to have student-centered learning that is hands-on and applicable to the student's home life (Colley, 2008). These scholars believed that knowledge should be constructed through practice and reflection while the student is responsible for their own learning (Colley, 2008). This responsibility can be used to provide differentiated activities for gifted students by encouraging them to follow their interests and pursue topics that require deeper knowledge of the content to solve the problems presented.

When preparing for a PBS lesson, teachers are encouraged to conduct an orientation session that introduces students to the process (Colley, 2008). In this orientation lesson, students should be informed about the expectations of the project, the importance of collaboration, and how they will be assessed (Colley, 2008). There is likely to be some resistance from the students as they change from receiver of content to an active participant in their learning. Some preemptive discussion of how the PBS process works and examples of completed projects increases the chances of successful implementation of the first PBS lesson (Colley, 2008). The orientation lesson is also an opportunity for the teacher to gain understanding about the students' prior knowledge and experience with the topic so that he or she is better able to ensure the learning activities are challenging for all students.

The projects of PBS are designed to teach science concepts and fundamental knowledge to students in a manner that is relevant to their lives. In the planning process, the student groups should spend considerable time thinking and discussing how the question is going to be investigated to ensure that they have the resources available to conduct the investigation. The students will also be responsible for creating a project plan that includes: the driving questions, purpose, procedures, tools, technology, time required for completion, and individual duties of each student within the group (Colley, 2008). The teachers assess these project plans for feasibility before the students begin to implement them. The students are also required to determine in advance what artifacts will be produced to demonstrate their learning.

The difficulty of a PBS lesson should be tailored to the ability level of the students. This allows a teacher to differentiate for the gifted learner in a manner similar to that of PBL in that advanced students have the freedom and flexibility to choose increasingly difficult tasks based on their interest and ability. There is also an opportunity for the teacher to plan activities of varying depth and breadth so that the teacher may scaffold tasks to suit the rigor of the creative assignment appropriately for the learner. While some students will be able to construct creative scientific explorations immediately, the teacher should plan some prompts to stimulate students' thinking as well as increase the rigor of the lesson.

The students are encouraged to work collaboratively throughout the project. Once they have collected and recorded data in the manner prescribed in the approved project plan, students will prepare reports that may undergo several revisions before the teacher accepts it. This repetition and revision of writing samples provides reinforcement of the students written communication skills and encourages students to reach for higher levels of achievement. This persuasive communication piece could be used as a point of collaboration between a science and an English class.

Once projects are completed, they are submitted to other groups in the class for peer review and presentation and the teacher will ask probing questions to ensure that the students understand the content. Students will be guided to reflect on the process of conducting the project as well as their learning during the process (Colley, 2008; Rivet & Krajcik, 2004).

Research on PBS has demonstrated improvements in students understanding of science content, inquiry skills, and their ability to draw relationships among science concepts (Rivet & Krajcik, 2004). This study also provided evidence on increased student engagement and meaningful investigations as posited in the description of PBS. In a comparison of two fourth grade classrooms, one using PBS and the other using a traditional lecture style delivery, Zumbach, Kumpf, and Koch (2004) found that students in the PBS class were more motivated and dedicated more of their time out of class pursuing solutions. In addition, Lee et al. (2005) found that third and fourth grade students showed considerable improvements on content knowledge and inquiry skills.

Enabling Standards

PBL and PBS instruction requires enabling standards, which involves extra time for teachers to prepare assignments and determine some examples of questions whose investigation will incorporate the content to be covered within the standard. These projects require a significant amount of detail in preparation for their successful execution and teachers must make ample plans for the variety of directions that student generated questions may lead (Colley, 2008; Rivet & Krajcik, 2004). In providing the time needed to prepare the lesson, the instructional leader may not need to increase the amount of time the teachers receive to plan, but ensure that the planning time they have throughout the day is uninterrupted.

Teachers will also need ample professional development to ensure they have an in-depth understanding of what it means to facilitate as opposed to instruct a lesson as well as ensuring the content is covered throughout the activities (Allen et al., 2011). For example, if the teacher is not aware of when to provide content-based instruction throughout the process, students may complete the research or project without gaining the scientific concepts the standard seeks to address. The instructional leader should make sure teachers have appropriate support in implementation, especially within the initial project (Allen et al., 2011; King, 1993; Larmer & Mergendoller, 2012).

Students may not be able to come up with sufficient questions on their own and it is the teachers' responsibility to ensure that the project will focus on the current science standards and prevent projects from drifting away from the instructional focus. The instructional leader should ensure that professional development is geared toward difficulties that teachers are experiencing implementing the PBL or PBS in their classroom (Allen et al., 2011; Colley, 2008).

The teachers will also need to be prepared for the confusion and frustration that will sometimes arise during a new style of learning experience. When undertaking a creative assignment, the students may become frustrated if they have a setback or determine that the path they would like to pursue is not going to provide the solutions they predicted. When there is a change in instructional practice, there should be some dedication to problem solving and reflection on the part of the instructor to ensure best fit of the learning strategy to the students. It would be the instructional leader's job to ensure that adequate support is in place for all parties involved so that success is attainable (Allen et al., 2011; Colley, 2008; King, 1993.

The classroom should be structured in a manner that allows for creative exploration in groups as well as individual settings. Problem-based learning and PBS lessons call for collaborative group production as well as individual accountability (Colley, 2008). In a classroom conducting a PBL or PBS lesson, there should be several organized stations that lend themselves toward the production of these tasks in small group settings. Once students have had opportunities for structured inquiry-based science experiences and have experienced success, the nature of science activities should shift toward increasing autonomy and more self-directed learning.

Independent Investigations

The true indicator of creative production in science is an originative research project. Independent science projects provide students with opportunities to find a problem of interest and focus on that one problem over an extended period of time. This type of activity is similar to the actual work of scientists, and the willingness to complete such a project is indicative of a student's ability to become a scientist (Berger, 1994). In his case studies, Wagner (2012) found that hands-on projects where students must solve a real-world problem and demonstrate mastery are essential elements of the education of future innovators. Science projects require students to move through the five parts of the scientific creative process described by Mansfield and Busse (1981). It is important to note that these skills can be developed with practice. Students who have the potential to be creatively productive in science may not be the students who earn the highest scores on intelligence or creativity tests (Brandwein, 1995; Mansfield & Busse, 1981; Neu, Baum, & Cooper, 2004). Teachers must look for students who display the predisposing factors for scientific creativity (Brandwein, 1995). Gifted students should engage in independent science projects throughout their schooling, however, students who come from disadvantaged backgrounds may require additional scaffolding and resources as they move toward autonomous learning. Students should be encouraged to ask questions and to discuss how the answers to such questions could be found out through inquiry or experimentation. Science students should learn about the famous experiments of those who have made creative contributions to science in ways that help these students to learn how to think like scientists. Examples of such science training programs can be found in schools such as the Bronx High School of Science that has produced greater numbers of winners of the Intel Science Talent Search than any other U.S. high school (Berger, 1994; Brandwein, 1995; Kopelman, Galasso, & Schmuckler, 1988). The Intel Science Talent Search was designed to recognize students who were creatively gifted in science and mathematics.

The program at Bronx Science has the main objective of developing the science talents of students to the point where they can conduct originative inquiry. A three-year sequence of coursework has been developed for this purpose. In the ninth grade, students choose biology, chemistry, or physics. The Socratic method is emphasized, in which students recognize problems and offer hypotheses. Many open-ended laboratory experiments are performed. The students practice identifying problems, offering hypotheses, planning experiments, analyzing data, testing hypotheses, and making conclusions. Honors science students read research articles and students are trained in how to read a body of literature and formulate a research question. Some students are selected for a biology research class in their junior year based on the display of characteristics associated with being creatively gifted in science. Those characteristics include: motivation, ability to work independently, curiosity, questioning, likes to solve problems, ability to see different approaches, and readily able to induce, deduce and make connections between related ideas (Kopelman et al., 1988). Teachers and/or mentors guide these students in their independent research projects. The success of this program in developing scientific creativity is evidenced by the number of students who complete independent science investigations and the number who have been recognized by the Intel Science Talent Search (Berger, 1994; Kopelman et al., 1988).

Training the Scientific Imagination

Scientific thinking goes beyond inductive and deductive reasoning. Many scientific discoveries have been related to the abilities of scientists to create analogies or use their imaginations. Holton (1995) describes how scientific creators used the visual and the metaphoric imagination. The visual imagination is important because much of the content that is described by modern science is invisible to the human eye. The progression of knowledge about the structure of the atom occurred through a series of different visualizations that were actually visual analogies. From Thompson's plum pudding model, to Rutherford's large positive nucleus, to Bohr's model that envisioned the atom as a sort of mini solar system, scientists used their visual imaginations to explain what they could not see by comparing it to what is already known. As additional experimental evidence accrued, scientists changed their models to account for the new evidence.

For example, the progression from the plum pudding model to the nuclear model occurred because Rutherford could not explain why the alpha particles in his gold foil experiment were being deflected at large angles. His explanation was the presence of a positively charged nucleus. What we now know about the atom is a result of scientists' interpretations of indirect evidence as a way to "see" subatomic events. Another example of the visual imagination is Einstein's work on the theory of special relativity that was largely derived from thought experiments that he conducted in his visual imagination (Einstein, 1922).

Scientists also use metaphoric imagination in the development of theories and models, although the use of such techniques may not be apparent in scientific publications. For example, Thomas Young made his claim that light was a wave based on an analogy he made between an interference pattern of light colors on a thin glass plate and the pattern of the frequencies of sounds made by organ pipes (Mathewson, 2005). This notion was contrary to the accepted scientific knowledge of his time, which was that light was composed of particles. Through analogy, Young saw that even though the two different phenomenon-sound and light-were different kinds of waves, that all waves have similar interference patterns. In his mind, the pattern of behavior that light exhibited was wave-like; therefore light must be a wave. Young made this intellectual leap based on his ability to imagine similarity between distal phenomena. Now, this analogy did not justify the publication of this scientific idea in a journal, but it did help the scientist to direct his further investigations. Young later supported his theory with evidence from his famous double slit experiment that beautifully showed that light exhibited wave properties (Young, 1802). Thus, creating metaphors and analogies in science are a somewhat risky activity that occurs in the early stages of the scientific imagination. These students may not document these activities in the final report of an experiment, but they are likely to be fundamental in generating the idea that led to the experiment. Modern scientists also use visual imagination to create new knowledge. Contemporary, mixed-method studies of scientists engaged in data analysis and hypothesis testing revealed extensive use of visualization and imagery abilities to construct mental models and make complex comparisons of these internal models to external models (Trickett & Trafton, 2007; Trafton, Trickett, & Mintz, 2005). Unfortunately, much of classroom instruction neglects the importance of visualization and creativity in gifted and science education (Adams, 2003; Andersen, 2014) even though many scientists solve problems through the creation of novel visualizations or mental model manipulation.

How can we help students to develop these skills? A scientist creates analogies by mapping the target, a concept, or problem that he or she is trying to solve (Dunbar & Fugelsang, 2005). A source is some other piece of knowledge that the scientist uses to explain the behavior of the target by mapping the features of the target onto those of the source. In this mapping process, new features of the target can be discovered or new concepts may be invented. Research on analogy has shown that training is required to develop the skills needed to create effective scientific analogies that focus on the underlying relations among the features of the source and the target instead of the sharing of superficial features (Dunbar & Fugelsang, 2005). Clement (1989) distinguishes *decorative* analogies from *scientific* analogies by their usefulness in making predictions and potential explanatory power. Both the structural similarities and the differences between the source and the target are important to the creation of a useful analogy.

The process of analogy making involves a number of steps that can be taught using *synectics* (Gordon, 1961). In synectics exercises, students learn to play with analogies and build skills in creating metaphoric analogies. These analogy-creating skills are then used in problem solving. Synectics provides a methodology to invent new perspectives and fresh ways to solve problems (Joyce, Weil, & Calhoun, 2009). This strategy leverages the similarities between domains, such as the arts and sciences, and demonstrates that creativity can be a conscious process. The objective of synectics is to use metaphor and analogy deliberately while harnessing the power of irrationality for constructive purposes. The strategy within synectics called "making the strange familiar" (Joyce et al., 2009, p. 233) teaches the process of mapping a target onto a source. To use this strategy, the teacher provides the students with information about a new topic. The teacher then provides a source for the students. Making a list of characteristics of the target and the corresponding or parallel properties of the source identifies the connections between the source and the target. After the list is completed, the students write a paragraph that describes the analogical connections they have made and to point out where the analogy fits and where it does not fit. The purpose of this is to provide a model for the students to create their own analogy.

The students are asked to create their own analogy for the target using a source of their own choosing. Divergent thinking is used to consider many possible sources and then convergent thinking skills are used to evaluate the chosen source. Students must carefully consider the properties of the target and the source to ensure that the source is a good fit for the target. Once a suitable source is determined, the students write about the connections they have created. The use of this teaching strategy helps students to connect two ideas and identify the connections between the ideas (Joyce et al., 2009). This mirrors the processes that scientists use when they employ visual or metaphoric imagination. However, care must be taken to show students how to create analogies that use the underlying relationship among features and not just the common superficial features. Science teachers should provide opportunities for students to practice creating their own analogies as well as provide examples of how analogies have been used in the development of scientific knowledge throughout history. In this way, the importance of creativity in science can be emphasized.

Creativity and Content Standards

The increase of accountability for content knowledge via standardized tests continues to be a driving force in the way that instruction is being delivered. Many teachers use the testing culture of public schools to justify their lack of attention to creative learning strategies. Teachers often believe that they do not have time to incorporate creative learning strategies even though these teaching styles can be used to deliver content in a way that connects with the students' lives outside the classroom. This will not require a complete renovation of teaching practices as much as a reconceptualization of the approach to delivering content.

The instructional strategies presented in this chapter are not only extensions of learning for gifted children; teachers can integrate them at every level of the teaching and learning process to engage gifted children on a regular basis in the classroom. Gifted creative students are more likely to retain content and be able to apply it in novel situations when content is learned in a manner that builds on and connects with their existing knowledge (King, 1993). This type of learning will require teachers to move from the position of content deliverer to the learning facilitator as described by King (1993) as moving from the "sage on the stage to guide on the side" (p. 1).

Ensuring Content Acquisition

Using these strategies alone will not automatically ensure that gifted students will learn the content; however, using these strategies will increase the engagement of gifted science students and can develop the creative skills of gifted science students. Teachers must take care to ensure that they create a clear connection to the science content within the lesson. When properly implemented, problem and inquiry based learning improves the meaningful understanding of science by students by increasing the real-world relevance of content (Kanter, 2009).

One method teachers can employ to ensure content is covered is carefully designing problem statements or driving questions that require the acquisition of content in order to answer them. An example provided by Kanter (2009) for a biology lesson would be to pose the question, "What will it take to redesign our school lunch choices to meet our bodies' needs? (p. 530)." Answering this question will require students to determine how to measure the energy their body receives from food, the amount of energy used in their daily activities, determine how much their current lunch provides them, and create a lunch that meets their bodies' nutritional needs (Kanter, 2009). By focusing the lesson in a real-world problem setting, students are able to explore a topic that directly relates to their lives while learning scientific content within a context that lends itself toward deeper learning.

Successful strategies for insuring curriculum coverage with the goal of content acquisition during a student centered creative learning assignment include ongoing lessons designed to guide students' exploration through the content (Kanter, 2009). Techniques like unpacking the task, highlighting incongruities, and try to apply, are activities that ensure students' do not drift into a lane leading to the creation of a product that circumvents the content in the process of completion. Those who design these lessons for students using the techniques espoused in this chapter should take significant care in ensuring the inclusion of the content within the projects and activities. The enhancement of creative skills in the sciences is needed to prepare students for the careers that will await them once they graduate. Educators must be diligent about incorporating connections to real-world situations if we are to properly educate our students for career success and lifelong learning.

REFERENCES

Adams, C. M. (2003). Nurturing talent development in science. In P. Olszewski-Kubilius, L. Limburg-Weber, & S. Pfeiffer (Eds.), *Early gifts: Recognizing and nurturing children's talents* (pp. 19–38). Waco, TX: Prufrock Press.

Alexakos, K. (2010). Teaching the practice of science, unteaching the "scientific method." *Science Scope, 33*(9), 74–80.

Allen, D. E., Donham, R. S., & Bernhardt, S. A. (2011). Problem-based learning. *New Directions for Teaching and Learning, 2011*(128), 21–29. doi:10.1002/tl.465

Andersen, L. (2014). Visual-spatial ability: Important to STEM, ignored in gifted education. *Roeper Review, 36*, 114–121. doi:10.1080/02783193.2014.884198

Andersen, L., & Schmidt, A. M. (2012, November). *Nanoscience! A problem-based learning approach.* Paper to be presented at the 59th Annual Convention of the National Association of Gifted Children, Denver, CO.

Bell, R. L. (2008). *Teaching the nature of science through process skills: Activities for grades 3–8.* Boston, MA: Pearson.

Beghetto, R. A. (2006). Creative self-efficacy: Correlates in middle and secondary students. *Creativity Research Journal, 18*, 447–457. doi: 10.1207/s15326934crj1804_4

Berger, J. (1994). *The young scientists: America's future and the winning of the Westinghouse.* Reading, MA: Addison-Wesley.

Bland, L. C., Coxon, S., Chandler, K., & VanTassel-Baska, J. (2010). Science in the city: Meeting the needs of urban gifted students with project Clarion. *Gifted Child Today, 33*(4), 48–57.

Brandwein, P. F. (1995). *Science talent in the young expressed within ecologies of achievement* (pp. 1–184). Storrs, CT: National Research on the Gifted and Talented, University of Connecticut.

Clement, J. (1989). Learning via model construction and criticism: Protocol evidence on sources of creativity in science. In J. Glover, R. Ronning, & C. Reynolds (Eds.), *Handbook of creativity: Assessment, theory and research* (pp. 341–381). New York, NY: Plenum.

Colley, K. (2008). Project-based science instruction: A primer—An introduction and learning cycle for implementing project-based science. *Science Teacher, 75*(8), 23–28.

Cross, J. R. (2014). *Summer program evaluation* (Unpublished report). Williamsburg, VA: Center for Gifted Education, College of William & Mary.

Csikszentmihalyi, M. (1996). *Creativity: Flow and the psychology of discovery and invention.* New York, NY: Harper Collins.
Dunbar, K., & Fugelsang, J. (2005). Scientific thinking and reasoning. In K. J. Holyoak & R. G. Morrison (Eds.), *The Cambridge handbook of thinking and reasoning* (pp. 705–726). Cambridge, NY: Cambridge University Press.
Einstein, A. (1922). How I created the theory of relativity (Y. A. Ono, Trans., 1982). *Physics Today, 35*(8), 45–47. doi: 10.1063.1.2915203
Gordon, W. J. J. (1961). *Synectics.* New York, NY: Harper & Row.
Holton, G. (1995). *Einstein, history, and other passions.* Woodbury, NY: AIP Press.
Joyce, B., Weil, M., & Calhoun, E. (2009). *Models of teaching* (9th ed.). Boston, MA: Pearson.
Kanter, D. E. (2009). Doing the project and learning the content: Designing project-based science curricula for meaningful understanding. *Science Education, 94*(3), 525–551. doi: 10.1002/sce.20381
Kim, K. H., VanTassel-Baska, J., Bracken, B. A., Feng, A., Stambaugh, T., & Bland, L. (2011). Project Clarion: Three years of science instruction in title I schools among K-Third grade students. *Research in Science Education, 42*(5), 813–829. doi:10.1007/s11165-011-9218-5
Kim, M., Bland, L. C., & Chandler, K. (2009). Reinventing the wheel. *Science and Children, 47*(3), 40–43.
King, A. (1993). From sage on the stage to guide on the side. *College Teaching, 41*(1), 30–35. doi: 10.1080/87567555
Kopelman, M., Galasso, V. G., & Schmuckler, M. (1988). A program for stimulating creativity in a citywide high school: The bronx high school of Science. In P. F. Brandwein & A. H. Passow (Eds.), *Gifted young in science: Potential through performance* (pp. 209–216). Washington, DC: National Science Teachers Association.
Larmer, J., & Mergendoller, J. R. (2010). 7 Essentials for project-based learning. *Educational Leadership, 68*(1), 34–37.
Lee, M., & Erdogan, I. (2007). The effect of science–technology–society teaching on students' attitudes toward science and certain aspects of creativity. *International Journal of Science Education, 29*(11), 1315–1327. doi:10.1080/09500690600972974
Mathewson, J. (2005). The visual core of science: Definition and applications to education. *International Journal of Science Education, 27,* 529–548. doi:10.1080/09500690500060417
Mansfield, R. S., & Busse, T. V. (1981). *The psychology of creativity and discovery: Scientists and their work.* Chicago, IL: Nelson-Hall.
Meador, K. (2003). Thinking creatively about science: Suggestions for primary teachers. *Gifted Child Today, 26,* 26–29.
Meyer, A. A. (2012). Teaching for creativity. *The Science Teacher, 79*(5), 54–56.
Neu, T. W., Baum, S. M., & Cooper, C. R. (2004). Talent development in science: A unique tale of one student's journey. *The Journal of Secondary Gifted Education, XVI*(1), 30–36.
Pecore, J. L. (2013). Beyond beliefs: Teachers adapting problem-based learning to preexisting systems of practice. *Interdisciplinary Journal of Problem-based Learning, 7*(2), 1–27.
Pecore, J. L., & Bohan, C. H. (2012). Problem-based learning: Teachers who flourish and flounder. *Curriculum and Teaching Dialogue, 14*(1), 123–126.
Rivet, A. E., & Krajcik, J. S. (2004). Achieving standards in urban systematic reform: An example of a sixth grade project-based science curriculum. *Journal of Research in Science Teaching, 41*(7), 669–692.
Scott, G., Leritz, L. E., & Mumford, M. D. (2004). Types of creativity training: Approaches and their effectiveness. *Journal of Creative Behavior, 38*(3), 149–179.
Starko, A. J. (2005). *Creativity in the classroom.* Mahwah, NJ: Lawrence Erlbaum.
Subotnik, R. F., Olszewski-Kubilius, P., & Worrell, F. C. (2011). Rethinking giftedness and gifted education: A proposed direction forward based on psychological science. *Psychological Science in the Public Interest, 12*(1), 3–54.
Trafton, J. G., Trickett, S. B., & Mintz, F. E. (2005). Connecting internal and external representations: Spatial transformations of scientific visualizations. *Foundations of Science, 10,* 89–106. doi:10.1007/s10699-005-3007-4

Trickett, S. B., & Trafton, J. G. (2007). What if…: The use of conceptual simulations in scientific reasoning. *Cognitive Science*, *31*(5), 843–875. doi:10.1080/03640210701530771

VanTassel-Baska, J. (1998). Planning science programs for high-ability learners. *ERIC EC Digest*, November.

VanTassel-Baska, J., & Brown, E. F. (2007). Toward best practice: An analysis of the efficacy of curriculum models in gifted education. *Gifted Child Quarterly*, *51*(4), 342–358. doi:10.1177/0016986207306323

VanTassel-Baska, J., Bass, G., Ries, R., Poland, D., & Avery, L. D. (1998). A national study of science curriculum effectiveness with high ability students. *Gifted Child Quarterly*, *42*(4), 200–211.

Wagner, T. (2012). *Creating innovators: The making of young people who will change the world.* New York, NY: Scribner.

Young, T. (1802). The Bakerian lecture: On the theory of light and colours. *Philosophical Transactions of the Royal Society of London*, *92*. doi:10.1098/rstl.1802.0004

Zumbach, J., Kumpf, D., & Koch, S. C. (2004). Using multimedia to enhance problem-based learning in elementary school. *Information Technology in Childhood Education Annual*, 25–37.

Anthony M. Washington
Richmond Public Schools
USA
and
William and Mary
USA

Lori Andersen
Achievement and Assessment Institute
University of Kansas
USA

ANGELA E. STOTT AND PAUL A. HOBDEN

6. A BELIEF SYSTEM AT THE CORE OF LEARNING SCIENCE

A Case Study of a Critical and Creative Gifted Learner

INTRODUCTION

This chapter describes the belief system which has been interpreted by the authors to be at the core of the self-regulated learning of science undergone by an individual high achieving, gifted learner. The case reported is of a learner referred to as André. He was observed to apply a number of cognitive and metacognitive strategies in his learning of science motivated by his beliefs of what the nature of learnt meaningful knowledge should be that is precise, elegant, and transferable. Most of the cognitive strategies André was observed to use had at their core critical and creative thinking. In this chapter we describe this belief system central to André's self-regulated learning of science and describe what he understood by knowledge which is precise, elegant and transferable and illustrate how these beliefs motivated André's critical and creative thinking.

This research proposes a set of beliefs which may be beneficial for science teachers of gifted learners to nurture in their efforts to enhance self-regulated learning. Self-regulation in learning is known to enhance achievement (Al-Khatib, 2010; Zimmerman, 1990), a goal highly valued in our school systems. Understanding self-regulation in the context of gifted learning may be a key to unlocking the potential of underachieving gifted and other learners. This research also supports the view that critical and creative thinking are closely related, and as evidenced in this study, are vital components of meaningful learning.

The claims made are based on observations over a period of three years in which the first author taught André on a one-on-one basis when he was aged 13–15. During this time written work was collected and interviews and lessons recorded. Data analysis was done in a grounded fashion through iterative inductive engagement with the data. The analysis and findings reported represented here are in response to the research question "What belief system drives this high achiever's effective learning of physical science?" Validity was ensured through long term engagement and use of member checks. Ethical considerations included obtaining consent from André and his parents, and the use of a pseudonym to protect his identity.

M. K. Demetrikopoulos & J. L. Pecore (Eds.), Interplay of Creativity and Giftedness in Science, 87–103.

CONCEPTUAL FRAMEWORK

André's learning can best be characterized as that of a gifted learner striving through self-regulation to meaningfully understand scientific knowledge through his critical and creative thinking. Consequently these concepts form the core of our interpretive framework in trying to understand André's learning and are discussed below.

Self-Regulated Learning

Self-regulated learning is understood to be learning which occurs under the metacognitive control and intrinsic motivation of the learner (Schraw, Crippen, & Hartley, 2006). Learners tend to be willing to engage deeply in learning for extended periods of time, make greater use of learning strategies, and achieve better, when they are undergoing self-regulated learning (Zimmerman, Bandura, & Martinez-Pons, 1992). During self-regulated learning, learners may enter a state of flow, in which they are deeply absorbed in their work to the extent that they almost become oblivious of their environments and of the passing of time (Fredricks, Alfeld, & Eccles, 2010). Flow is considered a particularly productive and gratifying state to be in, and highly conducive to the production of new knowledge through creative thought.

Internal and external factors can contribute to the development of, and triggering of, self-regulation in learning (Sinatra & Mason, 2008). Internal factors include possession of cognitive, metacognitive, and motivational skills (Schraw et al., 2006). Motivational skills may include self-efficacy and epistemological beliefs which are beliefs about the nature of knowledge and how it is obtained (Schraw et al., 2006). Sinatra and Pintrich (2003) show that people are more likely to engage in self-regulated learning if they have a constructivist and mastery-oriented epistemology. A person holding a constructivist epistemology believes that knowledge is obtained through individual and social construction, and can therefore be contested. This is in contrast to an absolutist epistemology, according to which knowledge consists of facts to be accepted unquestioningly from an authority. A person holding a mastery-oriented epistemology believes that the purpose of obtaining knowledge is to master a domain, rather than to achieve extrinsic rewards. Therefore these two epistemological beliefs which Sinatra and Pintrich link to engagement in self-regulated learning refer to the beliefs about how knowledge is formed and learned, its contestability and the purpose of learning new knowledge. What this list lacks, however, is the ideal properties of this knowledge. It seems reasonable that the perception of what constitutes meaningful knowledge, which is most conducive to inducing engagement in self-regulated learning, may be subject-specific. Therefore, in the context of physical science learning, we ask: "What epistemological belief, about the ideal properties of knowledge, drives self-regulated learning of physical science?" It is this gap in the literature which is addressed in this study.

Critical and Creative Thinking

Critical thinking is closely related to creative thinking, and both of these can be linked to meaningful learning (e.g. Lipman, 1989; Paul & Elder, 2008; Schraw et al., 2006). Schraw et al. list critical thinking as one of the cognitive strategies a learner uses during self-regulated learning while Lipman argues that critical thinking involves convergent thinking during which ideas are evaluated against criteria and monitored metacognitively. According to Paul and Elder (2008), effective learning always involves a series of creative acts, which are then evaluated against criteria. Paul and Elder argue that critical and creative thinking are inseparable in practice, although describing them as if they were separate may be useful. This conflicts with authorities, such as De Bono, who believe that creative thinking requires abandonment of the logic and standards of critical thought, and that knowledge in a domain can inhibit creativity in that domain (Bailin, 1987). It appears that the views of other authors on the topic (e.g. Bailin, 1987; DeHaan, 2009; Glassner & Schwarz, 2007; Novak, 2010) fall between these two extremes, that is they view critical and creative thinking as closely related, but separate.

Creative thinking involves divergent thinking in which new ideas are generated (called fluency), thinking switches between these ideas (flexibility) as they are evaluated, and some are selected for linkage (conceptual combination), and focus (selective mental attention) (Lubart & Zenasni, 2010; Mumford, Hester, & Robledo, 2010). The kind of creativity referred to here appears to correspond to what is sometimes called "mini c creativity" (DeHaan, 2009), or "petite creativity" (Schwartz, Varma, & Martin, 2008). This refers to generation of knowledge which is new to the particular learner, although not necessarily new to the domain as a whole. Creativity is also referred to as innovation, and occurs due to transfer of knowledge from one context to another (Schwartz et al., 2008). Adaptive expertise is required to perform such knowledge transfer. Transfer can also be encouraged by use of appropriate representational tools. However, adaptive expertise requires the individual to possess a highly structured knowledge base (DeHaan, 2009), resulting from having engaged in meaningful learning for an extended period of time (Novak, 2010). In this chapter creative thinking and creativity are understood in the broad sense of generation of new ideas or artifacts. This predominantly includes generation of ideas or artifacts new only to the learner, but could also include ideas and artifacts new to the domain as a whole.

Meaningful Learning

Ausubel contrasted meaningful learning with rote learning. Whereas rote learning focuses on the recall of isolated facts, meaningful learning involves linkage and subsumption of concepts to create a hierarchical, integrated knowledge structure. This is associated with positive affect and the development of expertise in the knowledge domain (Novak, 2010). The knowledge gained from rote learning tends to be inert (non-transferable to new contexts). In contrast, knowledge gained through meaningful learning is more likely to be active, and so enable innovation to occur as knowledge is transferred to new contexts (Schwartz et al., 2008). Ausubel's term *meaningful learning* corresponds to the terms learning for understanding, effective learning, and deep learning. All of these share the criterion of conceptual linkage resulting in a hierarchical, integrated knowledge structure. Such learning is steered by critical thinking. For example, the learner makes judgments about concept selection, accuracy and relevance of linkages between concepts, and appropriateness of assignment of concepts to relative hierarchical levels, and the learner monitors learning metacognitively (Paul & Elder, 2008). This discussion again draws attention to the link between critical and creative thinking, since critical thinking is required for meaningful learning to occur, and meaningful learning is required for the development of adaptive expertise, which is required for knowledge transfer, that is creativity.

Meaningful learning refers to formation of links between concepts, and organization of these concepts relative to one another. Therefore an understanding of meaningful learning should be informed by conceptual change theory (CCT). CCT has its origins within work by Piaget. As explained by Dykstra, Boyle, and Monarch (1992), according to Piaget, learners accept new information by assimilation (acceptance without the need for major modification in mental structure) if they consider this new information to be compatible with their existing knowledge. However, learners experience feelings of disequilibration when they consider that information they are presented with is incompatible with their existing conceptual frameworks. In such cases the learners have to undergo accommodation (major modification in mental structure) before they can learn the new information. Posner, Strike, Hewson, and Gertzog (1982) used the term conceptual change to refer to accommodation, and proposed that learners make judgements about whether to undergo conceptual change or not based on their perceptions of competing concepts' intelligibility, plausibility and fruitfulness. They suggested that exposing learners to discrepant events could induce a feeling of disequilibration, also called dissonance, which might cause the learners to undergo conceptual change.

More recent research, as reviewed by Vosniadou (2008), has provided some support for the role of dissonance in conceptual learning, particularly amongst gifted learners, but has also shown that learners often respond to dissonance by avoidance behavior, rather than by undergoing conceptual change. The social, contextual and motivational aspects of conceptual learning have also received greater attention in recent research. Sinatra and Mason (2008), and Sinatra and Pintrich (2003) are some of the leaders in this so-called warming movement of conceptual change. They focus on intentional conceptual change, which is a subcategory of self-directed learning, since both share the characteristics of being under the metacognitive and motivational control of the learner. Epistemological beliefs therefore drive the choices a learner makes during intentional conceptual change, as is the case in all self-directed learning. They also drive the choices the learner makes of whether to engage in intentional conceptual change or not, including whether to embrace or avoid dissonance. Other work on conceptual learning within a sociocultural perspective includes that by Tytler and colleagues on use of representations to socialize learners into a discipline (e.g. Hubber, Tytler, & Haslam, 2010), as well as work on use of analogical thinking in conceptual learning (e.g. Clement, 2008).

It is the authors' view that all learning, particularly meaningful learning, is constructivist in nature and is limited by the capacity of working memory. According to constructivist learning theory, learning cannot occur by passive absorption of knowledge, but only through active sense-making activity on the part of the learner (Dirks, 1998). According to the Information Processing Model (IPM) of learning, the small capacity of working memory is the greatest limitation to human learning (Jonassen, 2009). Working memory consists of whatever one is thinking about at a particular moment. The more individual items a learner tries to think about simultaneously, the greater discomfort, called cognitive load, the learner experiences. Both critical and creative thought present significant cognitive load to the learner, since they require the learner to think of multiple pieces of information simultaneously. The hierarchical, integrated knowledge structure of experts enables them to chunk knowledge elements. In this way they can represent more information in working memory within fewer individual items, thus reducing the limitations of cognitive load (Kirschner, 2009). Motivation appears to expand the size of working memory somewhat, and also make learners more prepared to persevere with their learning despite the discomfort afforded by cognitive load (Niaz & Logie, 1993). When intrinsic motivation is high, learners may engage passionately with their learning and be more disposed to engage in critical and creative learning.

Summary

Based on the above discussion of the core concepts, the following framework was used in the interpretation of André's learning. At times a learner may enter a state of self-regulated learning in which metacognitive control and intrinsic motivation play vital roles. During self-regulated learning, learners may become so intensely engaged in learning as to become almost unaware of their surroundings and the passing of time. Self-regulated learning, particularly when undergone in a state of flow, is very creative, is associated with strongly positive feelings, and is associated with high achievement. It is our view that gifted learners enter such states more frequently than other learners. Unfortunately, underachievement is well known amongst gifted learners and seen by some as very complex and an enigma (Reis & McCoach, 2000). Since self-regulation is known to improve achievement, and since gifted learners are known to be likely to benefit from instruction in learning strategies and styles, understanding beliefs that drive gifted learners may empower teachers to help underachieving gifted learners to reach their potential.

During self-regulation, learners make use of cognitive and metacognitive strategies under the control of their belief systems. One of the cognitive strategies learners engage in is critical thinking. This is closely related to creative thinking. Critical and creative thinking are therefore important components of self-regulated learning or for that matter any kind of meaningful learning. The learner's belief system includes epistemological beliefs which are beliefs about the nature of knowledge and how it should be acquired. Some belief systems are more conducive to promoting self-regulation in learning than others are. However, little is known about the particular beliefs of what constitutes meaningful knowledge in particular domains, such as school science and which promote self-regulated learning in that domain. These beliefs are the focus of this chapter.

RESEARCH DESIGN

The first author has known André since he was a baby, and became aware of his intellectual giftedness, particularly in the sciences, when he was very young, since he grew up in an isolated rural mission community in which the author also lived at the time. Although unrelated to André, the author did encounter him on a regular basis as a member of the mission community and as a teacher at the mission school that André attended. André, while still in primary school, would often visit the science laboratory as the author prepared demonstrations for lessons, or as he worked on his science fair projects. These encounters, together with reports from his teachers, confirmed and intensified impressions of André's giftedness and creative ability. His giftedness was confirmed by subsequent events. During his schooling, André won regional science fairs and the national science fair. In most of his projects he created electronic devices, such as a sonar positioning system and a soccer-playing robot. André also received numerous awards for science, mathematics and computer Olympiads on regional and national levels, and scored ten As in the grade 12 national examination. These subsequent achievements supported the idea that André exemplified a case of a gifted and creative learner from whom we could learn.

During André's grades 8 to 10 (age 13–15), the first author engaged in a case study on how André learned science, and the belief system which drove this learning (Stott, 2002). The author taught André science in an enriched and accelerated individual program when he first entered secondary school in grade 8. Detailed records of his learning were begun at this time. This formed the basis of a detailed case study (Yin, 2003) where the case was selected for its intrinsic interest (Stake, 1994), and for its potential to suggest ways in which less successful learners may be helped to improve their learning by observing the learning of more successful learners (Baron, 1987). This is similar to expert-novice research, in which the differences between experts and novices are examined in order to propose ways to help novices improve (see, for example, Kirschner, 2009). Within the context of studies on gifted education, this is also consistent with findings that the learning of lower-achieving gifted learners can be enhanced by explicit instruction in learning strategies observed in higher-achieving gifted learners (e.g. Lee, 2004; Scruggs, 1985; Sternberg, 1987).

Data collection occurred through participant observation, since a human instrument was considered best able to sense and examine the complexities of the case, and thus generate a rich, holistic description from which naturalistic interpretations and generalizations could be made (Merriam, 2009). By the end of the three-year study the data corpus was considerable and consisted of 35 detailed lesson reports, 21 detailed notes of critical incidents, 17 full audio-recorded one-on-one lessons, 12 interviews with André probing his learning, and 27 self-report notes written by André as he was going about learning. Data collection and analysis were focused by the research questions, such as "What belief system drives this high achiever's effective learning of physical science?" As patterns began to emerge in the data, these patterns were summarized in categories, which were used to code the data. Analysis of the data using these codes resulted in emergence of more patterns. This inductive, iterative process resulted in the creation, testing and refinement of an explanatory model, grounded in the data (Taber, 2000), to answer the research questions.

André and his parents gave consent for the conduction of this research and the publication of its results, under the chosen pseudonym. Long-term observation and multiple data sources were used to ensure validity (Merriam, 2009). Relevant sections of the research findings were shown to the people to which they refer, and appropriate adjustments were made to ensure valid representation. In all cases this only required minor changes. Rich descriptions have been given in Stott (2002), enabling readers to form interpretations and generalizations of their own, thus enhancing the validity of the study (Stake, 1994). The first author has maintained contact with André and his family to the present. Recent discussions with André, as well as André's confirmation of the contents of this chapter, provide additional support for the validity of the work presented here.

A CASE STUDY: ANDRÉ'S LEARNING

The authors interpret André's self-regulated learning of science as being motivated and directed by beliefs that meaningful scientific knowledge should be precise, elegant, and transferable. Therefore, André uses the criteria of precision, elegance, and transferability to test whether the scientific knowledge he possesses is acceptable, or whether he needs to continue to apply critical and creative thinking to transform the knowledge further until it does meet these criteria. The beliefs are defined, and their roles in André's use of the cognitive strategies of critical and creative thinking during learning are illustrated below.

Precision

The belief that meaningful scientific knowledge should be "precise" is defined to mean that the learner believes that conceptual boundaries and characteristics for conceptual abstraction must be clearly defined and logically coherent, so that they can be applied in an exact, consistent manner. This is illustrated by remarks and drawings André made (Figure 1) in response to the question, "How does a pendulum's length affect its frequency?" He said that if length determines a pendulum's frequency, then the two pendulums he drew in Part 1 of Figure 1 should have significantly different frequencies because their lengths are significantly different. But he doubted that this would be so. He said that adding a string of negligible mass below a pendulum bob would not change the position of the centre of mass significantly even though it would alter the length significantly. He reasoned that if it was the pendulum's length to centre of mass, rather than its length, that determined frequency, then the two pendulums drawn in Part 1 would swing with negligibly different frequencies. While asking whether this is universally applicable, he drew Part 2 and considered the effect of this situation on pendulum frequency. He said he intended the length to centre of mass to be the same as for the first pendulum, but the absolute length to differ in a different way to the middle picture (Field notes, 31/01/01).

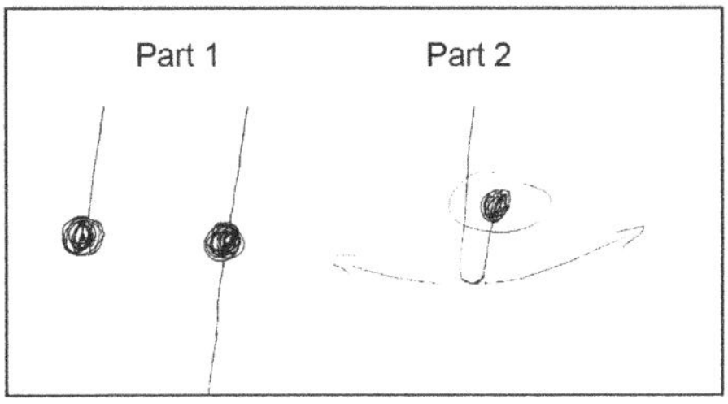

Figure 1. Pendulum variations André drew while exploring factors affecting pendulum frequency

André's behavior is interpreted here as searching for and testing a sharp (i.e. precise) conceptual line between a factor which determines pendulum frequency and a factor which does not. Notice that this searching and testing involved both critical and creative thinking. It involved creative thinking as André generated appropriate alternatives for testing his hypothesis, and it involved critical thinking as he made substantiated judgments about these alternatives. In another incident, André's comment that concepts must be mutually exclusive to prevent inappropriate generalization (Personal communication, 1/11/02), further supports the view that he values conceptual precision very highly for ensuring scientifically sound learning. In a more recent personal communication (04/12), André remarked that he thought that people's science learning could be improved if they could be shown that meaningful scientific knowledge is precise and well structured.

Learners experience a feeling of disequilibrium, also called dissonance, when new information is presented which they perceive not to fit satisfactorily into their existing knowledge structures, and this feeling may encourage them to undergo conceptual change (Posner et al., 1982). From this it follows that the more stringent a learner's requirements for the fit between knowledge elements to be acceptable, the more likely that discrepant information will challenge understandings and misconceptions, and possibly encourage the learner to undergo conceptual change when necessary. André's high regard for precision in knowledge causes him to set very stringent criteria for information to acceptably fit into his existing knowledge structure. This reduces the likelihood that he would undergo sloppy assimilation of incompatible information into his knowledge structure. It increases the likelihood that he is self-regulated to undergo conceptual change towards more scientifically accurate understandings whenever his existing knowledge structure differs even only slightly from a more scientifically acceptable structure. As pointed out by Howard (1987), the categorization of borderline instances puts conceptual boundaries to the test, thus clarifying the rules of conceptualization. Also, engagement with hypothetico-predictive reasoning helps learners to make predictions based on their hypotheses and test the validity of their understanding (Lavoie, 1995). However, it seems reasonable to expect that unless a learner values knowledge precision the activities of borderline categorization and hypothetico-predictive reasoning could be engaged in with little improvement in the learners' conceptual understanding.

Elegance

A belief that meaningful scientific knowledge should be elegant refers to a value for simple, compressed order in the organization of information. André defines information elegance as "explaining the most cases or the most observations, or thoughts, or whatever, in the least number of facts. Compressed data." (Interview, 09/10/02). André says that if an elegant outcome can be reached, then it is worth working at something beyond the point of understanding or functionality (Diary entry, 19/02/02).

André says he dislikes verbosity. He communicates minimalistically. He was frequently observed to search for patterns within information and to generate graphs and equations to aid communication and to consolidate learning. For example, Figure 2 shows a reproduction of André's summary of the effect of object distance from a lens on the position of the resultant image. This was an outcome of a lesson (Field notes, 06/03/01) in which he was taught about five discreet set-ups: the object being beyond 2F, at 2F, between 2F and F, at F, and between F and O, of the lens. At the start of the lesson he could not answer questions about where the image would be formed in each case, and no graph was drawn or referred to by the author during the lesson. It appears that André developed the graph to summarize learning which occurred in that lesson, rather than merely repeating something he had seen elsewhere. André produced the graph towards the end of the lesson in response to an instruction to repeat what he had been taught. His preference for drawing the graph over repeating the outcome of each of the five individual set-ups is interpreted as being due to a high regard for the graph's elegance.

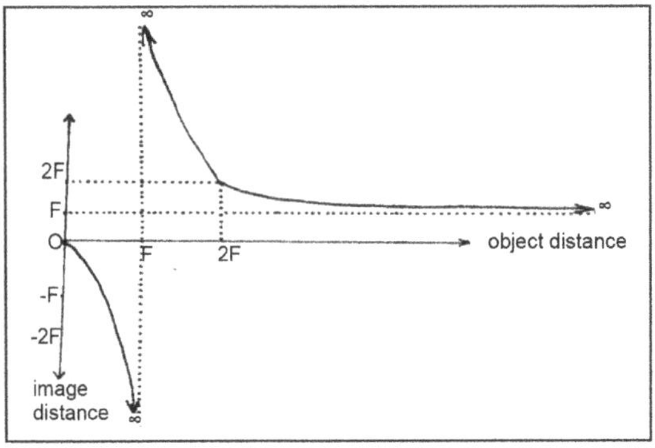

Figure 2. A reproduction of a graph André drew to summarize the relationship between object and image positions relative to a lens

André's value for elegance often motivated him to engage with information for a longer period than he would have otherwise, as he created successively more elegant representations. Therefore his value for elegance drove him to undergo creative thought as he generated representations, and critical thought as he evaluated these representations, particularly against the criterion of elegance. André is not alone in his high regard for elegance and the motivating influence this has on his creative activity. For example, Gooding (1982) maintains that Michael Faraday's belief in the elegance of nature drove his theory creation. Amongst more recent creative geniuses, Steve Jobs' love for elegance largely drove his product designs (Isaacson, 2011).

A learner can only link two concepts when both concepts are represented simultaneously in working memory, which is of very limited capacity (Jonassen, 2009). Greater knowledge elegance can be expected to result in lower cognitive load, despite representation of a larger amount of information in the working memory. Therefore it is reasonable that knowledge elegance should increase the likelihood that a learner can undergo beneficial and complex link formation. This is consistent with Bruner's view that effective learning must involve theory formation in order to avoid mental clutter (Bruner, 1971), since "Knowledge, to be useful, must be compact, accessible, and manipulative" (p. 106). It is also consistent with the finding that experts within a particular knowledge domain possess hierarchically organized and highly linked knowledge base, which includes abstracted levels of knowledge, all of which are necessary for elegant representation (Novak, 2010).

André's value for conceptualizing and representing knowledge elegantly drove him to self-direct extended engagement in creation and evaluation of representations. The contribution of this extended engagement with representations to André's effective learning can be understood in terms of dynamic transfer and the role representations play in coordinating aspects of conceptual understanding. According to theories of dynamic transfer, representations empower a learner to transfer learning to a new context through a series of steps (Schwartz et al., 2008). In other words, representations serve as tools to augment working memory. According to Pierce's triadic model, conceptual understanding involves coordination between a concept, its representation, and its referents (i.e. phenomena to which both the concept and representation refer) (Hubber et al., 2010). Creation, critical evaluation, and recreation of representations refine a learner's understanding of the coordination between these three elements, resulting in effective conceptual learning.

Transferability

By a belief that meaningful scientific knowledge should be transferable, we refer to the learner's value for aspects of understanding which enable knowledge to be used in new contexts, i.e. which make knowledge utilizable, manipulable and flexible (Bruner, 1971). André's value for being able to work something out rather than being limited to what has been learnt by repeated practice, in other words his high regard for the ability to transfer knowledge to new contexts, is shown in the way he works out formulae. This is illustrated in his learning of the formula $F = ma$. Every time André was observed to use this formula within a year of being introduced to it, he derived it from first principles. He would do this by slowly reasoning aloud through the direct and inverse relationships between the concepts. Once this verbal reasoning ceased to be observed he was asked if he was now recalling the formula from memory. He said this was how he was arriving at it:

> I think of a body that's decelerating because of a net force acting in the opposite direction to motion.... and obviously it will lose momentum, and the rate at which – momentum is actually force stored – and the rate at which it loses momentum is equal to the force it exerts on the resisting. (Interview, 29/01/02)

During another interview he remarked that thinking about the relationships between concepts was how he remembered most equations. When asked why this was so despite the fact that reproducing a memorized equation was less time consuming, he replied:

> In an application you're not going to get: "Here's this formula and now work that out". If you're working something out in a practical application, you need to work it out logically because it's not the same problem over and over." (Interview, 09/10/02)

In both these methods of deriving the formula, André is making use of links between concepts. He justifies using these conceptual links to derive the formula, rather than merely recalling the memorized equation, by saying that the derivation is more likely to be useful in a non-routine problem, that is, it is more transferable. This is consistent with work on knowledge transfer, such as that by Schwartz et al. (2008), which shows that multiple links between knowledge elements ensure flexibility and improves knowledge transferability. It is also consistent with expert-novice research, which lists the highly linked nature of an expert's knowledge system as a major reason why experts are able to apply knowledge to new contexts (Kirschner, 2009). The effectiveness of André's physical science learning, and his creative ability, can partially be understood as a result of the formation of multiple conceptual linkages in a manner which aids transfer between contexts. This is driven by the satisfaction André gets from knowledge linked in this way being utilizable in various contexts.

André would often test the transferability of his newly gained knowledge by using it to design various types of machines. This was usually done in a "playful" manner, and was accompanied by a lot of speaking to himself about "if this then that", and a general attitude of obvious enjoyment. This again illustrates André's use of both creative and critical thought to drive his deep learning style. André underwent creative thinking as he generated these designs, and critical thinking as he evaluated the transferability of his knowledge, based on his ability to use this knowledge flexibly within the new situations he had created.

CONCLUSION

Discussion

It appears that André's learning of science is driven by a motivating epistemological belief about the nature of meaningful science knowledge, namely that it should be precise, elegant, and transferable. This belief system drives André to undergo both critical and creative thinking during his self-regulated learning, in which he often enters a state of flow. A value for precision drives André to create test-cases and to stringently evaluate inclusion and exclusion of items into a conceptual category. This provides André with intrinsic motivation to undergo deep conceptual learning resulting in development of a highly accurate conceptual knowledge base. A value for elegance drives André to manipulate his knowledge as he creates and evaluates representations of this knowledge. He continues this manipulation and representation process until his representations are sufficiently concise, interlinked, structured and generalized to meet his criterion of knowledge elegance. This provides André with intrinsic motivation to undergo self-regulated learning resulting in a well-structured, integrated, hierarchical knowledge base, and also to develop powerful representational tools. A value for transferability of knowledge motivates André to manipulate his knowledge by applying it creatively to new contexts, linking memory elements within and between concepts, and so evaluating the compatibility of new learning with his existing knowledge.

These beliefs are powerful in motivating the gifted learner to engage in creative and critical thought in a self-regulated manner. It is not surprising to find that beliefs can so strongly influence learning. For example, Gooding (1982), in a study of the learning and creative process of the experimental and creative genius, Michael Faraday, also found epistemological beliefs, including a high value for elegance of knowledge, to drive Faraday's thinking. Also, Sternberg (1987) suggested that the most effective way to improve learning effectiveness, and even intelligence, is to alter a learner's belief system.

The value of a belief that meaningful scientific knowledge is precise can be understood in terms of conceptual change theory (Posner et al., 1982). A value for precision increases the learner's likelihood of undergoing dissonance and consequent conceptual change towards a more scientifically acceptable conception when exposed to appropriate discrepant events. The value for knowledge precision drove the learner to create test-cases and perform thought experiments, which contributed to him suspending judgment about his conceptions for a while. This was followed by him making conceptual decisions, using critical thinking, during which he underwent conceptual change. This is consistent with interpretations made by Gooding (1982) about Michael Faraday's process of conceptual change using thought experiments.

Much work on conceptual learning suggests that most learners do not experience conceptual learning as a revolutionary, gestalt-change, sudden, process in which a new conception suddenly clicks into place, mentally, immediately enabling the learner to apply it across contexts (e.g. Tao & Gunstone, 1999). Instead, learners tend to vacillate between variations of conceptions, and need explicit help in transferring newly obtained conceptions to new contexts (Schwartz et al., 2008). Interestingly, though, André describes his conceptual learning as involving a series of revolutionary "*clicks into place*", rather than an evolutionary process, and was surprised that this is not so for most people (Personal communication, 04/12). A possible explanation for this is that André's belief system drives him to undergo revolutionary conceptual change to a greater extent than most learners who lack such a belief system.

The value of a belief that meaningful scientific knowledge is elegant can be understood in terms of the information processing model of learning, and in the motivation this belief creates for representation production. Representation production can enable a learner to undergo dynamic transfer and can improve the quality of a learner's conceptual learning process. Greater knowledge elegance is accompanied by chunking, abstraction and hierarchical organisation of knowledge. These processes result in the formation of a knowledge structure characteristic of an expert. Such a knowledge structure enhances learning, problem solving and creativity by enabling more information to be represented and linked in working memory for a particular amount of cognitive load (Kirschner, 2009). Representation production aids the process of dynamic transfer, since representations serve as tools to augment the size of working memory (Schwartz et al., 2008). This enhances a learner's ability to undergo innovation, in other words creative thinking. Creation and evaluation of representations also enhance conceptual learning as learners coordinate their understandings of concepts, their representations and their referents (Hubber et al., 2010).

The value of a belief that meaningful scientific knowledge is transferable can be understood in terms of the motivation this provides for link formation between knowledge elements, and the flexibility and activation of a highly linked knowledge base (Lavoie, 1995). Experts are known to possess highly linked knowledge structures (Kirschner, 2009). In contrast, the knowledge of novices is often inert (cannot be activated within new contexts) due to the sparseness of links between novices' knowledge elements.

The belief system described in this chapter steers André's creative and critical thinking in effective ways. The high levels of motivation observed to propel André's self-directed learning are clearly closely linked to the opportunity for undergoing creative thinking. For example, participation in the national annual science fair particularly propelled his learning by providing a platform for his creative work. Undergoing self-directed learning driven by the motivation which creativity provided him, clearly enthused and invigorated him to learn science. André wrote that participation in the science fair "taught me to accept problems as part of any undertaking…. it imposed an annual rhythm of creative activity on me without which I would have probably run to a boring halt" (Written comment, 09/02). More recently, André remarked about the idea of flow and that he thinks that "almost 100% of real progress is made in that [flow] state, and the rest of the time is spent trying to reach it" (Written communication, 27/05/2011).

Limitations

The findings discussed above arose from a case study of a single gifted learner, who was learning Physical Science individually in a one-on-one relationship with a teacher, and who already possessed a belief system which steered his self-directed learning towards critical and creative thought. No claim can be made that these findings can be generalized to all gifted and creative learners. For a learner to undergo this form of self-regulated learning, a number of interacting internal and external factors are necessary. For example, it is widely accepted that optimal learning occurs with a mix of individual and social learning opportunities (Glassner & Schwarz, 2007). The context of the study described here did not allow observation of the influence of social contexts on the gifted learner's learning, and consequently this has not been addressed in this chapter. This does not mean, however, that socio-cultural perspectives on learning were seen as unimportant. This chapter has focused on the role of ontological and epistemological beliefs on critical and creative thought, and self-regulated learning. The gifted learner under study clearly possessed highly developed cognitive and metacognitive skills and high levels of self-efficacy. His learning also occurred within a stimulating and supportive environment. Clearly these factors are important, and without them it is unlikely that cultivation of the belief system described here would be possible, let alone effective. In addition the full study (Stott, 2002) described the learning strategies this gifted learner was observed to use, as well as the teaching strategies found to be effective in stimulating and supporting engagement with self-directed learning. Arising from the study it was suggested that teachers of gifted learners explicitly teach such learning strategies while nurturing, or encouraging the development of the belief system described here.

Implications

Despite the limitations discussed above, the insights gained from this study have implications for physical science education, particularly of gifted learners. It would be interesting to investigate the extent to which giftedness in the domain of physical science is linked to possession of this, or a similar, belief system, and what the effects of nurturing such a belief system amongst gifted learners would be. As Sternberg (1987) points out, altering a person's belief system is extremely difficult, and may, indeed, not even be possible. However, it is possible that many learners who are gifted in the sciences may naturally possess similar belief systems which need to be nurtured, possibly increasing the likelihood of their uptake of these principles. This study suggests that science teachers should value, model, and nurture an epistemological belief system which values knowledge which is precise, elegant, and transferable. For example, teachers should prompt learners to give more precise, rather than vague answers, encourage learners to refine representations to make them more concise and generic, and encourage learners to apply their knowledge to new contexts, and form links between memory elements and between concepts. The findings of this study suggest that such activity might encourage learners to become self-regulated as they use both creative and critical thinking to undergo effective learning. This should enhance the affective experience of learning, as well as enhance achievement levels (Novak, 2010). This study also reinforces our understanding that critical and creative thinking are very closely linked to one another and are at the core of meaningful and self-regulated learning.

REFERENCES

Al-Khatib, S. A. (2010). Meta-cognitive self-regulated learning and motivational beliefs as predictors of college students' performance. *International Journal for Research in Education, 27*, 57–72.

Bailin, S. (1987). Critical and creative thinking. *Informal Logic, 9*(1), 23 – 30.

Baron, J. (Ed.). (1987). *An hypothesis about the training of intelligence*. New Jersey, NJ: Lawrence Erlbaum Associates.

Bruner, J. S. (1971). *The relevance of education*. London, England: George Allen & Unwin.

Clement, J. (2008). The role of explanatory models in teaching for conceptual change. In S. Vosniadou (Ed.), *International handbook of research on conceptual change* (pp. 417–452). New York & London, UK: Routledge.

DeHaan, R. L. (2009). Teaching creativity and inventive problem solving in science. *CBE – Life Sciences Education, 8*, 172–183.

Dirks, A. L. (1998). *Constructivist pedagogy, critical thinking, and the role of authority*. Retrieved December 3, 2008, from http://webhost.bridgew.edu/adirks/ald/papers/constr.htm

Dykstra, D. I., Boyle, C. F., & Monarch, I. A. (1992). Studying conceptual change in learning physics. *Science Education, 76*(6), 615–652.

Fredricks, J. A., Alfeld, C., & Eccles, J. (2010). Developing and fostering passion in academic and nonacademic domains. *Gifted Child Quarterly, 54*(1), 18–30. doi: 10.1177/0016986209352683

Glassner, A., & Schwarz, B. B. (2007). What stands and develops between creative and critical thinking? Argumentation? *Thinking Skills and Creativity, 2*, 10–18.

Gooding, D. (1982). Empiricism in practice: Teleology, economy, and observation in Faraday's physics. *Isis, 73*(1), 46–67.

Howard, R. W. (1987). *Concepts and schemata: An introduction*. London, England: Cassel Education.

Hubber, P., Tytler, R., & Haslam, F. (2010). Teaching and learning about force with a representational focus: Pedagogy and teacher change. *Research in Science Education, 40*, 5–28. doi: 10.1007/s11165-009-9154-9

Isaacson, W. (2011). *Steve Jobs*. New York, NY: Simon & Schuster.

Jonassen, D. (2009). Reconciling a human cognitive architecture. In S. Tobias & T. M. Duffy (Eds.), *Constructivist instruction: Success or failure?* (pp. 13–33). New York, NY: Routledge.

Kirschner, P. A. (2009). Epistemology or pedagogy, that is the question. In S. Tobias & T. M. Duffy (Eds.), *Constructivist instruction: Success or failure?* (pp. 144–157). New York, NY: Routledge.

Lavoie, D. R. (1995). The cognitive-processing nature of hypothetico-predictive reasoning. In D. R. Lavoie (Ed.), *Toward a cognitive-science perspective for scientific problem solving: A monograph of the national association for research in science teaching, number six* (pp. 13–49). Manhattan, KS: Ag Press.

Lee, R. L. (2004). *The impact of cognitive task analysis on performance: A meta analysis of comparative studies* (Ed.D.). University of South California, Los Angeles, CA.

Lipman, M. (1989). *Misconceptions in teaching for critical thinking* (Resource publication, Series 2 No. 3). Upper Montclair, NJ: Institute for Critical Thinking, Montclair State College.

Lubart, T., & Zenasni, F. (2010). A new look at creative giftedness. *Gifted and Talented International, 25*(1), 53–58.

Merriam, S. B. (2009). *Qualitative research: A guide to design and implementation*. San Fransisco, CA: Jossey-Bass.

Mumford, M. D., Hester, K. S., & Robledo, I. C. (2010). Scientific creativity: Idealism versus pragmatism. *Gifted and Talented International, 25*(1), 59–64.

Niaz, M., & Logie, R. H. (1993). Working memory, mental capacity and science education: Towards an understanding of the "working memory overload hypothesis". *Oxford Review of Education, 19*(4), 511–521.

Novak, J. D. (2010). Learning, creating, and using knowledge: Concept maps as facilitative tools in schools and corporations. *Journal of e-Learning and Knowledge Society, 6*(3), 21–30.

Paul, R., & Elder, L. (2008). *Critical & creative thinking*. Dillon Beach, CA: The Foundation for Critical Thinking.

Posner, G. J., Strike, K. A., Hewson, P. W., & Gertzog, W. A. (1982). Accommodation of a scientific conception: Towards a theory of conceptual change. *Science Education, 66*(2), 211–227.

Reiss, S. M., & McCoach, D. B. (2000). The underachievement of gifted students: What do we know and where do we go? *Gifted Child Quarterly, 44*(3), 152–170.

Schraw, G., Crippen, K. J., & Hartley, K. (2006). Promoting self-regulation in science education: Metacognitition as part of a broader perspective on learning. *Research in Science Education, 36*, 111–139.

Schwartz, D. L., Varma, S., & Martin, L. (2008). Dynamic transfer and innovation. In S. Vosniadou (Ed.), *International handbook of research on conceptual change* (pp. 479–508). New York, NY & London, England: Routledge.

Scruggs, T. E. (1985). Maximizing what gifted students can learn: Recent findings of learning strategy research. *The Gifted Child Quarterly, 29*(4), 181–185.

Sinatra, G. M., & Mason, L. (2008). Beyond knowledge: Learner characteristics influencing conceptual change. In S. Vosniadou (Ed.), *International handbook of research on conceptual change* (pp. 560–582). New York, NY & London, England: Routledge.

Sinatra, G. M., & Pintrich, P. R. (2003). The role of intentions in conceptual change learning. In G. M. Sinatra & P. R. Pintrich (Eds.), *Intentional conceptual change*. Mahwah, NJ & London, England: Lawrence Erlbaum Associates.

Stake, R. E. (1994). Case studies. In N. K. Denzin & Y. S. Lincoln (Eds.), Handbook of qualitative research (pp. 236–247). Thousand Oaks, CA: SAGE.

Sternberg, R. J. (1987). Teaching intelligence: A triarchic model. In D. N. Perkins, J. Lochhead, & J. Bishop (Eds.), *Thinking: The second international conference* (pp. 53–60). New Jersey, NJ: Lawrence Erlbaum Associates.

Stott, A. (2002). *A case study of a high achiever's learning of physical science* (Unpublished masters in education). University of Natal, Durban, South Africa.

Stott, A., & Hobden, P. (2006). *Intrinsic factors driving a high achiever's learning of physical science*. Paper presented at the 14th annual meeting of SAARMSTE, University of Pretoria, Pretoria, South Africa.

Taber, K. S. (2000). Case studies and generalizability: Grounded theory and research in science education. *International Journal of Science Education, 22*(5), 469–487.

Tao, P.-K., & Gunstone, R. F. (1999). The process of conceptual change in force and motion during computer-supported physics instruction. *Journal of Research in Science Teaching, 36*(7), 859–882.

Vosniadou, S. (2008). Conceptual change research: An introduction. In S. Vosniadou (Ed.), *International handbook of research on conceptual change* (pp. xii–xxvii). New York, NY & London, England: Routledge.

Yin, R. K. (2003). *Case study research: Design and methods* (3rd ed.). Thousand Oaks, CA: Sage.

Zimmerman, B. J. (1990). Self-regulated learning and academic achievement: An overview. *Educational Psychologist, 25*(1), 3–17.

Zimmerman, B. J., Bandura, A., & Martinez-Pons, M. (1992). Self-motivation for academic attainment: The role of self-efficacy beliefs and personal goal setting. *American Educational Research Journal, 29*(3), 663–676.

Angela E. Stott
Schools Partnership Project
University of the Free State
South Africa

Paul A. Hobden
School of Education
University of KwaZulu-Natal
South Africa

SECTION 3

DEVELOPING GIFTED AND CREATIVE LEARNERS IN SCIENCE EDUCATION CLASSROOM

CLAIRE E. HUGHES AND TIMOTHY A. GOODALE

7. MIND YOUR P'S AND E'S

Developing Creativity in the Science Classroom

INTRODUCTION

In its report, "Preparing the Next Generation of STEM Innovators: Identifying and Developing our Nation's Human Capital", the National Science Board (2010) of the National Science Foundation clearly states that "the U.S. education system too frequently fails to identify and develop our most talented and motivated students who will become the next generation of innovators" (p. 2). Too often, students with tremendous potential and high levels of creativity and talent go unrecognized and undeveloped because they lack the opportunities needed to develop their untapped potential. There are many successful programs focused on advanced learners in which today's top scientists have participated. Students who are encouraged to develop their abilities and who participate in activities related to science during their youth generate significant numbers of patents, win more Nobel Prizes, and are more likely to hold tenured academic faculty positions at top universities (Lubinski & Benbow, 2006).

Creativity is one of the required cognitive attributes for success in the 21st century, according to Howard Gardner, stating that "Everything that can become automated, will be", (Sparks, 2011, p. 1). The question of creativity as a domain-specific skill is a hotly contested one. While it would appear that there are some general skills that are "creative" (Mayer, 2010), it cannot be assumed that by promoting arts education, for example, that students will transfer those skills over to the domain of science. Research has found that "[t]hese transfer claims have been posited without any particular mechanism; there's a lot of magical thinking going on," states Ellen Winner, a long-time creativity researcher (Sparks, 2011, p. 3). If one wants to develop creativity in science, it is critically important to do so within the field of science, and not assume that such skills in other content areas will transfer.

Creativity is highly dependent upon the context. A study, done by Mumford et al. in 2010, found differences in how creativity is demonstrated between scientific fields irrespective of experience within the field. Health scientists demonstrated stronger performance at problem definition and solution appraisal, while biological scientists were stronger at information gathering and idea generation. Social scientists were found to be stronger in idea generation and conceptual combinations. Such differences found across expertise levels appear to be more related to the structure of the discipline, rather than expertise in the related skills.

M. K. Demetrikopoulos & J. L. Pecore (Eds.), Interplay of Creativity and Giftedness in Science, 107–126.

Because of its dependence on expression within the structure of a discipline, creativity is a construct that many claim to recognize, and yet few can define. In addition, there are cultural differences of the understanding of creativity, with East Asians being more likely to view creativity as an outside demand or experience but one with internal rewards, and Americans more likely to perceive creativity as an internal personality trait that results in innovative external products (Lubart, 2011; Paletz, Peng, & Li, 2011). There are also heated debates about how to measure creativity, whether it is rare or common, and how to study it. The tension between defining creativity as a "property of people, products, or processes" (Mayer, 2010, p. 450) is one that is not resolved in this chapter, nor is the concept of creativity as an individualistic characteristic or a result of social and cultural environments. For the sake of this chapter, which focuses on practitioners within a science educational environment, we leave most of the theoretical arguments to researchers and will focus on multiple ways of developing creativity within the science classroom. To develop creativity in a practical, content-based manner, we will examine the integration of the four elements of creativity, known as the 4Ps, in conjunction with the content-focused process of the 5Es, as the organizing method.

Creativity is often described as the 4Ps, or the combination of people, products, and processes that occurs within a given place (Mayer, 2010, Kozbelt, Beghetto, & Runco, 2011). Given this four-sided aspect of creativity, it is worthwhile to explore these aspects as separate components to be developed and nurtured within a classroom.

Table 1. 4Ps of creativity and teacher actions

Element of Creativity	*Teacher Responsibilities*
Place	Establish context, support and underlying classroom culture; identify materials and resources
Person	Support personal creative qualities; encourage lack of conformity and questioning
Process	Link explorations to concepts; support innovative combinations and experimentation; provide models of divergent thinking
Product	Evaluate the usefulness of a product; determine quality of end result

Likewise, in science education, there is an organizing method or process to teach inquiry and discovery of most concepts. Science education has similar goals and objectives to fostering creativity as an inquiry-based approach that recognizes individualized elements of discovery and critical problem solving. Although, probably not recognized on a grand scale as novel products or processes, the outcomes in a true inquiry-based approach to science education do foster creative elements because the findings or discoveries are original to the student and authentic learning has taken place. Most science educators recognize that the learning cycle is an effective approach to an inquiry-based classroom. The learning cycle was developed in the 1960's by Karplus and Their (1967) and had three distinct phases of instruction: (1) Exploration, (2) Concept Introduction, and (3) Concept Application. Since the introduction of the original learning cycle, many revisions and alterations have taken place, but the most popular and widely recognized version now is the 5E model: Engagement, Exploration, Explanation, Elaboration, and Evaluation (Bybee, 1997). The 5E model incorporates the original three phases of the learning cycle but adds two critical elements: Engagement and Evaluation.

The 5E model and its subsequent impact on student achievement has been extensively evaluated and its appropriate use has resulted in greater student academic achievement in science, higher retention rates of scientific concepts and improved reasoning ability and process skills (Hanuscin & Lee, 2008). As a process for planning and executing instruction the 5E model allows teachers to sequence and organize a range of activities and applications and avoid randomness and lack of connection to the curriculum. The table below depicts the each of the 5E phases and subsequent teacher responsibilities as adapted from Abell and Volkmann, 2006 along with an additional E phase.

Table 2. 6E (5E+1) Process and teacher actions (adapted from Abell and Volkmann, 2006)

Phase of Instruction	*Teacher Responsibilities*
Engagement	Establish context; motivate; identify misconceptions
Exploration	Provide common experiences; determine student conceptual understanding
Explanation	Link explorations to concepts; introduce formal content
Elaboration	Expand or apply student knowledge; provide extension activities
Evaluation	Assess student understanding of formal content; determine revisions
Experience	Provide context and preparation; model play and creativity

In the subsequent sections each of the standard 5 phases are more thoroughly explained and placed within context of the 4P's. In addition, a sixth "E" is proposed. "Experience" provides context and preparation for the entire encounter of a science-based lesson that fosters creativity and intertwines the two research-based strategies found separately in science and gifted/creativity investigations. "Experience" as a step in lesson design is meant in both the noun and the verb form; teachers themselves have to model "play" and "experience" creativity themselves in order to teach it (Starko, 2005), and they have to plan for the experience by preparing an array of materials and resources that may be only indirectly related to the nature of the problem, but that promote curiosity- which directly leads to the next "E" of "Engagement".

P'S AND E'S INSTRUCTIONAL MODEL

It is worthwhile pursuing a model of instruction that combines the process of creativity development with the inquiry-based process of science instruction. Such a model provides a "road map" for teachers to explore the development of creativity and science talent. Such development cannot be left up to talented and gifted programs alone. According to Kim, Cramond, and VanTassel-Baska, (2010), there is a high level of correlation between creativity and intelligence scores, up to an IQ score of 120. However, above 120, there does not appear to be any relationship. Thus, the most creative individuals may be merely bright, whereas the most academically gifted students may not be highly creative. It is important to recognize that developing creativity in science is not an activity reserved only for gifted students or gifted programs, and in fact, may be very appropriate for students with learning differences, who often think "outside of the box" (Hughes-Lynch, 2010). Teachers must plan to incorporate the process of creativity within the process of science instruction for all students.

Experience

It is essential that the science teacher purposefully plans and prepares the experience of the classroom in order to develop creativity (Egan, 2005). A teacher cannot merely take a set of educational standards and hope that preparing a strong science lesson will consequently produce creative scientists. Creativity has to be a separate goal; one that is supported and fostered, despite the innate tensions and challenges that emerge as a result. Technical scientific knowledge is not the goal; creative "messes" using the tools of science are, and the teacher has to be prepared to establish that experience.

Place. The teacher has to focus on the classroom, and even the school, as an environment that promotes and encourages creativity. Therefore, the teacher has to plan the experience as one that is promoting, rewarding and encouraging continued development of creative outputs. Csiksentmihalyi (1999), for example, stated that a "set of rules must be transmitted from the domain to the individual, the individual must then produce a novel variation in the context of the domain… and then variation must then be selected by the field for inclusion in the domain" (p. 315). The classroom has to have materials, space and time that encourage "messing around", rather than specific, content-based, goal-oriented opportunities.

Table 3. Ps and Es instructional model

	Place	*Person*	*Process*	*Product*
Experience	Knowing the rules of the field; selecting which areas to explore; providing openness of experience within limits of content	Risk taking and ability issues; classroom management and discipline	Thinking skills; questions to develop open-ended responses	Teacher knowledge of content; understanding of what constitutes originality
Engage	Devising an activity and materials that will generate questions	Facilitating question-asking, rather than answer-providing	Providing challenging, interesting and relevant questions to solve; allowing time for processing	Asking for a non-predetermined product that solves a relevant and interesting problem.
Explore	Providing materials, time, and problems for open discovery and exploration	Encouraging risk taking and opportunities for mistakes	Noticing obscure information and shifting, dynamic problems	Individualized and varied outcomes or products
Explain	Develop an open forum for student explanation with facilitation of formal content	Reinforcing honesty and ethical behavior; not rushing to judgment; comfort with lack of closure	Guided didactic conversation of formal content	General student understanding of formal content; evolving criteria
Elaborate	Provide a place to expand and apply basic conceptual knowledge	Facilitate approaches that go into deeper detail and deeper thought	Allowing students to elaborate and revise hypotheses	General and refined outcomes or products of formal content
Evaluate	Determine where evaluation data will be collected	Decide level of data (individual or group)	Formative/ Summative or both	Real time data that can be used to reform future lessons.; Quantity and originality of ideas

Person. The teacher must establish a tone of risk-taking within the classroom; a tone that often is difficult to maintain. In an interview with renowned psychologist Robert Sternberg, he noted that university students who took more risks got higher marks for creativity in a drawing contest. But when they took controversial stands in other content areas, the raters often scored them down. He stated that "the raters were saying 'I want you to be creative, and be sure you agree with me'" (Sparks, 2010, p. 6). Classroom management of behavior is a necessary element; a classroom culture that encourages respect without close-mindedness; questioning without anarchy. The teacher can provide modeling for this behavior by keeping the intellectual curiosity on the subject matter and away from the interpersonal.

Process. There are a number of commercially available programs to develop creativity. These began with Osborn's "brainstorming" techniques in the 1940's, (Lehrer, 2012), and evolved to include Gordon's "synectics" (1961) Edward de Bono's "Six Thinking Hats" (1999), von Oech's Roles and "Whacks" (1998), and Cameron's "The Artist's Way" (2002). Despite lack of research in their effectiveness, and even some research that shows that they may inhibit creativity (Lehrer, 2012), these programs have been amazingly popular. As the teacher prepares to develop creativity, it would be worthwhile to examine some of these programs and think about how to incorporate brainstorming coupled with individual time for pondering; to think about new combinations of materials while maintaining integrity of previous scientific knowledge.

Product. A teacher's effectiveness hinges on an understanding of both content and the learning process. In order to be highly effective a teacher needs to have a rich, coherent conceptual map of their discipline; an understanding of why a subject is important; and an understanding of how to communicate knowledge of that subject to others (Darling-Hammond & Baratz-Snowden, 2005). It is not enough for a teacher to know the content. An effective teacher can draw relevant connections and provide real world examples within their subject area. A teacher of creativity can recognize and appreciate creative responses from a solid understanding of what constitutes originality within a subject.

In planning the experience, it is important that teacher prepare for, understand, and appreciate both the field in which they are teaching and the creative process. Prior research has explored whether teachers' knowledge and ability are associated with student learning in the classroom. In short, major studies have found that students learn more from mathematics teachers who majored (earned a four-year degree) in mathematics than from teachers who did not (Goldhaber & Brewer, 1997). Similarly, students learned at higher levels from mathematics and science teachers (with a major) who studied teaching methods in the subject they teach than from those who did not (Monk, 1994). In summation, in the established experience, it is important that teachers have a strong conceptual understanding of what is to be taught and be prepared to provide concrete real world examples while allowing students to become engaged with the content themselves, rather than merely providing information.

Engage

The "Engagement" task accesses the learners' prior knowledge and helps them become engaged in a new concept through the use of short activities that promote curiosity and elicit prior knowledge. The activity should make connections between past and present learning experiences, expose prior conceptions, and organize students' thinking toward the learning outcomes of current activities. (Bybee et al., 2006). The goals of the engagement phase are to invite learner's consideration, encourage interest and spur them to unearth their prior experiences with the formal concepts to be studied (Tanner, 2010).

Place. The first opportunity a teacher has with a student is to provide him with a place that invites questions. The teacher has to provide materials and questions that engage a student. Having a variety of complex experiences with which to engage, allows students to develop more cognitive flexibility (Ritter et al., 2012). By providing an atmosphere of challenge and encouraging students to be curious, adventurous and take risks, teachers can provide opportunities for increased creativity (Sparks, 2011).

Person. Students who are engaged consider their work to be more play than effort. In order to engage, they must see the activity as something that involves a degree of risk-taking and related to their interests (Sparks, 2011). Interestingly enough, a recent longitudinal study found that it is often students from lower socio-economic, disadvantaged background who are most willing to engage in scholarly "play" and risk-taking necessary for creativity, especially if they perceive that the purpose of the work is to enrich and assist their local communities (Heath, 2011). If the teacher establishes the classroom in such a manner that questions are asked, student questions are encouraged, and demonstrations of creativity are not only welcome but desired, schools today become training grounds for laboratories of tomorrow (Deo, Wei, & Daunert, 2012).

Process. The process of engagement comes about from the complexity of the problem that is presented to the student. In order to more fully engage the student's creative processes, the teacher must present unusual and unexpected events, or "schema violations" (p. 962) that require involvement in order to solve (Ritter et al., 2012). However, it is important to note that in order to be creative in a content area, prior knowledge contributes significantly to the degree of creativity possible (Kyung-Nam, Moon, & French, 2011). It is difficult for students to be creative in science, if they do not have a good general knowledge of science. Creativity must build from knowledge, but it is knowledge for a purpose. In other words, teachers cannot justify rote learning as preparation for creativity. Students must perceive a need for knowledge in order to solve complex problems. They then seek the knowledge in order for the second step of creative problem solving. It is the problem that drives the knowledge and the resultant creativity in a two-step process.

Product. There is a fairly agreed-upon relationship between creativity and usefulness- with several scholars positing that creativity occurs only if person goes through a process that results in an original and useful product (Gruber, 2012; Mayer, 2011). The initial engagement for students has to have a purpose, an end in mind. It is this mindful "solution" that has to be at the forefront of student thought in order to develop creativity. Creativity is inspired by a "usefulness" rather than mere cognitive thought- there is a desire to come up with a conclusion that serves a purpose (Gruber, 2012).

Explore

Exploration experiences provide students with a common base of activities within which current concepts (particularly misconceptions), processes, and skills are identified and conceptual change is facilitated. Learners may complete lab activities that help them use prior knowledge to generate new ideas, explore questions and possibilities, and design and conduct a preliminary investigation. (Bybee et al., 2006). Typically, teachers will devise activities in which students work alone or in groups to develop an understanding of the content, process or phenomenon. Students often encounter confusion, conflicting ideas and unanswered questions during their exploration (Tanner, 2010). This is why it is critical to foster an environment that does not punish "mistakes" and use the variations that students develop as teachable moments to better understand the concepts being taught.

Place. During the process of creative exploration, it is necessary to have relevant, and possibly irrelevant information and materials handy so that students can test the limits of their explorations. In this process, described by Eisner (2004) as the thinking within and through the limits of the material, students have access to resources that allow extensions of thought and ideas. Similarly, students should be provided interesting and hypothetical connections between and within a field to solve. It is during the exploration phase that essential knowledge can be sought to solve the original problem, but only if students know that such knowledge exists and have the skills to manipulate it.

Person. A person who is willing to explore a scientific concept is one who is willing to take risks and to play (Sparks, 2011). However, the teacher can encourage the element of persistence that is necessary to exploring a topic, or what Duckworth et al. (2007) call "grit". With persistence and task commitment, students move beyond engagement to an exploration of inter-connected topics. During the process of exploration, care should be taken that students do not rush to early judgments and decisions, but are encouraged to take the time to fully explore the questions. Early foreclosure of understanding will limit the depth of creativity.

Process. In the development of scientific creativity, the teacher cannot establish a goal of a single "static" answer but an evolving problem that continues to shift and to change. This "dynamic" nature of the exploration, in which a student explores solutions, and shifts their understanding as the situation shifts and changes is critical to developing creativity (Gruber, 2012). Teachers can train students to notice obscure information that can shift a problem's solution and create innovative approaches. In a study of problem-solving, students trained to notice and look for obscure information were able to solve 67% more problems than an untrained group (McCaffrey, 2012).

Product. In today's educational climate of single–construct educational standards, the development of multiple responses to a single set of problems or questions is one that can be problematic. However, it is necessary that while basic information can be used to solve problems, there must be a comfort with ambiguous, numerous, and evolving solutions. Defining a product as complete is not part of the Exploration stage- testing and proving is.

Explain

The explanation phase focuses students' attention on a particular aspect of their engagement and exploration experiences and provides opportunities to demonstrate their conceptual understanding, process skills, or behaviors. This phase also provides opportunities for teachers to directly introduce a concept, process, or skill. Learners explain their understanding of the concept. An explanation from the teacher or the curriculum may guide them toward a deeper understanding, which is a critical part of this phase. (Bybee et al., 2006).

Place. Optimally, the explanation phase involves active participation by both teacher and student (Tanner, 2010). All too often, this phase is dominated by the teacher and the lesson becomes one-sided and lecture-based. In a classroom that seeks to develop creativity, the discussion follows an intellectual coaching model, in which the teacher guides student thinking, but allows the students to do the explanation for themselves. There is less reliance on power point slides and static information and more dependence on student inquiry and student-generated need for information.

Person. During the explanation phase of instruction, the issues of scientific vocabulary and knowledge become problematic. Without access to the language of science, students will have difficulty explaining what it is that they are questioning. Great care, however, should be taken to ensure that scientific vocabulary is not taught out of isolation, but within the setting of the questions that are posited. It is also perhaps noteworthy to emphasize the importance of honest results. It has been found that creative people are more likely to cheat and to justify their unethical behavior (Gino & Ariely, 2012). In the pursuit of scientific creativity, the issues of ethics and reliable information are ones that must be dealt with on a personal and individual level.

Process. A recommended approach during this phase is that of Holmberg's guided didactic conversation. Essentially, there should be a constant interaction ('conversation') between the teacher and students that are stimulated through the students' interaction during the explore phase that is inherently linked to the formal content. There are five of the six basic characteristics of true, guided didactic conversation that Holmberg (1983) outlines that are pertinent:

1. Easily accessible (readability and complexity) presentations of content
2. Explicit advice and suggestions to the student as to what to do and what to avoid
3. Invitations to an exchange of views, to questions
4. Involve the student emotionally to take a personal interest in the subject
5. Personal style including the use of the personal and possessive pronouns.

Product. As students seek to explain their results, they are constructing and testing multiple ideas and products with an evolving set of criteria. Students should be encouraged to identify and define the criteria for their products, and encouraged to change these definitions as the problem changes. As conclusions do or do not fit the established criteria, students must be encouraged to reject their initial ideas or products and continue working.

Elaborate

Teachers challenge and extend students' conceptual understanding and skills. Through new experiences, the students develop deeper and broader understanding, more information, and adequate skills. Students apply their understanding of the concept by conducting additional activities. (Bybee et al., 2006). During the elaboration phase, teachers should explicitly guide students in the application of presented content. In essence, the elaboration phase should let students try out their new understandings established in the explanation phases (Tanner, 2010).

Place. As in so much of creativity development, there is time needed for students to deepen their understanding and to allow the creative process to occur. Restricted time is one of the greatest limitations to developing creativity (Sternberg & Kaufman, 2010). However, in today's classrooms where there is a given scope and sequence of content, time is a luxury. In schools such as Thomas Jefferson High School in Arlington, Virginia, time has been "bought" by combining courses in a focused effort. For example, they have a course called "IBET"- or Integrated Biology, English and Technology, where students are encouraged to seek technological solutions to biological challenges and write up the results. Such combining of creativity with other subjects allows students and teachers to experience biology in a richer, more holistic manner.

Person. Perhaps one the strongest ties can be drawn between "failure" and creativity in that creativity is often spurred by a perceived failure, an awareness of what other choices could be made, and a willingness to try again (Rodgers, 2012). There must be a genuine relationship between teachers and students in order for a student to feel free to elaborate on their responses and to feel free to make mistakes. Mistakes must be seen as necessary steps towards a deeper solution and that by examining mistakes, students can learn. This growth-set mental orientation (Dweck, 2006), is one that encourages students to perceive learning not as a set outcome, but a dynamic and reiterative process.

Process. There are many methods to facilitate elaboration of content, which can consist of cooperative learning or discussion, lab or activity extensions or even deeper discussion. It is in the scientific processes of organization, dis-organization, and re-organization, that creativity occurs (Barker, 2012).

Product. The outcome of the problem or question should be explored well enough to flesh out the details and to make connections to other learning and information. The criteria for the product should explore these inter-connections and students need to more completely describe how their product meets the solution or criteria.

Evaluate

It is in the evaluation stage that all of the previous stages are given value and measure. The evaluation phase encourages students to assess their understanding and abilities and provides opportunities for teachers to evaluate student progress toward achieving the educational objectives. (Bybee et al., 2006). The evaluation phase is one that most teaches are probably familiar with but complete in many different manners. Basically, the added "Experience" planning phase allows an instructor to define the level and depth of the evaluation and the types of assessment data that are required. In its simplest form, any evaluation data (formative or summative) should be used to drive instructional decisions and lesson reforms.

Place. Creativity is unique among psychological traits in that it is dependent upon dual evaluation of the person and an outside audience (Sternberg & Kaufman, 2010); both the creator and the audience have to decide that something is "creative" for it to fall under that construct. When students perceive that their attempts at creativity are going to be noticed by a teacher, they are more likely to be creative (Randel, Jaussi, & Wu, 2011). The evaluation in the classroom should be constructed so that both teachers and students are seeking creative results as an outcome, and creative output should be valued by class and teachers alike.

Person. It is important for teachers and students to place a student's creativity in the context of instruction. According to Ellen Winner, a respected researcher, there is a difference between "revolutionary creativity" in which a new style of insight is developed, and more general creativity, such as adding on to existing work. "It's not at all clear to me that this [revolutionary] kind of creativity can be cultivated, though perhaps it can be asphyxiated," she said (Sparks, 2011, p. 8). It cannot be expected, although it can be encouraged, to develop creativity that produces results original to the child, and not necessarily original to the field.

Process. The issue of grading is a significant one. When people know that they are being evaluated, such as for a grade or by a judge, they demonstrate less creativity (Collins & Amabile, 2010). However, when students are in competition with each other, creativity- and resultant stress- tends to increase (Eisenberg & Thompson, 2011). To determine a process of evaluation, both formative evaluations of the process of creating, and a summative evaluation of the usefulness of a product should be considered.

Product. Perhaps the hallmark of creative thought is divergent thinking (Sternberg & Kaufman, 2010). Certainly E. Paul Torrance (1981) in his classic measure of creativity provides guidelines for evaluating the creativity of products, including:

- Flexibility
- Originality
- Fluency
- Elaboration

These measures extend the concept of evaluation beyond that of mastery learning to one of increasing ideas, extending concepts and explaining interconnections. While the nature of the content should drive the evaluation of the usefulness of the product in terms of scientific merit, the creativity of a product should feature in its evaluation as well.

There is an assumption among working scientists that younger scholars have it easier these days than scientists of earlier times. Because of the emphasis on STEM programs in schools and the concern for developing creativity, the context of education has provided a nurturing environment for young scientists to grow. Needed is "an environment to foster creativity and facilitate research performance" (Deo, Wei, & Daunert, 2012, p. 2065). Yet, the mechanisms for such development are not clear, nor have there been extensive models of instruction developed that combine creativity with content. This chapter develops a model of instruction that directly promotes creativity within the content field of science through the integration of the 4Ps of creativity and the 5Es of inquiry-based science education.

SAMPLE COMPARATIVE LESSON PLANS

In Practice, a 5E lesson is quite similar to most traditional approaches to effective teaching and learning. In essence, two steps within the instructional process are switched, the change recognized in the 5E model allows for student discovery and group connection to the content. Table 4 outlines the two distinct models and basic teacher tasks for each. Following is a sample lesson plan comparing the two models for this process, along with questions and evaluation components that incorporate creativity with introductory science content.

Table 4. 5E versus traditional approaches to teaching and learning

Traditional		*5E*	
Step	*Action*	*Step*	*Action*
Warm-Up	Preview Content/ Standards	Engage	Connect to Content/ Standard
Direct Instruction	Lecture	Explore	Facilitate Student Centered Activities
Indirect Instruction	Facilitate Student Centered Activities	Explain	Lecture, Q & A, Didactic Conversation
Connection/Extension	Real World Connections	Elaborate	Real World Connections
Evaluate	Summative/Formative Evidence	Evaluate	Summative/Formative Evidence

Plainly, in the 5E model the process of direct and indirect instruction are switched compared to most traditional models of teaching and learning. Varied nuances exist within the inquiry-based approach associated with the 5E model. Nonetheless, in its simplest form the 5E allows for individual student and whole class "experience" of the phenomenon or content under investigation. Specifically, by allowing "Exploration" prior to direct instruction allows for common explorations and actions that all students participate within. This mitigates the problem that teachers often face of guessing student prior knowledge of a subject. A generalized example could center on the use of a roller coaster (conceptually or physically) to explain kinetic and potential energy. Traditionally, a teacher may start the class during the "warm-up" phase by asking: "Who in class has ever taken a ride on a roller coaster?" In most diverse classroom settings this preface is a crap shoot. A teacher may get several hands of students that have had this experience and focus on them solely to help explain concepts. The problem with this is that the other students are left out of the dialogue and still fail to connect to the concepts about to be taught. In contrast, in a

5E lesson the teacher's job at the start of the class is to "hook" the class, so they may show a clip of an extreme roller coaster or retell a story of riding a roller coaster and have students imagine themselves within that story. Again, in traditional instruction a teacher may begin the next phase of class by lecturing about the concepts of kinetic and potential energy and how they relate back roller coasters. Again, the students that have not had this experience still have nothing to relate to and the content presented is abstract. The most powerful component of the 5E model lies within the "explore" phase which can serve as the "great equalizer" since it puts students on a somewhat common footing with relation to the concepts about to be taught. In the above example, the 5E teacher in the next phase of class would allow students to manipulate a model coaster on a track and observe actions during incline, rest and decline. A technology based lesson, could have the students create an animation of a coaster and have the students again observe the same actions. Lastly, in a resource poor classroom, teachers could have students physically replicate the motion of a coaster by forming a line and going uphill, resting on top and going down the hill. Nonetheless, what hold true in all the examples is whole class participation to preview concepts that are going to be covered. In the next phase of the 5E, when the teacher wants to provide definitions and concepts in direct instruction he or she can cite examples from the activity that everyone just participated within. The 5E is in no way a silver bullet for instructional issues, but its simplicity and power to provide a common experience is an element that any teacher would find beneficial to reach a greater spectrum of students.

Below is a step by step basic lesson of creating simple circuits with a battery, insulated wire and a mini light bulb. The first description is of the traditional model and approach to teaching and learning that is followed by the 5E (+1)model. Differences are then highlighted and at the conclusion of this section the 4P process and its integration are explained.

Step 1

Traditional: Warm-up. Teachers typically preview the content standards associated with circuits and electricity. In addition, questions may be asked about prior knowledge or reviewed from previous classes. The structure of the day's activities is previewed.

5E: Experience and engage. The teacher has to have been exposed to the concept of "playing" with circuits. In addition, to play the experience, the teacher has to think of the materials that students might want to work with, such as different forms of energy and different conductors. One enterprising teacher brought in dog poop for the methane as a source of energy in addition to a battery for students to experiment with. Teachers could also model the concept of circuits by rapidly turning on and off the lights and then asking students if they know how that process works. Students may illustrate their preconceptions of how a light turns on and off. In an obligatory sense, content standards associated with circuits and electricity is previewed without being rote.

Comparison. All too often, traditional K-12 classrooms are started with a standardized procedure of a warm up that focuses on examining the content standard to be previewed and many times homework is reviewed or questions are presented on content that has not been deeply explored. In contrast, the engage phase is suppose to spark interest, by breaking from the norm and flashing the lights and having students provide preconceptions without fear of being wrong such that students can fully engage within the classroom processes about to be undertaken.

Step 2

Traditional: Direct instruction. Transitioning from the warm-up phase, often in traditional instruction a teacher will move to a mode of direct instruction that involves a mixture of lecture and note taking and some question and answering. The motive behind this practice is to build a base of knowledge through verbal communication of concepts that can later be drawn upon. In essence, students are supposed to build a library of factual knowledge that they can later apply in discussion and extensions.

5E: Explore. The explore phase takes a leap of faith by the teacher to allow students to manipulate and investigate the phenomenon. In this example, the students would be provided the simple materials, given the task to light the bulb and draw how they did it, if and when successful. This will set the stage for the next phase by providing a common basis of student experience. The students are encouraged to explore the relative materials in order to pursue a common goal- lighting the bulb.

Comparison. The goal of both processes is to develop a foundation of student knowledge. The difference is that the traditional approach aims to build a basis of factual knowledge first whereas the 5E approach is more of an experiential foundation. Going forward, teachers within both models would attempt to draw upon either facts (traditional) or experiences (5E) in applications and elaborations in the content.

Step 3

Traditional: Indirect instruction. Progressing from the elements of direct instruction, teachers often establish time for students to practice or extend upon the content that was projected in the lecture. This can take many forms based on the content. Examples include, cooperative and collaborative learning and discussion, exploration and or investigation. Using the example above, this is where students would build circuits that are most likely a part of a cookie-cutter lab activity. The purpose of this technique is to allow students to apply or extend their knowledge formed from the prior stage.

5E: Explain. Transitioning from the explore phase, teachers utilizing a 5E model often move to a more traditional mode of direct instruction. The caveat within this phase is that the teacher will utilize a more didactic mode of conversation that draws upon the experiences of the students while tying them to the standards and concepts related to the content. The talking head mode of direct instruction is avoided while students are allowed to provide examples that the teacher can use to explain the traditional concrete concepts of electricity and circuits.

Comparison. The main difference between these phases lies within how they were preceded within the lesson. The opportunity for discovery and exploration is lacking in the traditional model. Students were exposed to what constitutes a circuit and how a light bulb works, this may make the student centered learning portion more of a rote exercise of practice rather a time of creativity, discovery and exploration. In addition, the direct instruction model becomes more robust with the implementation of student perspectives and the ability to draw upon preconceptions. Engagement in this process is increased compared to the traditional talking head mode of direct instruction.

Step 4

Traditional: Connections & extensions. In a traditional model of instruction teachers typically want to establish time to review and make/draw extensions to the student centered indirect instruction. In the example above, a teacher would typically review the lab worksheet and try to review the multiple ways that were discovered to light the bulb but reiterate the key elements. The teacher may then preview complex circuits or relate it to house or building wiring systems.

5E: Elaborate. The elaborate phase is similar to the traditional approach of connections and extensions. Simply, teachers want to extend knowledge and preview and connect related content. Based on student understanding teachers may be able to allow students to explore creating more complex electrical circuits. But at minimum, teachers utilize this time to gauge student understanding and establish critical understanding of concepts.

Comparison. The key differences among this phase may depend on how well students understand concepts based on prior steps. The goal or outcomes of this phase is the same, but the depth of understanding may differ, based mainly on the personal connection that could have been established within the 5E model. In any case, both models need to complete a cyclical nature of reviewing the lessons objectives and outcomes.

Step 5

Evaluate. Any effective instruction will incorporate both formative and summative assessment within lessons and units. In the above example, the traditional model may have students create a summative statement about circuits as "ticket out the door" to end the class. The teacher can later evaluate these statements for accuracy and misconceptions. In the 5E model, the teacher may have the students revisit their preconception and expand upon their initial thoughts. Fluency of ideas can be measured, and originality determined. Teachers can evaluate this work for growth and understanding. Both models could utilize a lab element that grades student participation in creating and documenting successful and unsuccessful models of circuits.

Connection with 4ps

Throughout this lesson, the creative person, place, process and products were supported by teacher actions, using the 5E (+1) model as the structure. In a more traditional lesson, the student is led inexorably to the final conclusion, with little to no input from the student. There is a "wrong" and a "right" set of knowledge and skill development sequence. In the development of creativity within the science classroom, students are encouraged to actively participate within the process, developing the knowledge and skills as a result of the need to solve the problem, rather than the problem supporting the acquisition of limited knowledge and skills.

CONNECTING SCIENCE WITH CREATIVITY

In 1957, spurred by the launch of Sputnik, United States schools launched an effort to recruit the best and the brightest American minds to form a new generation of leaders and innovators in science and engineering. It was an effort that ended too quickly. By 1983, the Nation at Risk report noted that the ideal of academic excellence as the primary goal of schooling had faded across the board in American education. The next 25 years has not changed the essential nature of schools.

Recent reports warn that our world cannot progress with a work force that has mastered only minimum competencies. Reiterating a nation's interest in developing creative youth, Florida (2005) notes that it is the creative graduate who is the most highly sought-after commodity and valuable resource pursued by global economies. The National Science Board recommends that today's educational programs 1) Provide opportunities for excellence, 2) cast a wider net, and 3) foster a supportive ecosystem (NSB, 2010). "Tough Choices or Tough Times", a report from the

National Center on Education and the Economy (2006) noted that students of the future "will have to be:

- comfortable with ideas and abstractions,
- good at both analysis and synthesis,
- creative and innovative,
- self-disciplined and well organized,
- able to learn very quickly, and
- work well as a member of a team, and
- have the flexibility to adapt quickly to frequent changes in the labor market as the shifts in the economy become ever faster and more dramatic" (p. 8).

Today's world problems can only be solved using new strategies. Yet, our school systems are often so focused on bringing all students to a single point of instruction or developing a single set of skills, that the process of learning has been reduced to a single method of instruction, contained within a teacher's manual. Recent teacher surveys indicate that 65% of teachers have never received any information or training about how to develop creativity (Farkas & Duffet, 2008). The P's and E's model integrates two distinctly unique processes of instruction: science content instruction with the development of creativity. It is our hope and goal that teachers use this model to design a variety of science lessons and units that will allow students to explore, create, and ultimately, learn how to find and solve problems- all with a spirit of deep engagement and appreciation for the joy and wonder of exploring.

REFERENCES

Abell, S. K., & Volkmann, M. J. (2006). *Seamless assessment in science*. Portsmouth, NH: Heinemann.

Barker, T. S. (2012). Becoming organ-ized: The creativity of organization, dis-organization, and re-organization in scientific and artistic experiments. *Leonardo, 45*(3), 263–268.

Bybee, R. W. (1997). *Achieving scientific literacy: From purposes to practices*. Portsmouth, NH: Heinemann.

Bybee, R. W., Taylor, J. A., Gardner, A., Van Scotter, P., Powell, J. C., Westbrook, A., & Landes, N. (2006). *The BSCS 5E instructional model: Origins, effectiveness, and applications*. Colorado Springs, CO: BSCS.

Cameron, J. (2002). *The artist's way*. New York, NY: Tarcher.

Collins, M. A., & Amabile, T. M. (2011). Motivation and creativity. In R. J. Sternberg (Ed.), *Handbook of creativity* (pp. 297–312). New York, NY: Cambridge University Press.

Csikszentmihalyi, M. (1999). Implications of a systems perspective for the study of creativity. In R. Sternberg (Ed.), *Handbook of creativity* (pp. 313–335). Cambridge, NY: Cambridge University Press.

Darling Hammond, L., & Baratz-Snowden, J. (2005). *A good teacher in every classroom: Preparing the highly qualified teachers our children deserve*. San Francisco, CA: Jossey-Bass.

deBono, E. (1999). *Six thinking hats*. New York, NY: MICA Management Resources, Inc.

Deo, S., Wei, Y., & Daunert, S. (2012). Probing a myth: Does the younger generation of scientists have it easier? *Analytical and Bioanalytical Chemistry, 403*(8), 2065–2067.

Duckworth, A. L., Peterson, C., Matthews, M. D., & Kelly, D. R. (2007). Grit: Perseverance and passion for long-term goals. *Journal of Personality and Social Psychology, 92*(6), 1087–1101.

Dweck, C. (2006). *Mindset: The new psychology of success*. New York, NY: Random House.

Egan, K. (2005). *An imaginative approach to teaching*. San Francisco, CA: Jossey-Bass.

Eisenberg, J., & Thompson, W. F. (2011). The effects of competition on improvisers' motivation, stress and creative performance. *Creativity Research Journal, 23*(2), 129–136.
Eisner, E. (2004). *Arts and the creation of the mind.* New Haven, CT: Yale University Press.
Farkas, S., & Duffet, A. (2008). Results from a national teacher survey. In Thomas B. Fordham Institute (Ed.), *High achievement students in the era of NCLB* (pp. 50–82). Washington, DC: Author.
Florida, R. (2005). *Flight of the creative class: The new global competition for talent.* New York, NY: HarperBusiness.
Gino, F., & Ariely, D. (2012). The dark side of creativity: Original thinkers can be more dishonest. *Journal of Personality and Social Psychology, 102*(3), 445–459.
Goldhaber, D. D., & Brewer, D. J. (1997). Evaluating the effect of teacher degree level on educational performance. In W. J. Fowler (Ed.), *Developments in school finance* (pp. 197–210). Washington, DC: National Center for Education Statistics, U.S. Department of Education.
Gordon, W. J. J. (1961). *Synectics: The development of creative capacity.* New York, NY: Harper and Brothers.
Gruber, C. (2012). What is the most creative thing you have ever seen? "Useful" is the new "valuable" and dynamic systems are the key! *Integrative Psychological and Behavioral Science, 46*(2), 243–248.
Hanuscin, D. L., & Lee, M. H. (2008). Using the learning cycle as a model for teaching the learning cycle to pre-service elementary teachers. *Journal of Elementary Science Education, 20*(2), 51–66.
Heath, S. B. (2011). *Risk-taking and creativity.* Presentation at the National Academic of Education annual conference, October 29. Washington, DC.
Holmberg, B. (1983). Guided didactic conversation in distance education. In D. Sewart, D. Keegan, & B. Holmberg (Eds.), *Distance education: International perspectives* (pp. 114–122). London, England: Croom Helm.
Hughes-Lynch, C. E. (2011). *Teaching children with high-functioning autism.* Austin, TX: Prufrock Press.
Karplus, R., & Thier, H. D. (1967). *A new look at elementary school science.* Chicago, IL: Rand McNally.
Kim, K. H., Cramond, B., & VanTassel-Baska, J. (2010). The relationship between creativity and intelligence. In J. C. Kaufman & R. J. Sternberg (Eds.), *Cambridge handbook on creativity* (pp. 395–412). New York, NY: Cambridge University Press.
Kozbelt, A., & Beghetto, R. A. (2010) *The Cambridge handbook of creativity.* Cambridge, England: Cambridge University Press.
Kyung-nam, J., Moon, S. M., & French, B. (2011). Differential effects of divergent thinking domain knowledge and interest on creative performance in art and math. *Creativity Research, 23*(2), 61–70.
Lehrer, J. (2012, January 30). Groupthink: The brainstorming myth. *The New Yorker.* Retrieved from http://www.newyorker.com/reporting/2012/01/30/120130fa_fact_lehrer
Lubart, T. I. (2011). Creativity across cultures. In R. J. Sternberg (Ed.), *Handbook of creativity* (pp. 339–350). New York, NY: Cambridge University Press.
Lubinski, D., & Benbow, C. P. (2006). Study of mathematically precocious youth after 35 years: Uncovering antecedents for the development of math-science expertise. *Perspectives on Psychological Science, 1*, 316–345.
McCaffrey, T. (2012). Innovation relies on the obscure: A key to overcoming the classic problem of functional fixedness. *Psychological Science, 23*(3), 215–218.
Mayer, R. E. (2010). Fifty years of creativity research. In R. L. Sternberg (Ed.), *Handbook of creativity* (pp. 449–460). New York, NY: Cambridge, University Press.
Monk, D. H. (1994). Subject area preparation of secondary mathematics and science teachers and student achievement. *Economics of Education Review, 13*(2), 125–145.
Mumford, M. D., Antes, A. L., Caughron, J. J., Connelly, S., & Beller, C. (2010). Cross-field differences in creative problem-solving skills: A comparison is health, biological and social sciences. *Creativity Research Journal, 22*(1), 14–26.
National Center on Education and the Economy. (2006). *Tough choices or tough times: The new commission on the skills of the American workforce.* Washington, DC: NCEE.
National Science Board. (2010). *Preparing the next generation of STEM innovators: Identifying and developing our nation's human capital.* Retrieved from http://www.nagc.org/uploadedFiles/Information_and_Resources/Hot_Topics/NSB%20-%20Stem%20innovators.pdf

Paletz, S. B. F., Peng, K., & Li, S. (2011). In the world or in the head: External and internal implicit theories of creativity. *Creativity Research Journal, 23*(2), 83–98.
Randel, A. E., Jaussi, K. S., & Wu, A. (2011). When does being creative lead to being rated as creative? The moderating role of perceived probability of successfully bringing ideas to a supervisor's attention. *Creativity Research Journal, 23*(1), 1–8.
Ritter, S. M., Damian, R. I., Simonton, D. K., van Baaren, R. B., Strick, M., Derks, J., & Dijksterhuis, A. (2012). Diversifying experiences enhance cognitive flexibility. *Journal of Experimental Social Psychology, 48*(4), 961–964.
Rodgers, T. (2012). How art and research inform one another, or choose your own adventure. *Canadian Journal of Communication, 37*(1), 155–161.
Sparks, S. (2011). Science looks at how to inspire creativity. *Education Week, 31*(14), 1–16.
Starko, A. J. (2005). *Creativity in the classroom* (3rd ed.). Mahwah, NJ: Lawrence Erlbaum.
Sternberg, R. J., & Kaufman, J. C. (2010). Constraints of creativity. In J. C. Kaufman & R. J. Sternberg (Eds.), *Cambridge handbook of creativity* (pp. 467–481). New York, NY: Cambridge University Press.
Tanner, K. D. (2010) Order matters: Using the 5E model to align teaching with how people learn. *CBE Life Science Education, 9*, 159–164.
Torrance, E. P. (1981). Empirical validation of criterion-referenced indicators of creative ability through a longitudinal study. *Creative Child and Adult Quarterly, 6*, 136–140.
Von Oech, R. (1998). *A whack on the side of the head: How you can be more creative*. New York, NY: Warner Books Inc.

Claire E. Hughes
Department of Education and Teacher Preparation
College of Coastal Georgia
USA

Timothy A. Goodale
College of Education
North Carolina State University
USA

LINA SOARES

8. SCIENCING

Creative, Scientific Learning in the Constructivist Classroom

INTRODUCTION

Scientific thinking and discovery is not restricted to scientists. Children are natural born scientists. From the time children are born, they inquisitively touch their toes, shake and rattle their toys, and throw unfamiliar objects as they seek to understand their world. Zeece (1999) offered that actions such as these suggest that young children engage in scientific thinking long before they begin the formal study of science in school. Unfortunately for many children when they begin their formal science education, they are introduced to the study of science as a mastery of scientific facts. For gifted children, this method of rote memorization in science education stifles their need for creativity and can be disastrous (Mackin, Macaroglu, & Russell, 1996).

Intellectually gifted students acquire knowledge at a rapid pace. They perform skills at exceedingly high levels of achievement in comparison to students of their age. Further, the intellectually gifted have exceptional thinking abilities that can be easily applied to scientific concepts. However, in an effort to push the intellectually gifted to reach their optimal levels of academic performance, teachers often stifle gifted students' creative talents by an overemphasis on content (Reis & McCoach, 2000). The problem in science education is that many teachers do not view science as a set of methods that can be applied to discover new ideas. Consequently, this narrow view of science as content knowledge is particularly problematic for talented students who have the capacity and curiosity to engage in scientific discovery.

Sciencing is a term that means active engagement through hands-on involvement whereby students learn about the scientific world through first-hand investigative experiences. Chaille and Britain (2003) posit that sciencing is both hands-on and minds-on because to do science, students need to be mentally and physically involved as they question, probe, formulate an hypothesis, and experiment as a means to understand their world. Because the intellectually gifted have an insatiable quest for knowledge, sciencing is one method that holds promise for gifted learners to satisfy their intellectual curiosity and engage in scientific discovery. Subsequently, for sciencing to thrive as an effective method for doing science, teachers of the gifted must possess certain dispositions conducive to learning that will allow their students to co-construct knowledge through collaborative, active engagement in the

M. K. Demetrikopoulos & J. L. Pecore (Eds.), Interplay of Creativity and Giftedness in Science, 127–151.

creative processes of scientific meaning-making. Such an approach to learning is grounded in the constructivist theory to learning; an approach that is often viewed as contradictory to the more traditional approach to science education whereby the teacher is directly responsible for transmitting scientific facts to students. In the constructivist classroom, the teacher's role shifts to that of a facilitator (Chaille & Britain, 2003); a shift that has become increasingly accepted in both the scientific and the gifted community when addressing the needs of today's young adolescent students (Duschl, Schweingruber, & Shouse, 2007).

Central to the orientation of this chapter is a view that learning and teaching is an interconnected social practice in middle grade science classrooms. To allow young adolescent intellectually gifted students to use science content in creative ways to produce novel and valuable ideas, middle grade science teachers need to construct student-centered learning environments that will foster inquiry, exploration, and discovery. The consistent theme throughout this chapter emphasizes how gifted middle grade students can be encouraged to think more creatively and apply scientific reasoning in the context of science classrooms through the process of sciencing – the act of doing science. Particular emphasis is given to the constructivist epistemological view that embraces the personal and interpersonal construction of knowledge, while recognizing that scientific knowledge is a product of social processes and learning science requires students to be active participants in this scientific culture.

Setting the Stage

The vision brought to this chapter is a focused discussion that forefronts the careful integration of research, theory, and practice related to the intersection of creativity with giftedness for young adolescent learners in the science classroom. In short, the purpose of this chapter is to enhance the science education of intellectually gifted adolescent students by using creative methods for scientific discovery.

To set the stage, the chapter first examines the sad plight of science education for young gifted adolescents that currently exists in today's middle schools. Current reform initiatives for K-12 public schools have done little to improve the quality of science education for intellectually gifted students.

The chapter then briefly describes the nature of young gifted adolescents in terms of both their intellectual and creative characteristics and their relevance to science education since truly effective science instruction of gifted adolescent learners must take into account these future science leaders.

The next part of the chapter establishes the view that learning and teaching is one interconnected social practice in middle grade science classrooms and that knowledge is constructed as a product of the social interactions. Such a view to learning is the social constructivist approach and is central to the theoretical orientation of this chapter.

The role of the middle grade teacher in a constructivist classroom is examined with particular emphasis given to the special knowledge and skills needed to foster creative scientific thinking for intellectually gifted adolescent learners. The focus then moves to a hands-on and minds-on science approach that middle grade science teachers can implement in their classrooms whereby gifted adolescent students can learn about the scientific world through first-hand investigative experiences; the approach is known as sciencing. With this approach, science is taught as science is done through active exploration and scientific discovery.

To illustrate theory into practice, the chapter offers research-based instructional methods and classroom examples for middle grade science teachers to engage their intellectually gifted students in doing science with creativity through problem-based learning, project-based learning, scenario learning, and service learning projects. Each method featured in this section permits young adolescent, intellectually gifted learners to pursue their intellectual curiosity, engage their creative abilities, and apply scientific thinking in the process of doing science.

In summary, the chapter serves a dual purpose. First, it is a text for pre-service middle grade science teachers who will one day teach intellectually gifted students in science education. Secondly, the chapter offers current middle grade teachers with the knowledge and pedagogy to raise the bar and develop science content using creative processes to meet the unique needs of gifted adolescent learners.

TOUGH TIMES IN GIFTED EDUCATION

It's a tough time to raise, teach or be a highly gifted child. Schools are to extraordinarily intelligent children what zoos are to cheetahs. Every organism has an internal drive to fulfill its biological design. The same is true for unusually bright children. From time to time the bars need be removed, the enclosures broadened. Zoo Chow, easy and cheap as it is, must give way, at least some of the time, to lively, challenging mental prey. – Stephanie Tolan (1996, pp. 6–7)

The metaphorical comparison of gifted children to cheetahs should serve as a reminder that to constrain intellectually gifted students is antithetical to the concept of giftedness. Unfortunately, many intellectually gifted students are ignored in today's schools. Across the educational landscape, the role of gifted education is at a difficult crossroads. Since the passage of the No Child Left Behind Act (NCLB) (2001), the field is criticized for grouping practices seen as counter to the current interest in inclusion (Van Tassel-Baska & Stambaugh, 2005). Furthermore, gifted education is considered irrelevant by some critics because reform initiatives promote critical thinking, interdisciplinary curriculum, and project work for all students (Van Tassel-Baska, 1998). Today, many regular classroom teachers are faced with the dilemma of how to meet the needs of their diverse learners. Given time constraints and working in an era of high-stakes accountability, it has been reported that many classroom teachers now focus their attention on low-performing students (Farkas & Duffett, 2008). As a result, many students who are identified as gifted and talented are not challenged in classrooms today, and schools that do offer specific program services for high ability students are often organized in ways that fail to translate into talent development for advanced learners (Van Tassel-Baska, 1998). Davis, Rimm, and Siegle (2011) posit that gifted learners "are silently paying a price" (p. 2) as a result of the current unfavorable circumstances, and the "price is lost academic growth; lost creative potential; and sometimes, lost enthusiasm for educational success, eventual professional achievement, and substantial contributions to society" (p. 2). What must not be overlooked is the fact that young adolescent gifted science students are involved in this turmoil. Consequently, too many intellectually gifted students who are sitting in classrooms today are denied the opportunity to learn advanced science process skills and develop their cognitive and creative talents; a dismal outlook exists for future science leaders.

As concerns continue to mount about the status of today's gifted education programs, the development of the intellectually gifted middle school students' science talent becomes the central concern. To understand this significance, it becomes necessary to take a look at the characteristics of gifted students.

CHARACTERISTICS OF YOUNG GIFTED ADOLESCENT LEARNERS

Creativity is an elusive factor in its relationship to giftedness – Van Tassel-Baska (2004, p. 1).

When one begins to equate the words *gifted*, *creative*, and *scientific* to an individual, perhaps the most obvious historical figure who captures the interplay of gifted, science and creativity is Leonardo da Vinci. In his life's work, his creative expressions and scientific thinking exemplified how the intersection of creativity and science can support one another (Deckert, 2001; Potter, 2006). First and foremost, he was a renowned artist of the Renaissance, but he was also a gifted scientist in many areas of science, such as botany, civil engineering, hydrology and anatomy (Nicholl, 2004). Just as Leonardo da Vinci demonstrated how science and creativity can interact in productive ways during his life's undertakings, gifted adolescent students possess extraordinary abilities to be future da Vincis when given the opportunity to engage in the creative processes of scientific meaning-making.

An examination of gifted characteristics has been addressed by many scholars in the field (Colangelo & Assouline, 2000; Van Tassel-Baska, 2003; Winebrenner, 2000), but one thing for certain is there is no one defining characteristic to address the "typical" gifted student (Gallagher & Gallagher, 1994). However, many of the characteristics of gifted learners have been addressed repeatedly such that a summary is important because this knowledge can help teachers plan innovative instructional approaches for doing science in their classrooms.

Intellectual characteristics. Most intellectually gifted students are classified as precocious; they are advanced learners and can master new material rapidly ahead of other learners. They have the ability to memorize and learn rapidly, maintain a wide information base, and enjoy multiple interests (VanTassel-Baska, 2003). In addition to precocity, gifted students are characterized as intense learners. The intensity is often manifested in both the affective and cognitive domains through heightened emotions and superior reasoning (Clark, 2002). Furthermore, intensity is often demonstrated through the ability to focus and concentrate for long periods of time with complex concepts. Csikszentmihalyi (1990) characterized this ability to exercise high degrees of concentration as a natural "flow" that intellectually gifted students use to tackle new experiences. In addition, high-ability students enjoy working on multiple levels simultaneously, such as problem-solving complex real-world problems that have many parts and perspectives to study (Feldhusen, 1993; Reis, 1990; Renzulli & Reis, 1997). As a result, young gifted adolescents are likely to absorb extraordinary quantities of information, possess a high degree of retentiveness and advanced comprehension, and possess thought processes that move at an accelerated pace (Clark; VanTassel-Baska & Stambaugh, 2005).

Creative characteristics. Creativity in and of itself means many things to different people, but the obvious question to address is to what extent is creativity linked to the concept of giftedness? For purposes of this chapter, creativity will be defined as "a mental process by which an individual creates new ideas or products, or recombines existing ideas and products, in a fashion that is novel to him or her" (Gallagher & Gallagher, 1994, p. 319). Based on this premise, one can presume that the operative words, 'creates new ideas or products,' suggests an association does exist between creativity and intelligence since it has been determined that intellectually gifted adolescent students are inquisitive and have a fascination for discovery and experimentation (Clark, 2002). In fact, Renzulli went as far in 1992 to theorize that gifted students must possess degrees of creative ability to explain why high ability students are given to extraordinary productive capabilities. Renzulli referred to this characteristic that is common to most intellectually gifted students as creative productivity. However, seminal research studies have shown that the relationship between high intelligence and high creative ability are not one and the same (Getzels & Jackson, 1962; Wallach & Kogan, 1965). In other words, teachers cannot expect highly intelligent students to be the most creative students in the classroom.

Today, it is widely accepted that a dynamic interaction among cognitive and personality attributes and certain environmental components are good indicators of creatively gifted students (Davis, Rimm, & Siegle, 2011). Such indicators used to describe the creatively gifted are imaginative, intuitive, prefers the abstract, logical, fascination for discovery, task-committed, and thinks outside the box. In addition, Davis (1999) adds motivation and risk taking to the list of important personality traits that must be taken into account when referring to highly creative students.

Scientific thinkers. To understand how gifted adolescent learners have the capacity for scientific ways of thinking, it is noteworthy to highlight some of the important habits of mind, or the cognitive abilities and affective attitudes (Brandwein, 1986; Elder & Paul, 2007; Gallagher & Gallagher, 1994; Van Tassel-Baska, Gallagher, Bailey, & Sher, 1993) that are critical to the practice of science (see Table 1).

Table 1. The scientific thinker

Habits of Mind
• Know, uses, and interprets scientific explanations of the natural world
• Skeptical; raises vital scientific questions and problems; develops a method for probing
• Thinks open-mindedly; curious; questions theory; uses a self-correcting method
• Generates and assesses scientific evidence
• Makes logical scientific conclusions and communicates

The scientific habits of mind that are shown in Table 1 are a sample of some of the distinguishing characteristics of all scientists. It is interesting to note that the important work of a scientist involves both a set of cognitive abilities and processes that are applied to the exploration of scientific knowledge. What is not listed, but is the impetus that prompts scientists to understand the unknown is motivation (Gallagher & Gallagher, 1994). Scientists are motivated to discover and perhaps Judson (1980) explained it best by coining the fascination to discover the "rage of the unknown."

Like scientists, intellectually gifted adolescent students have a passion to understand their world (Brown & Knowles, 2007). They are curious, intuitive, open minded, and have a keen interest in investigating scientific phenomena. Moreover, they have the ability to direct their own thought processes, to be persistent, and to plan, monitor, question, reflect, and evaluate their own work (Freeman, 2003; Neber & Atkins, 2003). In short, gifted adolescent learners possess all the essential scientific habits of mind to pursue the unknown.

VanTassel-Baska, Gallagher, Bailey, and Sher (1993) offered that young adolescent gifted students can grow in the development of valuable habits of mind found among scientists when given the opportunity to inquire, explore, negotiate, and share ideas. This opportunity is made possible when learning and teaching is viewed as a joint enterprise or an interconnected practice of co-constructing meaning. Such an approach to learning is grounded in the constructivist theory to learning.

SCIENTIFIC LEARNING IN THE CONSTRUCTIVIST CLASSROOM

To support the interplay of science education and creativity for early adolescent gifted learners, the theory of constructivism provides the epistemic engagement for both teachers and students. The constructivist view of learning is that scientific knowledge is individually acquired through social processes and learning science requires students to be active participants (Crotty, 1998), who draw upon prior knowledge and past experiences to engage in new learning (Driver, Asoko, Leach, Mortimer, & Scott, 1994; Piaget, 1952; Schulte, 1996).

In the middle grade science classroom, constructivism means reconstituting classrooms as communities of inquiry (Wenger, 1998). The social interactions that occur during collaborative inquiry play a pivotal role in shaping students' co-construction of meaning because the context provides the problem-solving environment in which learners can draw from prior experiences and then probe, challenge, and work together in the scientific meaning-making process (Savery & Duffy, 1995). This means that middle grade science teachers and their intellectually gifted students pose questions, explore, and evaluate previous ideas by using what they already know or altering existing knowledge to fit with the new understandings before drawing conclusions (Llewellyn, 2002). In other words, students' scientific thinking skills are the tools to learn scientific concepts as students interact with their environment (Hassard, 1992). For teachers who adopt the constructivist theory of learning, it does not mean they dismiss their role as knowledge experts, but rather, they view their role as facilitators who guide their students in the construction of scientific knowledge (Hemlo-Silver & Barrows, (2006). To do so, constructivist teachers use their expert knowledge to ask the guiding questions to propel student growth, and they provide the learning resources and the strategies to problem solve (Schulte, 1996). Further, constructivist teachers establish the inquiry activities so their intellectually gifted students can engage in the creative processes of scientific meaning-making and formatively assess students on an on-going basis. Constructivist teachers understand the important role of collaboration for their intellectually gifted students and the time required to share and reflect. They provide real world issues to drive learning in a nurturing environment as students probe, challenge, and work together to problem solve. By grounding learning activities in an authentic real-world context, constructivism stimulates intellectually gifted students to apply their natural curiosity in creative ways to produce novel and valuable ideas (Trumbull, 1999). This development begins with specialized teachers who understand the value in challenging their young adolescent intellectually gifted students to actively think and negotiate meaning in a socially accepted forum.

MENTORING SCIENTIFIC MINDS AND CREATIVITY

A common myth embedded in many teachers' minds is the belief that gifted students will naturally excel in science education because of their extraordinary intellectual abilities. On the contrary, researchers (Croft, 2003; Renzulli, 1968; Sisk, 1989) have advocated that gifted students as a group are more susceptible to their teachers' attitudes and actions than are other students because they have different cognitive, affective, physical, intuitive, and societal characteristics compared to their peers (Karnes & Bean, 2001). Because of these unique characteristics, Colangelo and Davis (1997) have long advocated that intellectually gifted learners need the tutelage of specialized teachers who know how to stimulate creative and scientific thinking. Thus, the question then becomes, Who are the specialized science teachers?

The Role of the Middle Grade Science Teacher

The role of middle grade science teachers of intellectually gifted students is critical due to today's concerns about the quality of science education for these special students. Research has shown that science teachers must have strong knowledge of subject matter content, pedagogical content, and knowledge of gifted students (Clark, 1997; Feldhausen, 1997; Gallagher, 2000; Van Tassel Baska, 1998). While the teacher's knowledge is paramount to raise the quality and the sophistication of science education for intellectually gifted adolescent learners, middle grade science teachers of intellectually gifted adolescents must also possess certain dispositions conducive to learning that will allow their students to co-construct knowledge through collaborative, active engagement in the creative processes of scientific meaning-making. Additionally, their unique role requires them to design effective pedagogical strategies to implement challenging science content and establish student-centered learning environments that will foster inquiry, exploration, and discovery. When middle grade science teachers adopt appropriate dispositions and create nurturing environments, they demonstrate their actions are focused on the needs of their intellectually gifted adolescents.

Middle grade science teachers of intellectually gifted adolescent students should emphasize concepts, principles, relationships, and generalizations, not isolated facts (Van Tassel-Baska, 1989). They should also encourage the study of real world problems and interdisciplinary themes (Johnson, Boyce, & Van Tassel-Baska, 1995). Middle grade science teachers need to ask challenging "How," "Why," and "What if" questions to engage their students to make connections to past knowledge, to make their students think and hypothesize, and to enable their students to apply and transfer knowledge to other disciplines. At the same time they need to pose questions such as, What might have caused this? Why do you believe this occurred? and How can we find out? to stimulate creative and critical thinking. Divergent questions such as these provide a multitude of student responses (Lawson, 1995), and the middle grade science teacher's role is to guide students to explore and analyze the problem, collect and interpret evidence, and then draw appropriate conclusions based on the evidence. In addition, middle grade science teachers who teach young adolescent gifted students in science should provide multiple opportunities for students to use technology, not only for research purposes, but to connect students in classrooms around the world to form a global science community (Van Tassel-Baska, 1998). In short, middle grade teachers of intellectually gifted students need to view science education as a set of methods that can be applied to discover new ideas through investigative processes. Investigative processes in this context means teaching science as science is done – through inquiry that demands mental and physical activity.

When outstanding middle grade science teachers view science as inquiry and their intellectually gifted students as constructive learners, they understand the efficacy of the student-centered nature of inquiry instruction as a mode to differentiate instruction. Additionally, they welcome their students' ideas and encourage their students to examine their suppositions. In short, they practice a philosophy that science is a way of knowing about the world. Van Tassel-Baska and Stambaugh (2005) explicated that no other curriculum area better challenges "the natural curiosity and intellectual spirit of gifted students than does science" (p. 159). To achieve knowledge about the world, intellectually gifted adolescent learners must be free to inquire, explore, and channel their natural curiosity. One sensible solution is to introduce students to sciencing (Chaille & Britain, 2003).

SCIENCING FOR EARLY ADOLESCENT GIFTED LEARNERS

Sciencing is a process of active exploration that begins with inquiry from one's personal interest or a question that has been posed. Dunkhase (2003) offers that inquiry is a means to bring together content and process, since inquiry "focuses on content knowledge in the context of the process of developing scientific understanding" (p. 11). Sciencing in the middle grade science classroom for intellectually gifted adolescent learners does not mean searching for one possible solution to a problem. Rather, it involves a certain amount of risk taking and experimenting to investigate phenomena (Llewellyn, 2002). Scientific inquiry consists of actions that allow for multiple results.

Sciencing is an appropriate discovery process for intellectually gifted adolescent students in the science classroom because "the focus is on the active search for knowledge or understanding to satisfy students' curiosity" (Lind, 1999, p. 79). It is not a process that focuses on memorization; it emphasizes exploring questions and how the questions might be answered. From this perspective, sciencing, as a process of inquiry, is a good fit for intellectually gifted adolescent learners because they are intuitive and have a desire to understand the world around them. Additionally, they are naturally driven by their own curiosity to engage in scientific discovery, and the process of sciencing provides the means to step into the role of a scientist to observe, to develop theories, conduct experiments to test theories, draw logical scientific conclusions, and effectively communicate scientific findings (Llewellyn, 2002). Further, sciencing can enhance intellectually gifted adolescent learners' capacity to think, both critically and creatively, and to promote collaborative discussions for sharing new ideas and solving scientific problems. It is a process that permits intellectually gifted adolescents to work with like-minded peers to clarify, elaborate, debate, and negotiate meaning (Van Tassel-Baska, 2000). From an epistemic perspective, sciencing promotes a culture that scientific knowledge is a product of social processes and learning science requires students to be active participants in the scientific culture.

SCIENCING IN ACTION

In 2005, the United States Congress commissioned the Committee on Prospering in the Global Economy of the 21st Century (CPGE) to define specific science and technology enterprise actions and strategies to maintain the United States' competitive presence in the global community. The committee's final report identified the need to increase the number of United States citizens who enter college prepared to earn a science, technology, engineering, or mathematics (STEM) bachelor's degree as a critical component in their action plan to regain the United States' competitive edge. In response to STEM initiatives, appropriate science curriculum that promotes high quality learning for all students, including intellectually gifted adolescent learners, has been recognized by leading science organizations (National Science Board, 2007; National Science Foundation, 2010). The end goal is to increase all students' achievement in science and to prepare learners to be active participants in the ever-expanding global scientific community. These reports have far reaching implications for intellectually gifted students sitting in science classrooms today and have provided the needed catalyst for US school systems to begin to meet the needs of these unique learners.

In an attempt to nurture intellectually gifted adolescent students' scientific habits of mind, VanTassel-Baska and Bracken (2004) recommend multiple opportunities to learn science content by integrating the development of conceptual and content understanding through the use of scientific inquiry and investigative experiences. Specifically, these opportunities must allow for innovative methods of science instruction that permit intellectually gifted adolescent learners to pursue their intellectual curiosity, engage their creative abilities, and apply scientific thinking in the process of doing science. Framed in this manner, sciencing can be viewed as an appropriate process for intellectually gifted adolescent learners to gain scientific knowledge using the opportunities provided by problem-based learning, project-based learning, scenario-learning, and service learning projects.

Problem-Based Learning

Problem-based learning (PBL) was first introduced in medical education at McMaster University in Canada (Birch, 1986). PBL is a constructivist method of instruction characterized as "focused, experiential learning (minds-on, hands-on) organized around the investigation and resolution of messy, real-world problems" (Torp & Sage, 2002, p. 15). The focus is on experiential learning and is viewed as a manner of inquiry (Chiappetta & Koballa, 2006) whereby students gather and process information to construct a resolution, rather than a solution to the problem (Gallagher & Gallagher, 1994). For intellectually gifted adolescent students in science, this type of learning has been shown to be an effective process for students to apply both creative and critical thinking skills as they generate ideas, analyze, and discover a solution to the problem (Gallagher, Stepien, & Rosenthal, 1992). For middle grade science teachers who use problem-based learning as an instructional method, they serve more as metacognitive facilitators who help their students analyze their reasoning and to think about their thinking while confronting the problem (Barrows, 1988).

Borrowing from Torp and Sage (1998) and expanded by Pecore (2012), the classical problem-based learning process involves permanent groups of students who work on a new case every three class meetings. During PBL instruction, students are led through a process that involves objectives, problems, research experiences, solution development activities, and assessments (Torp & Sage). Students work in groups and are presented with a problem on the first day. They are asked to analyze preliminary data. With instructor assistance, the group determines the issues to research on the second day. On the third day, groups then share their research with the class, receive additional information and/or conduct an exploratory activity, and continue researching the problem. For PBL assessment purposes, groups pull together their knowledge and prepare a final solution to the problem (Pecore, 2012). The following PBL steps can serve as a guide for implementation:

- A student reads the problem aloud to their group.
- Students identify the facts, "What they know" from reading the problem.
- Students identify learning issues, "What they don't know."
- Students identify what could be going on, their ideas to move them forward in exploration.
- Students make decision about how to proceed.
- Students acquire new information through research or additional resources.
- Students test their ideas against new knowledge, re-rank ideas as needed.
- Students continue to acquire new information and integrate it with what they know.
- Students arrive at most viable and defendable hypothesis/solution (Torp & Sage, 1998).

To understand sciencing in action through problem-based learning, a sample of cases have been selected from Creating Active Student Engagement in the Sciences (CASES). Developed by Emory University (www.cse.emory.edu/case), CASES is a free online repository that requires users to register and is an excellent source of real-world problem-based learning activities. While the two cases briefly described in Table 2 were constructed for a classroom of middle grade science students, each case exemplifies how problem-based learning can enhance intellectually gifted adolescent learners' capacity to think, both critically and creatively, and to promote collaborative discussions for sharing new ideas and solving scientific problems.

To summarize, problem-based learning is one constructivist instructional strategy for middle grade science teachers to teach intellectually gifted adolescents sophisticated science content regarding real-world problems. Scientific creativity and thinking is manifest as gifted adolescents become active inquiring scientists who explore and seek solutions to problems they see in their world. However, sciencing occurs in other cases. In the next section, our attention turns to project-based learning. As students investigate and seek resolutions to problems, they acquire a better understanding of key scientific concepts and principles from which to then construct scientific artifacts – the projects to symbolize what have been learned (Moore, Sherwood, Bateman, Bransford, & Goldman, 1996).

Table 2. Problem-based learning cases

Activity	*Description*
A Specular Find	This problem-based learning activity (Turner & Shamsid-Deen, 2008) is an earth science lesson designed for middle grade students. The problem is established when Chinese palaeontologists discover a human skull in a region of Southwestern China nearby where dinosaur fossils have been uncovered. In preparation for a CNN interview, the palaeontologists gather all the necessary data to determine if humans and dinosaurs did coexist.
A Friday in September	This problem-based learning activity (Wade, Sumrall, Webb, Chatraw, & Embree, 2011) is a life science lesson designed for middle grade students. The problem is established during a field trip when a group of Ecology students get lost in the woods from taking a wrong path. The group eventually finds its way back to the awaiting members in the Ecology class, but one of the lost students becomes very ill and has to be hospitalized. The class decides they need to better understand the disease that required their classmate to be hospitalized for their class project.

Project-Based Learning

According to the Buck Institute of Education (BIE), project-based learning is "a systematic teaching method that engages students in learning knowledge and skills through an extended inquiry process structured around complex, authentic questions and carefully designed products and tasks" (para 1). Project-based learning is an appropriate process for intellectually gifted adolescent learners to do science because the process requires the use of creative, critical, and information skills, multiple applications of technology, and constructive investigations to perform cognitively complex tasks. Investigations may occur in many forms such as, design, decision-making, problem-finding, problem-solving, discovery, or model representations, but all investigations must focus on student transformation and construction of knowledge represented in the final artifact (Thomas, 2000). Van Tassel-Baska and Stambaugh (2005) remind us that this form of learning is especially appealing to intellectually gifted adolescent learners who relish working with authentic problems that require their expert problem solving skills. In addition, the teamwork involved in project-based learning permits all members within the learning group to apply their existing knowledge to real world issues in a nurturing environment as they probe, challenge, and collaboratively work together to construct an authentic real-world project (Diehl, Grobe, Lopez, & Cabral, 1999).

To begin project-based learning for gifted adolescent learners, middle grade science teachers can use the Creative Problem-Solving (CPS) model. CPS serves as an appropriate sciencing approach for middle grade science teachers to implement in their classrooms to engage their gifted adolescent students in problem solving experiences. The CPS model was first developed by Osborn (1963) and later addressed by Parnes (1981) as a useful process to afford intellectually gifted learners the opportunity to tackle a problem in creative ways through real action (see Figure 1). Since its inception, further changes have been made (Treffinger, Isaksen, & Dorval, 1994) to the CPS model that has evolved into a six-step process and organized into three phases. The process allows students to first activate divergent thinking skills to fact find, analyze, and evaluate criteria, then use convergent thinking skills to come to a consensus by selecting only the most promising ideas to solve the problem.

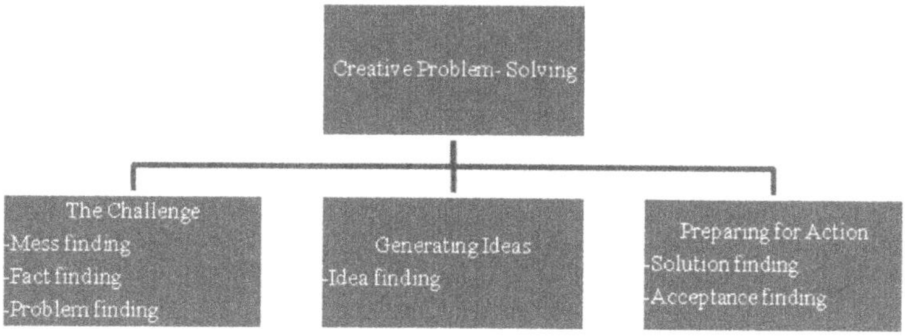

Figure 1. Creative problem-solving model

Generally, the first step begins with locating a problem (divergent thinking) or "*mess finding*" (Parnes, 1981; Treffinger et al., 1994) from which to initiate the model for problem-solving. Once a problem is located, the second step involves the *fact finding* stage in which students look at many sources of data and list all the things they know about the problem to find the main focus of the challenge. Parnes (1981) suggested that students use *who*, *what*, *when*, *where*, *why*, and *how* questions in this stage to generate many facts about the problem for consideration. From the list of generated facts, the students then collaborate (convergent thinking) to narrow the list in order to focus on only the most important pieces of data that pertain to the problem. Following this stage, students engage in *problem finding* to discover the underlying or the most important problem of the challenge and then craft a problem statement that will guide the creative problem-solving process. This stage permits students to consider alternative definitions for the problem because knowing how the problem is defined can determine the solution or how it will be solved (Davis, Rimm, & Siegle, 2011). Next, students engage in the *idea finding* stage to generate many different ways to solve the problem. This is a brain storming stage whereby students freely list their ideas. Following this stage, students then begin the *solution finding* stage which requires them to analyze and evaluate all criteria in order to determine the best possible solution. Depending on the problem, criteria evaluation may include cost, resources, time, legality, space, safety, quality, and feasibility. In the final stage, students engage in *acceptance finding* to plan and implement the solution to solve the problem. This stage requires students to identify the *assisters* or key players who will lend support for solving the problem and the *resisters* or those people who will impede the process (Treffinger, 1995). The students then identify an action plan to resolve the problem and assign roles to perform various tasks to implement the action plan.

To wrap-up a learning-based project in science, middle grade teachers of intellectually gifted adolescents encourage their students to present their findings through multiple venues and formats. Van Tassel-Baska (1989) posits that the use of product differentiation for intellectually gifted adolescents permits students to demonstrate what they have learned in a variety of forms that will reflect their content knowledge and their ability to connect their knowledge within and across disciplines. For assessment purposes, Goodnough and Cashion (2003) recommend that middle grade science teachers use a rubric with a point value for each criterion. Such a rubric would assign point values for defining the problem, gathering appropriate resources, presenting a solution, and then convincing others the problem.

Newmann, Bryk, and Nagaoka (2001), the goal of project-based learning is to engage students in authentic intellectual work that the authors describe as "construction of knowledge through disciplined inquiry in order to produce products that have value beyond school" (p. 14). For our readers, a small sample of learning activities has been selected to illustrate sciencing in action through project-based learning. As in the previous section on problem-based learning, the following activities were designed to engage all science learners. However, each activity highlights the advanced content and complex ideas needed to allow intellectually gifted adolescent science students to consider multiple perspectives, research in-depth, and determine the problem (Feldhusen, 1993) before constructing a project to

Table 3. Project-based learning units

Learning Units
• *Beat the Heat* This project-based learning activity is an earth science unit designed for middle grade students to investigate environmental issues and to explore the impact of global warming. Intellectually gifted adolescent learners assume the role of budding scientists in order to gather all the data needed to determine the causes of climate change and to construct a project. The question for consideration in this study is: *What effects do our choices have on the world around us*?
• *Designer Genes: One Size Fits All*? This project-based learning activity is a life science unit designed for middle grade students to explore and understand the concept of genetic engineering. Young adolescent learners assume the role of biologists to research the pros and cons of altering agricultural products and then construct an artifact to demonstrate their understanding. The lesson is appropriate for intellectually gifted adolescent students in science who have a keen desire to understand the world around them and engage in ethical dilemmas. The question for consideration in this study is: *Just because we can, should we?*

extend their knowledge base. The following two examples were selected from Intel's Designing Effective Projects: Project-Based Units to Engage Students (http://www.intel.com).

Programs for Creative Problem-Solving

Middle grade science teachers will find there are a plethora of programs they can use for creative problem-solving with their intellectually gifted adolescent students. Some of the more prominent programs featured in Table 4 permit young adolescent, intellectually gifted learners to pursue their intellectual curiosity, engage their creative abilities, and apply scientific thinking in the process of doing science through competitive venues.

Table 4. Creative problem-solving programs

Programs
• *Future Problem-Solving Program International* (FPSPI) (http://www.fpspi.org) is a creative problem-solving program that middle grade science teachers can use for enrichment. Founded by E. Paul Torrance (1977), FPSPI encourages students to use critical and creative thinking skills and develop a vision for the future. FPSPI provides competitive opportunities for gifted students at the state, national, and international level.
• *Destination ImagiNation* (http://www.destinationimagination.org) is another competitive program that is an excellent program for teaching creative thinking and problem-solving. The program permits intellectually gifted adolescents to develop teamwork and leadership skills as students collaborate to research and solve real-world problems. In the actual competition, students are given a challenge on the spot and must find a solution with very little teacher input who serves as a facilitator.
• *Odyssey of the Mind* (OM) (http://www.odysseyofthemind.com) was developed by Ted Gourley (1981) to develop students' minds through mental games that require critical and creative thinking skills and problem-solving abilities. Students work in teams over the school year to apply their creativity to solve one of five problems and then take their solution to regional competitions in May. Winners compete in a national competition with an eye on competing in an international competition.

Sciencing comes alive in project-based learning when middle grade science teachers implement the CPS model (Parnes, 1981) to engage their gifted adolescent students in problem solving experiences. The project-based learning approach engages intellectually gifted adolescent students in the creative enterprises of science learning: discovery, decision-making, problem-finding, problem-solving, and artifact construction. One additional way to involve intellectually gifted adolescents in both a hands-on and minds-on scientific inquiry is through solving problems that focus on a specific situation. Challenging new situations invite intellectually gifted adolescent students to question, investigate, and to build theories.

Scenario-Based Learning

Today's young adolescent gifted science students desire multiple opportunities to problem solve and to engage in inquiry. Scenario-based learning (SBL) is a constructivist approach that ensures young adolescent gifted learners can play an active role in their own learning and construct knowledge through investigation, creative thinking, and metacognition (Savery, 2006). It is grounded in the principles of Situated Learning Theory (Brown, Collins, & Druid, 1989; Lave & Wenger, 1991) as the learning context involves the simulation of complex real-world scientific problems and knowledge can be achieved within this authentic context.

SBL is a problem solving method that engages students in active learning using authentic contexts. Students are presented with problems and accompanying choices that must be made to reach an outcome (Errington, 2003). Just as in real life, each scenario puts students in a situation where the decisions they make can significantly affect or change the outcomes. Scenario-based learning is particularly important for intellectually gifted adolescent students in science because they have the ability to organize information around key scientific concepts, identify patterns, and implement complex cognitive procedures that allow different pieces of knowledge to be tapped into and then applied to new situations (Van Tassel-Baska, 2005). Subsequently, the process further allows gifted adolescent learners to work on multiple levels simultaneously (Feldhusen, 1993; Reis, 1990; Renzulli & Reis, 1997) as they interact in a learning community (Duffy & Cunningham, 1996) to make meaning. Knowledge is co-constructed as each member in a learning community has a role to fulfill in order to complete the scenario activity. In addition, simulations expedite the complex learning processes that are often encountered in the real world, and therefore, students can explore and manipulate phenomena that might be too time-consuming or even dangerous. Bransford (2000) posited that simulations can "engage learners as active participants in their learning by focusing their attention on critical elements, encouraging abstraction of common themes (principles), and evaluating their own progress toward understanding" (p. 68).

While there are many different forms of simulation that are used to enact scenario-based learning, game-based learning is one simulation tool that can capture the interest of intellectually gifted adolescent learners to engage in real-world learning that is situated in real-world context (May, 1997). Essentially, students engage problem-solving processes within a game context that uses science content. According to Ketelhut (2007), game-based scenario learning is a much needed improvement for the isolated problem-solving format found in textbooks to challenge intellectually gifted learners. Intellectually gifted adolescent students will find the virtual game environment (Spires, Rowe, Mott, & Lester, 2011) to be a fascinating 21st century tool. The virtual game environment permits complex decision-making while engaging in scientific problem-solving that is prompted by scientific inquiry.

The simulation models featured in Table 5 deal with advanced content and complex ideas to allow intellectually gifted adolescent science students to interpret scientific explanations of the natural world, assess scientific evidence, and actively participate in the co-construction of scientific knowledge. Additionally, the simulation models featured in this section hold promise for productive scenario-based learning that affords young adolescent intellectually gifted students to engage in the process of sciencing by using science content in creative ways to produce novel and valuable ideas.

Table 5. Simulation models

Programs
• *River City* (http://muve.gse.harvard.edu/rivercityproject) is a multi-user virtual environment (MUVE) that is designed to engage middle grade students in the exploration and solution finding process for problems that occur in real-world environments. Specifically, River City is a simulation that uses the virtual world and actual museum exhibits from the Smithsonian to engage middle grade science students in scientific inquiry. The simulation features an unidentified town set in the late 19th century where citizens are falling ill – a "messy" situation exists and a solution must be found. To play the game, students take on the role of 21st century scientists who travel back in time to help the mayor find the cause of the illness. The game requires students to pose a problem question, such as *Why are the citizens of River City getting sick?* Then, students collaboratively investigate the health problems in the city, identify the problem, and then find the solution to the citizens' ills. To do so, students must study the complex terrain of River City where water run-off into the city allows many insects to propagate, and industry along the river presents another health concern. River City is an excellent scenario-based learning game for intellectually gifted students because the problem presented requires learners to collect water samples, conduct water quality experiments, interview the towns' inhabitants, research hospital records to study patients' symptoms, and collaboratively work together to determine the solution. The game concludes when a final report if presented to the mayor with the prescribed solution and recommendations.
• *The Case of Crystal Island* (www.intellimedia.ncsu.edu/projects.html) is a virtual game environment that targets scientific learning related to microbiology. To play the game, students are situated on an unidentified tropical island and assume the role of Alyx, who is the central character of the research team, and other camp members on the island. Several members of his research team have fallen ill and it is up to Alyx and healthy members on his team to determine the cause of the outbreak. A possible question for students to consider is *What bacteria or virus has infected the camp researchers?* This scenario then requires classroom teams of students to study the make-up of pathogens, such as bacteria and viruses, research the sick camp members' symptoms, noting and recording the details, formulating an hypothesis, and testing the hypothesis. When students have completed their investigations, determined the problem, and the solution, a final report is given to the camp nurse that consists of the cause and the treatment for the illness.

In the constructivist learning environment, scenario-based learning stimulates intellectually gifted students to apply their natural curiosity in creative ways to co-construct scientific knowledge when confronted with new situations. It is a way to involve intellectually gifted adolescents in the investigative nature of science. In our final section, intellectually gifted adolescents can further engage in sciencing through service activities that stimulate critical and creative thinking skills and problem-solving abilities.

Service Learning

Dewey (1938) believed that communities play an integral role in students' educational experiences. While Dewey's philosophy did not speak directly about service learning, it is well-known that he believed that education should be about experiential learning and represent students' daily lives (Kunin, 1997). Service learning is an integration of civic engagement and traditional academic curricula. More specifically, service learning is a form of education whereby students engage in real-world service that is related to their academic studies. The intended goals of all service learning projects (Thomsen, 2006) are for students to extend learning beyond the classroom, to acquire a sense of civic responsibility, and to advance their own personal growth through meaningful service experiences and self-reflection. According to Waterman (1997), service-learning is a means for students to become actively engaged in the process of their own learning, rather than simply following classroom instruction because the time spent in a service learning project provides the necessary ingredients for hands-on community action.

In keeping with the focus of this chapter, service learning is one more creative approach to engage intellectually gifted adolescent students in the process of sciencing - doing science that is connected to real-life experiences. The process is initiated when the service is connected to science learning and involves mental engagement and physical activity. In recent years, service learning has become increasingly popular for intellectually gifted programs because intellectually gifted students are generally perceived to be more socially and morally mature than their non-gifted classmates (Krystal 1999). Their advanced cognitive and affective development prepares them to be highly in tune to issues of social justice, fairness, and doing what is right and good for others (Lee & Olszewski-Kubilius, 2006). Researchers and educators both find that service learning projects are appropriate for intellectually gifted adolescents because these students possess high degrees of responsibility, self-confidence, and strong leadership abilities (Ablard, 1997; Chan, 1988; Davis & Rimm, 1998). Service learning further meets the curricula demands of advancement, depth, and complexity that are highly recommended for gifted learners (VanTassel-Baska, 2003). Open-ended learning experiences, opportunities to focus on the most important problems or problems in need of attention, and opportunities to implement solutions through community action all provide a differentiated foundation to nurture creativity and scientific reasoning for gifted adolescent students.

Young intellectually gifted adolescent students are engaged in service learning projects each and every day throughout the US. In Minnesota, a group of middle grade students (http://www.nylc.org/resources/projects/preventing-west-nile) took action when their community became infected with the West Nile Virus. As part of their service-learning project, they conducted an Internet research and reviewed vital information from the State Department of Public Health and the State River Basin Commission to gather all the facts they could before they narrowed down to the one problem they would explore. From their research, they learned that there is no cure for West Nile Virus, but the students elected to conduct a community-wide awareness program on the preventive measures the public could take as the solution to the problem. To inform the public, some students created a brochure detailing methods to eliminate mosquitoes around their homes and then distributed their brochures at a nearby mall. Other students created PowerPoint presentations that were shown to multiple groups throughout the community and presented to participants at a state-wide environmental conference. Finally, other students created a public service video that was aired by the local cable company to build public awareness and actions that could be taken to prevent the spread of the West Nile Virus.

While studying about the impact of human actions on the natural environment, eighth grade students in Maryland learned about the harmful effects from marine debris on marine ecology and wildlife. From the study, the students decided to develop a service-learning project that evolved into a beach clean-up at Assateague Island State Park. To extend their academic learning in the classroom, the eighth graders planned, organized, and implemented the beach clean-up to stop pollution in the coastal area (http://www.marylandpublicschools.org/MSDE).

Middle grade students in California participate in a state-wide partnership with the US Forest Service to rehabilitate degraded watersheds in their local communities (http://www.calstem.org/documents/calservebrochure.pdf). The state-wide service learning opportunity is funded by the California Department of Education's CalServe Iniative. As part of the middle school curriculum on environmental sensitivity issues, students conduct in-depth studies on the importance of watersheds to local communities as sources of drinking water and wildlife habitats. Their academic focus in the classroom requires them to understand what constitutes a healthy watershed and to research the various factors that have devastating impacts, such as the removal of vegetation to make room for roads and buildings and increased water runoff that carries harmful pollutants. Through service learning activities, students work with environmental engineers, biologists, and hydrologists, to restore degraded watersheds by replacing vegetation and repopulating the natural wildlife to create an ecological balance for the preservation of watersheds.

One final approach for young adolescent gifted students to get involved in service learning is to engage in Citizen Science. According to Cohn (2008), Citizen Science "... refers to volunteers who participate as field assistants in scientific studies" (p. 193). Working in conjunction with scientists, volunteers collect and analyze data in many scientific fields such as the environment, wildlife, forestry, ornithology, and ecology to name a few. Citizen Science has grown in popularity in elementary, middle, and high school classrooms because the scientific projects engage students in doing real science that address real world scientific needs. Table 6 provides a sampling of Citizen Science projects that will foster young adolescent gifted students' scientific habits of mind and inquiry skills.

Table 6. Citizen service projects

Programs
• World Water Monitoring Day (www.worldwateringmonitoringday.org) is a Citizen Science project that seeks to teach middle schoolers how humans can impact the water supply. Students participate by monitoring local bodies of water using test kits to measure pH, acidity, oxygen levels, and temperature. • Galaxy Zoo (http://www.galaxyzoo.org) is a Citizen Science project that teaches students to examine images of galaxies taken from the Hubble Telescope. Students participate in solar system research by learning to record observed data and classify galaxies by shape and size. • S'COOL (http://scool.larc.nasa.gov) is a Citizen Science project that is designed to teach students how clouds directly impact the earth's climate and weather systems. Students participate by observing clouds at various times and reporting their data online regarding cloud formation, types, height, and thickness.

From the scientific inquiries that first began with academic content, multiple service learning projects have been successfully implemented by intellectually gifted adolescents. Only a few examples have been cited here, but young adolescent gifted learners are using their classroom content knowledge in science to make important contributions to their communities across the US and beyond. The opportunities for intellectually gifted adolescent learners to work on authentic real-world problems and build their creative thinking and problem-solving skills in active collaboration with the community in which they live, not only makes their education relevant, but communities benefit as well.

CONCLUSION

Children are naturally curious about their world and have insatiable appetites to know all they can so as to understand their world. Intellectually gifted adolescent students in particular are driven to discover. Many want to be scientists. They want to question and challenge the unknown, and they do not like answers and solutions dropped in their laps or voluminous amounts of information to memorize. These preferences to learning should shape the structure of science education for intellectually gifted adolescent students. Constructivist middle grade science teachers of the intellectually gifted recognize these wants and provide the opportunities for their students to satisfy their intellectual curiosity and engage in scientific discovery. They understand that gifted adolescent students can learn about the scientific world through hands-on and minds-on investigative experiences. From this perspective, it should be clear that sciencing is an approach that will foster curiosity and stimulate creative ideas. Sciencing can transform middle grade classrooms into sites of inquiry for young intellectually gifted adolescents where creative thinking is evident and students can behave as the scientific explorers as they were born to be.

REFERENCES

Ablard, K. E. (1997). Self-perceptions and needs as a function of type of academic ability and gender. *Roeper Review, 20*, 110–115.

Barrows, H. S. (1988). *The tutorial process*. Springfield, IL: Southern Illinois University School of Medicine.

Birch, W. (1986). Towards a model for problem-based learning. *Studies in Higher Education, 11*, 73–82.

Brandwein, P. (1986). A portrait of gifted young with science talent. *Roeper Review, 8*(4), 235–243.

Bransford, J. (2000). *How people learn: Brain, mind, experience, and school.* Washington, DC: National Academy Press.

Brown, J. S., Collins, A., & Duguid, P. (1989). Situated cognition and the culture of learning. *Educational Researcher, 18*(1), 32–42.

Brown, D. F., & Knowles, T. (2007). *What every middle school teacher should know* (2nd ed.). Portsmouth, NH: Heinemann.

Buck Institute for Education. (2014). *Project-based learning.* Retrieved from http://www.bie.org/index.php/site/PBL/pbl_handbook_introduction/#history

Chaille, C., & Britain, L. (2003). *The young child as scientist* (3rd ed.). Boston, MA: Allyn & Bacon.

Chan, L. K. S. (1988). The perceived competence of intellectually talented students. *Gifted Child Quarterly, 32*, 310–314.

Chiappetta, E. L., & Koballa Jr., T. R. (2006). *Science instruction in the middle and secondary schools: Developing fundamental knowledge and skills for teaching* (6th ed.). Upper Saddle River, NJ: Pearson Merrill Prentice Hall.

Clark, B. (1997). Social ideologies and gifted education in today's schools. *Peabody Journal of Education, 72*(3–4), 81–100.

Clark, B. (2002). *Growing up gifted: Developing the potential of children at home and school.* Columbus, OH: Prentice Hall.

Cohn, J. P. (2008). Citizen science: Can volunteers do real research? *Bioscience, 58*(3), 192–197.

Colangelo, N., & Assouline, S. G. (2000). Counseling gifted students. In K. A. Heller, E. J. Mönks, R. J. Sternberg, & R. F. Subotnik (Eds.), *Handbook of gifted education* (3rd ed., pp. 572–581). Boston, MA: Allyn & Bacon.

Colangelo, N., & Davis, G. A. (1997). Introduction and overview. In N. Colangelo & G. A. Davis (Eds.), *Handbook of gifted education* (2nd ed., pp. 3–23). Boston, MA: Allyn & Bacon.

Committee on Prospering in the Global Economy of the 21st Century: An Agenda for American Science and Technology, National Academy of Sciences, National Academy of Engineering, Institute of Medicine. (2007). *Rising above the gathering storm: Energizing and employing America for a brighter economic future.* Washington, DC: National Academies Press.

Croft, L. J. (2003). Teachers of the gifted: Gifted teachers. In N. Colangelo & G. A. Davis (Eds.), *Handbook of gifted education* (3rd ed., pp. 558–571). Boston, MA: Pearson Education.

Crotty, M. (1998). *The foundations of social research: Meaning and perspective in theresearch process.* Thousand Oaks, CA: Sage.

Csikszentmihalyi, M. (1990). *Flow: The psychology of optimal experience.* New York, NY: HarperCollins.

Davis, G. A. (1999). Barriers to creativity and creative attitudes. In M. A. Runco & S. R. Pritzker (Eds.), *Encyclopedia of creativity* (Vol. 1, pp. 165–174). New York, NY: Academic Press.

Davis, G. A., & Rimm, S. B. (1998). *Education of the gifted and talented* (4th ed.). Needham Heights, MA: Allyn & Bacon.

Davis, G. A., Rimm, S. B., & Siegle, D. (2011). *Education of the gifted and talented* (6th ed.). Boston, MA: Pearson.

Deckert, D. (2001). Science and art: Lessons from Leonardo da Vinci? In G. Burnaford, A. Aprill, & C. Weiss (Eds.), *Renaissance in the classroom: Arts integration and meaningful learning* (pp.125–139). Mahwah, NJ: Lawrence Erlbaum.

Diehl, W., Grobe, T., Lopez, H., & Cabral, C. (1999). *Project-based learning: A strategy for teaching and learning*. Boston, MA: Center for Youth Development and Education, Corporation for Business, Work, and Learning.

Dewey, J. (1938). *Experience and education*. New York, NY: Collier Books.

Driver, R., Asoko, H., Leach, J., Mortimer, E., & Scott, P. (1994). Constructing scientific knowledge in the classroom. *Educational Researcher, 23*(7), 5–12.

Duffy, T., & Cunningham, D. (1996). Constructivism: Implications for the design and delivery of instruction. In D. Jonassen (Ed.), *Handbook of research on educational communications and technology* (pp. 170–198). New York, NY: Simon &Schuster.

Dunkhase, J. A. (2003). The coupled-inquiry cycle: A teacher concerns-based model for effective student inquiry. *Science Educator, 12*(1), 10–15.

Duschl, S., Schweingruber, H. A., & Shouse, A. W. (2007). *Taking science to school: Learning and teaching science in grades K-8*. Washington, DC: National Academic Press.

Elder, L., & Paul. R. (2007). *The miniature guide to the human mind* (3rd ed.). Dillon Beach, CA: Foundation for Critical nation's faculty, Thinking.

Ellis, K. (Producer, Writer, Director). (2009). *Project-based learning: An overview* [Video]. United States: Edutopia.

Errington, E. P., (2003). *Developing scenario-based learning: Practical insights for tertiary educators*. Palmerston North, NZ: Dunmore Press.

Farkas, S., & Duffett, A. (2008). Findings from a national teacher survey. In Thomas B. Fordham Institute (Ed.), *High achieving students in an era of NCLB* (pp. 49–74). Washington, DC: Author.

Feldhusen, H. J. (1993). Individualized teaching of the gifted in regular classrooms. In C. J. Maker (Ed.), *Critical issues of gifted education: Programs for the gifted in regular classrooms* (Vol. 4, pp. 263–273). Austin, TX: PRO-ED.

Feldhausen, H. J. (1997). Secondary services, opportunities, and activities for talented youth. In N. Colangelo & G. A. Davis (Eds.), *Handbook of gifted education* (2nd ed., pp. 189–197). Boston, MA: Allyn & Bacon.

Freeman, J. (2003). Scientific thinking in gifted children. In P. Csermely & L. Lederman (Eds.), *Science education: Talent recruitment and public understanding* (pp. 17–30). Amsterdam, The Netherlands: IOS Press with NATO Scientific Affairs Division. Retrieved from www.chaperone.sote.hu

Gallagher, J. J. (2000). Unthinkable thoughts: Education of gifted students. *Gifted Child Quarterly, 44*(1), 5–12.

Gallagher, J. J., & Gallagher, S. A. (1994). *Teaching the gifted child* (4th ed.). Boston, MA: Allyn & Bacon.

Gallagher, S. A., Stepien, W. J., & Rosenthal, H. (1992). The effects of problem-based learning on problem solving. *Gifted Child Quarterly, 36*(4), 195–200.

Getzels, J. W., & Jackson, P. W. (1962). *Creativity and intelligence*. New York, NY: Wiley.

Goodnough, K., & Cashion, M. (2003). Fostering inquiry through problem-based learning. *The Science Teacher, 70*(9), 21–25.

Gourley, T. J. (1981). Adapting the varsity sports model to nonpsychomotor gifted students. *Gifted Child Quarterly, 25*, 164–161.

Hassard, J. (2000). *Science as inquiry: Active learning, project-based, web-assisted, and active assessment strategies to enhance student learning*. Parsippany, NJ: Good Year Books.

Helm, J. H. (2004). Projects that power young minds. *Educational Leadership, 62*(1), 58–62.

Hemlo-Silver, C. E., & Barrows, H. (2006). Goals and strategies of a problem-based learning facilitator. *Interdisciplinary Journal of Problem-baaed Learning, 1*(1), 21–39.

Johnson, D., Boyce, L., & VanTassel-Baska, J. (1995). Evaluating curriculum materials in science. *Gifted Child Quarterly, 89*(1), 35–43.

Jones, B. F., Rasmussen, C. M., & Moffitt, M. C. (1997). *Real-life problem solving: A collaborative approach to interdisciplinary learning*. Washington, DC: American Psychological Association.

Judson, H. F. (1980). *The search for solutions*. New York, NY: Holt, Rinehart, and Winston.

Karnes, F. A., Bean, S. M. (Eds.). (2001). *Methods and materials for teaching the gifted*. Waco, TX: Prufrock Press.

Kohn, A. (1993). Choices for children: Why and how to let students decide. *Phi Delta Kappan, 75*(1), 8–16, 18–21.

Krystal, S. (1999). The nurturing potential of service learning. *Educational Leadership, 56*(4), 58–61.

Kunin, M. M. (1997). Service learning and improved academic achievement: The national scene. In J. Schine (Ed.), *Service learning: Ninety-sixth yearbook of the national society of the study of education*. Chicago, IL: The University of Chicago Press.

Lave, J., & Wenger, E. (1991). *Situated learning: Legitimate peripheral participation*. Cambridge, England: Cambridge University Press.

Lawson, A. E. (1995). *Science teaching and the development of thinking*. Belmont, CA: Wadsworth.

Lee, S.-Y., & Olszewski-Kubilius, P. (2006). The emotional intelligence, moral judgment, and leadership of academically gifted adolescents. *Journal for the Education of the Gifted, 30*, 29–67.

Lind, K. (1999). Science in early childhood: Developing and acquiring fundamental concepts and skills. In *American association for the advancement of science (AAAS): Dialogue on early childhood science, mathematics, and technology education* (pp. 73–83). Washington, DC: AAAS.

Llewellyn, D. (2002). *Inquire within: Implementing inquiry-based science standards*. Thousand Oaks, CA: Corwin Press.

Mackin, J., Macaroglu, E., & Russell, K. (1996). Science seminar: Providing the opportunity to go beyond traditional curricula. *Gifted Child Today, 19*(3), 16–20, 49.

May, D. G. (1997). Simulations: Active learning for gifted students. *Gifted Child Magazine, 20*(2), 28–35.

Moore, A., Sherwood, R., Bateman, H., Bransford, J. D., & Goldman, S. R. (1996). Using problem-based learning to prepare for project-based learning. In J. D. Bransford (Chair) (Ed.), *Enhancing project-based learning: Lessons from research and development*. Symposium conducted at the 1996 Annual meeting of the American Educational Research Association, New York, NY.

National Science Board. (2007). *A national action plan for addressing the critical needs of the US Science, Technology, Engineering, and Mathematics education system*. Arlington, VA: National Science Foundation.

National Science Foundation. (2010). *Chapter 3: Science and engineering labor force*. Retrieved from http://nsf.gov/statistics/seind04/pdf/c03.pdf

Neber, H., & Atkins, M. S. (2002). Self-regulated science learning with highly gifted students: The role of cognitive, motivational, epistemological, and environmental variables. *High Ability Studies, 13*(1), 59–74.

Newmann, F. M., Bryk, A. S., & Nagaoka, J. K. (2001). *Authentic intellectual work and standardized tests: Conflict or coexistence?* Chicago, IL: Consortium on Chicago School Research.

Nicholl, C. (2004). *Leonardo da Vinci: The flights of the mind*. London, England: Allen Lane.

No Child Left Behind of 2001. (2001). *Public Law*, 107–110.

Osborn, A. F. (1963). *Applied imagination: Principles and procedures of creative problem-solving* (2nd ed.). New York, NY: Scribner's.

Parnes, S. J. (1981). *The magic of your mind*. Buffalo, NY: Bearly Limited.

Paul, R., & Elder, L. (2005). *The miniature guide to critical thinking concepts and tools*. Dillon Beach, CA: Foundation for Critical Thinking.

Pecore, J. L. (2012). Beyond beliefs: Teachers adapting problem-based learning to preexisting systems of practice. *Interdisciplinary Journal of Problem-based Learning, 7*(2). Retrieved from http://dx.doi.org/10.7771/1541-5015.1359

Piaget, J. (1952). *The origins of intelligence in children*. New York, NY: International University Press.

Potter, P. (2006). *Art, science, and life's enigmas*. Retrieved from http://dx.doi.org/10.3201/eid1208.AC1208

Reis, S. M. (1990). What to teach, when to teach it. *Learning*, 44–45.

Reis, S. M., & McCoach, D. B. (2000). The underachievement of gifted students: What do we know and where do we go? *Gifted Child Quarterly, 44*, 152–170.

Renzulli, J. S. (1968). Identifying key features in programs for the gifted. *Exceptional Children, 35*, 217–221.
Renzulli, J. S. (1992). A general theory for the development of creative productivity through the pursuit of ideal acts of learning. *Gifted Child Quarterly, 36*(4), 170–182.
Renzulli, J. S., & Reis, S. M. (1997). *School-wide enrichment model: A how-to-guide for educational excellence* (2nd ed.). Mansfield Center, CT: Creative Learning Press.
Savery, J. R. (2006). Overview of problem-based learning: Definitions and distinctions. *The Interdisciplinary Journal of Problem-based Learning, 1*(1), 9–20.
Savery, J. R., & Duffy, T. M. (1995). Problem based learning: An instructional model and its constructivist framework. *Educational Technology, 45*(1), 31–38.
Schulte, P. L. (1996). A definition of constructivism. *Science Scope, 20*(3), 25–27.
Sisk, D. (1989). *Creative teaching of the gifted.* New York, NY: McGraw-Hill.
Spires, H., Rowe, J. P., Mott, B. W., & Lester, J. C. (2011). Problem solving and game-based learning: Effects of middle grade students' hypothesis testing strategies on learning outcomes. *Journal of Educational Computing Research, 44*(4), 453–472.
Stepien, W. J., Gallagher, S. A., & Workman, D. (1993). Problem-based learning for traditional and interdisciplinary classroom. *Journal for the Education of the Gifted, 16*(4), 338–357.
Tolan, S. (1996). *Is it a cheetah?* Ghanna, OH: Ohio Association for Gifted Children Review.
Thomas, J. W. (2000). *A review of research on project-based learning.* Report prepared for The Autodesk Foundation, San Rafael, CA. Retrieved from http://www.bie.org/index.php/site/RE/pbl_research/29
Thomsen, K. (2006). *Servicing learning in grades k-8: Experiential learning that builds character and motivation.* Thousand Oaks, CA: Sage.
Torp, L., & Sage, S. (1998). *Problems as possibilities for K-12 education.* Alexandria, VA: Association for Supervision and Curriculum Development.
Torp, L., & Sage, S. (2002). *Problems as possibilities for K-16 education* (2nd ed.). Alexandria, VA: Association for Supervision and Curriculum Development.
Torrance, E. P. (1977). *Creativity in the classroom.* Washington, DC: National Education Association.
Treffinger, D. J. (1995). Creative problem solving: Overview and educational implications. *Educational Psychology Review, 7*, 301–312.
Treffinger, D. J., Isaksen, S. G., & Dorval, K. B. (1994). Creative problem solving: An overview. In M. Treffinger & A. Runco (Ed.), *Problem finding, problem solving, and creativity* (pp. 223–236). Norwood, NJ: Ablex.
Trumbull, D. J. (1999). *The new science teacher: Cultivating good practice.* New York, NY: Teachers College Press.
Turner, B. L., & Shamsid-Deen, K. K. (2008). *A spectacular find.* Retrieved July 21, 2012, from Emory University, CASES Online, http://www.cse.emory.edu/casedisplay.cfm?case_id=170
VanTassel-Baska, J. (1989). Appropriate curriculum for the gifted. In J. Feldhusen, J. VanTassel-Baska, & K. Seeley (Eds.), *Excellence in educating the gifted* (pp. 175–191). Denver, CO: Love.
VanTassel-Baska, J. (1993). Theory and research on curriculum development for the gifted. In K. A. Heller, F. J. Mönks, & H. Passow (Eds.), *International handbook of research and development of giftedness and talent* (pp. 365–386). Oxford, England: Pergamon Press.
VanTassel-Baska, J. (1998). The development of academic talent: A mandate for educational best practice. *Roeper Review, 18*(2), 98–102.
Van Tassel-Baska, J. (2000). *Planning science programs for high ability learners.* Retrieved from http://www.kidsource.com/education/science.gifted.html
VanTassel-Baska, J. (2003). *Curriculum planning and instructional design for gifted learners.* Denver, CO: Love.
VanTassel-Baska, J. (2004). Creativity as an elusive factor in giftedness. *Update, the Electronic Magazine of the School of Education at William, and Mary*, 1–7.
VanTassel-Baska, J., & Bracken, B. (2004). *Project Clarion: An integrative curriculum scale-up to promote scientific conceptual understanding in promising young children.* Williamsburg, VA: The College of William and Mary, Center for Gifted Education.

VanTassel-Baska, J., & Stambaugh, T. (2005). Challenges and possibilities for serving gifted learners in the regular classroom. *Theory into Practice*, *44*(3), 211–217.

VanTassel-Baska, J., Gallagher, S., Bailey, J., & Sher, B. (1993). Scientific experimentation. *Gifted Child Today*, *16*(5), 42–46.

VanTassel-Baska, J., Bass, G., Ries, R., Poland, D., & Avery, L. (1998). A national study of science curriculum effectiveness with high ability students. *Gifted Child Quarterly*, *42*(4), 200–211.

Wade, A., Sumrall, S., Webb, A. L., Chawtraw, J. H., & Embree, M. (2011). *A Friday in September* [Online]. Emory University: CASES. Retrieved from Emory University, CASES Online http://www.cse.emory.edu/cases/casedisplay.cfm?case_id=1684

Wallach, M. A., & Kogan, N. (1965). *Modes of thinking in young children*. Westport, CT: Greenwood Press.

Waterman, A. S. (1997). An overview of service learning and the role of research and evaluation in service learning programs. In A. S. Waterman (Ed.), *Service learning: Applications from the research* (pp. 1–11). Mahwah, NJ: Lawrence Erlbaum.

Wenger, E. (1998). *Communities of practice: Learning, meaning, and identity*. New York, NY: Cambridge University Press.

Winebrenner, S. (2000, September). Gifted students need an education, too. *Educational Leadership*, 52–56.

Zeece, P. D. (1999). Things of nature and the nature of things: Natural science-based literature for young children. *Early Childhood Education Journal*, *26*(3), 161–166.

Lina Soares
Department of Teaching and Learning
Georgia Southern University
USA

ADELE L. SCHMIDT

9. QUANTIFYING THE EFFECTS OF PERSONALIZED ASSESSMENT TASKS IN SECONDARY SCIENCE TEACHING

INTRODUCTION

Tensions between the perception and reality of scientific practice have produced significant problems, including the fact that high proportions of students do not view science as a creative endeavor. The resultant, systemic devaluation of science has significant implications for scientific research, and science education.

Throughout the course of human history, social and cultural change has always been effected by, and reflected in, changes to systems of education. Current interest in teaching creativity within the field of science can be traced to a pervading belief that contemporary individuals and nations are living through a period of transition from old-world forms of work based on physical labor, to more intellectually intense, knowledge-based modes of operation.

This perception of a rapidly changing world, and a consequent need for new education and training paradigms (Calhoun, 2009; Coates & Goedegebuure, 2012; Douglass, Thomson, & Zhao, 2012; Frodeman, 2011; Hayden & Lam, 2007; Kitagawa & Oba, 2010; Lam, 2010; Obamba & Mwema, 2009; Oprescu, 2012; Ramoniene & Lanskoronskis, 2011; Whitchurch, 2012) emerges from interplay between personal and political conceptualizations of what it is, and what it means, to be creative and can be interpreted through reference to four distinct, but overlapping, discourses of creativity (Schmidt, 2011a, 2011b):

- *A developmental discourse*, which assumes that all individuals are capable of a degree of creativity that is commensurate with their level of cognitive development.
- *A psychometric discourse*, which is concerned with the interaction of internal and external traits, characteristics and events that can be measured, manipulated or exploited to predict, calculate or control creative output.
- *A sociocultural discourse*, which is concerned with the social, cultural and economic factors that stimulate, refine and sustain interest in creativity in the first instance; and the ways that these might generate or erode social and economic inequity at the level of individuals, communities and nations.
- *An entrepreneurial discourse,* which is concerned almost exclusively with the economic and commercial value of creative products.

M. K. Demetrikopoulos & J. L. Pecore (Eds.), Interplay of Creativity and Giftedness in Science, 153–186.

Rhetoric surrounding reforms to science education tends to focus on reversing a trend of declining enrolments in science subjects, the need to generate a technologically competent, scientifically literate workforce and the economic, environmental and social benefits associated with initiation and development of novel technologies and industries (Harris, 2012; Kessels, Rau, & Hannover, 2006; McWilliam, Poronnik, & Taylor, 2008; Universities Australia, 2012). This is consistent with sociocultural and entrepreneurial discourses, but lack of connectivity to concrete teaching and learning practices frustrates educators (Newton & Newton, 2009; Settlage, 2007) and failure to recognize the importance of moral and ethical frameworks in academic and educational settings poses a significant threat to quality and originality of intellectual output, particularly at the postgraduate and professional levels (Clegg, 2008; Schmidt, 2011a).

Explicit attention to pedagogy is a relatively new phenomenon in the tertiary education sector (Krause, 2012; Shay, 2012), but primary and secondary educators have a long history of translating developmental and psychometric theory to teaching and learning practice. In science education, it is widely recognized that development of key skills and knowledge is facilitated by well-planned and skillfully implemented learning programs that incorporate inquiry and argumentation activities (Barrow, 2006, 2010; Nadelson, 2009; Nancy Butler, Hee-Sun, & Scott, 2003; Nowak, 2007; Taylor, Jones, Broadwell, & Oppewal, 2008; William, 2005; Windschitl, Thompson, & Braaten, 2008).

To deploy inquiry methods in ways that develop creativity, it has been suggested (Schmidt, 2010, 2011a) that learning programs should incorporate opportunities for students at all levels to: 1) Acquire a high level of domain-specific knowledge; 2) Practice application of that knowledge across a gradient of difficulty and; 3) Link knowledge of science to knowledge of other fields in order to solve problems with personal relevance.

That the initial acquisition of domain-specific knowledge is highly dependent on fundamental (e.g. language, literacy and numeracy) skills is both consistent with a developmental approach and supported by empirical evidence. Prior academic performance is a significant predictor of achievement in secondary (Hogrebe & Tate, 2010) and tertiary science students (Universities Australia, 2012) and the ability to generate creative output is linked to above-average cognitive development/ability (Runco & Chand, 1995; Runco & Okuda, 1988; Sweller, 2009; Wu & Chiou, 2008). There is however, also strong evidence that those who go beyond knowledge accumulation and generate creative output display complex, and highly variable, combinations of social, psychological and intellectual characteristics (Boden, 2001; Christine & Glenn, 2007; Miller, 2000; Simonton, 2003a).

Studies of gifted and talented primary students highlight the importance of factors other than foundation skills. Dispositional elements such as emotional intelligence (Agnoli et al., 2012) and willingness to engage with, and respond flexibly to, challenge (Klavir & Gorodetsky, 2011) are strongly correlated with performance on academic tasks, but can be highly developed in students who would not be recognized as gifted in tests that examine academic skills alone (Klavir & Gorodetsky, 2011; Tzuriel, Bengio, & Kashy-Rosenbaum, 2011).

The significance of emotional-motivational factors also appears to increase as students progress through the education system. Studies of Italian students in the latter years of secondary schooling show that grade point average is strongly influenced by the extent to which students are able to manage emotions (DiFabio & Palazzeschi, 2009). Further, a study of Spanish students has shown that high teacher expectations and a positive learning environment in secondary school are some of the most powerful predictors of successful transition to post-compulsory education (Martín, Martínez-Arias, Marches, & Pérez, 2008).

An individual's early experience of schooling is therefore significant not only in terms of enabling or limiting access to further education and development opportunities and determining socioeconomic status, but also in shaping psychosocial orientations to self and others. Individuals' attitudes, beliefs and perceptions in relation to their own academic ability are strongly correlated with scholastic performance (Areepattamannil & Freeman, 2008; Griffin, Chavous, Cogburn, Branch, & Sellers, 2012) and studies of tertiary students from disadvantaged and/or non-dominant backgrounds show that interventions focused on resolution of intrapersonal tensions are more likely to result in program completion than those focused on content alone (Griffin et al., 2012; Reinheimer & McKenzie, 2011). Establishing and maintaining a positive, constructive orientation to learning may even be a particular requirement for success in science, as specific measures of emotional intelligence appear to be elevated in Bachelor of Science students, when compared to their Bachelor of Arts counterparts (Aslam & Ahmad, 2010).

To design and implement education programs that support and facilitate conversion of creative potential to creative output, educators must recognize the need for a more holistic approach to teaching and learning. Awareness of this is a driving force behind calls for greater personalization of learning experiences (Milliband, 2004; Verpoorten, Renson, Westera, & Specht, 2009). In a tertiary context, personalization has become synonymous with use of information and communications technology [ICT] (e.g. Beres, Magyar, & Turcsanyi-Szabo, 2012; Peter, Bacon, & Dastbaz, 2010; Sampson & Karagianidis, 2002; Tu, Sujo-Montes, Yen, Chan, & Blocher, 2012). In primary and secondary settings however, personalization is more accurately aligned with the concept of differentiation.

In recognizing that individuals within any given cohort of same-age students will differ in their life circumstances, past experiences, and readiness to learn (Tomlinson, 2000), proponents of differentiation advocate a dynamic, flexible approach to teaching and learning where teachers engage in on going adjustment of content, process, and products to ensure that all students are challenged to work slightly above what they can do independently (Rock, Gregg, Ellis, & Gable, 2008; Tomlinson, 1999).

In primary schools, attention to personal needs through small group instruction is up to four times as effective as undifferentiated, whole-class instruction

(Connor et al., 2010). For educators working with students at higher levels of education however, attempts to differentiate must overcome significant challenges. The first of these is low teacher-student ratios. In tertiary settings, these may realistically lie in the vicinity of one lecturer to several hundred students, which is one likely reason why ICT-mediated instruction has become so prevalent.

At a secondary level, there is greater recognition of the need for interpersonal connection and teacher-student ratios are more favorable. In this setting however, the challenge is not simply providing pathways from generic language and literacy skills to domain-specific proficiency, but doing so in a way that navigates sociocultural terrain characterized by challenges associated with access to material resources, relationships, identity, power and control, cultural adherence, social justice and personal cohesion (Fondacaro et al., 2006; Garbrecht, 2006).

The aim of this study is to investigate the impact of personalization on student learning in senior secondary students undertaking a two-year, tertiary preparation course in Chemistry. As in other countries, Australian secondary education is in a period of transition to national curriculum, but the study was undertaken in an environment where course content remained the mandate of The State of Queensland. The syllabus stipulates a requirement for context-based units, defined as provision of opportunities for students to learn "…in circumstances that are relevant and interesting to them…" with knowledge and understanding "… developed, consolidated and refined in, about and through the context" (Queensland Studies Authority, 2007, p. 45), but teachers in each school retain responsibility for writing and marking assessment tasks. To ensure that these comply with content and delivery requirements, folios of student work are reviewed by regional panels of experienced teachers at the end of the first (moderation) and second (verification) years of study (Queensland Studies Authority, 2007). The system is not without fault, but the approach is consistent with findings from targeted studies of high-performing schools and educators, which show that locally developed solutions to local issues and problems are a hallmark of quality education (Hargreaves & Shirley, 2009).

Skepticism about the utility of more generic, national testing regimes arises from evidence that test scores often map more accurately to sociocultural and socioeconomic status than to student ability (Cheng, Fox, & Zheng, 2007; Grodsky, Warren, & Felts, 2008; Hogrebe & Tate, 2010; Rubin, 2008). Despite significant correlations between performance on national and classroom tests, a majority of teachers believe that classroom assessment provides superior insights into student learning (Leighton, Gokiert, Cor, & Heffernan, 2010; McBride, Ysseldyke, Milone, & Stickney, 2010). Kyriakides (2004) has argued that one of the key reasons for this is that distancing classroom teachers from assessment processes constrains connectivity with interpersonal knowledge of the individuals within the classroom.

The core aim of this study was to examine the contribution that interpersonal knowledge makes to student learning. In particular, the aim was to determine whether personalization of assessment tasks delivers quantifiable improvements in student performance that are independent of general academic ability. The findings are then discussed in terms of the role personalized learning and assessment tasks might play in the education of gifted and talented students.

DATA COLLECTION AND ANALYSIS

The study population consisted of 79 (39 females, 40 males) 15 to 18 year-old Chemistry students from four cohorts (graduating years 2010, 2011, 2012 and 2013) attending a government funded, secondary school in Queensland, Australia.

To qualify for tertiary entrance in the Queensland system, students must complete four semesters (two years) of study in a minimum of five authority subjects that contribute to a tertiary ranking (Overall Position or OP) score. Chemistry is an authority subject and most students in the study were enrolled in five or six authority subjects (Chemistry plus four or five others) in total (Table 1). A small number (n=3) however, were electing not to apply for university and were also undertaking studies in non-authority subjects such as English Communication and Pre-Vocational Mathematics. As these students had all transferred after achieving non-pass grades in an authority subject (English or Mathematics A), further analysis used only results from the authority subject (to avoid distortion).

Complete (four semester) data were available for 26 students (2010 and 2011 cohorts). Less than four semesters of data were available for the remaining 53 students, because they were partway through the program of study (n = 29), transferred to other subjects (n = 21) or transferred to or from another school (n = 3).

Analysis was therefore based on average-to-date (fullest and latest) data for 53 (67%) students (2012 and 2013 cohorts) and complete (four semester) data for the remaining 26 (33%) (2010 and 2011 cohorts). All statistical analyses were undertaken using PASW® 18.0 software (SPSS Inc. 2009).

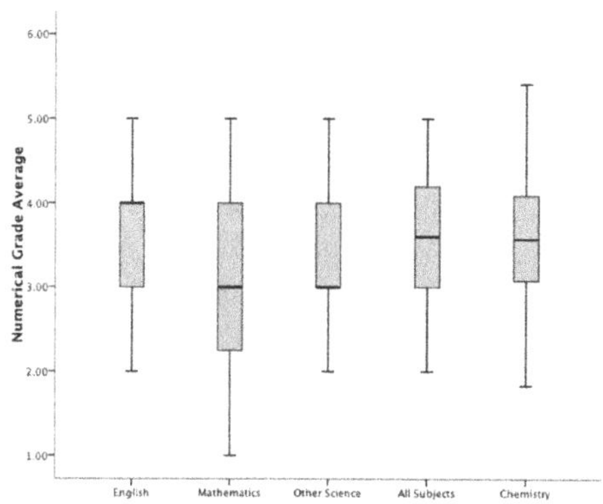

Figure 1. Academic performance across subjects
Deviations from normal distribution were detected for English and Other Science (Physics and/ or Biology). Chemistry grades were not significantly different to grades in other subjects

Table 1. Summary of academic achievement across subjects

Subject	*n*	*Mean* ± *S.E.*	*Skewness* ± *S.E.*	*Kurtosis* ± *S.E.*
English	75	3.52 ± 0.10	–0.12 ± 0.27	–0.52 ± 0.55
Mathematics A	20	3.38 ± 0.23	0.37 ± 0.51	–0.87 ± 0.99
Mathematics B	63	3.19 ± 0.14	0.01 ± 0.30	–0.57 ± 0.60
Mathematics C	16	3.69 ± 0.26	–0.11 ± 0.56	–1.09 ± 1.09
Physics	32	3.45 ± 0.18	0.06 ± 0.41	–1.00 ± 0.81
Biology	40	3.43 ± 0.12	0.17 ± 0.37	-0.11 ± 0.73
Legal Studies	5	3.00 ± 0.00		
History		3.74 ± 0.18	–0.33 ± 0.52	–0.22 ± 1.01
Ancient	9			
Modern	8			
Geography	1	3.00*		
Health & Physical Education	8	4.12 ± 0.23	–0.07 ± 0.75	0.74 ± 1.48
ICT/Business	13	3.38 ± 0.23	–0.28 ± 0.50	0.11 ± 0.97
Language		4.33 ± 0.14	0.57 ± 0.50	–0.26 ± 0.97
Japanese	10			
German	11			
Art	7	3.71 ± 0.42	–0.25 ± 0.79	–0.94 ± 1.59
Graphics	6	4.67 ± 0.21	–0.97 ± 0.85	–1.88 ± 1.74
Technology Studies	11	3.95 ± 0.26	–1.17 ± 0.66	2.12 ± 1.28
Music	11	4.09 ± 0.24	0.01 ± 0.66	–1.57 ± 1.28

**Average, skewness and kurtosis statistics were not calculated for Geography (n = 1).*

Academic achievement. To generate comparable measures of achievement for each cohort, A to E grades were converted to numerical variables (A = 5, B = 4, C = 3, D = 2, E = 1) and averaged across subject groups (Figure 1). Single-subject averages were calculated for English, Mathematics and Other Sciences (Physics and Biology), but due to relatively small numbers of students in other subjects (Table 1), a single All Subject Average was considered more viable than separation into discipline groups (e.g. combining History and Art to give an index of achievement in the Humanities).

This produced four new variables summarizing achievement in English, Mathematics, Other Sciences (Biology and Chemistry) and All Subjects. Deviations from normal distribution were detected for the English (S-W statistic = 0.889, df = 75, $p < 0.001$) and Other Science (S-W statistic = 0.919, df = 64, $p < 0.001$) variables

(Figure 1), but as the dataset was transformed prior to further analysis (see below), this was not problematic.

Variable reduction. Subject-specific data were converted to a more general index of academic capacity through principal components analysis. Given the high inter-correlation of grades across subjects, and an increase in sample size when the dataset was not limited to students studying an additional science (Physics or Biology), only the English, Mathematics and All Subjects indices were subjected to further analysis.

The data reduction procedure used the matrix of covariance, with pairwise elimination of cases missing data for one or more of the original subject-specific indices (n = 74) and varimax rotation. This analysis generated a single variance component with an eigenvalue greater than one (2.492), which explained 83% of the covariance. Subject-specific loadings were high for all three of the subject specific indices (English = 0.831; Mathematics = 0.937; All Subjects = 0.962). Individual scores were saved and used as an Academic Performance Index (API) in further analysis.

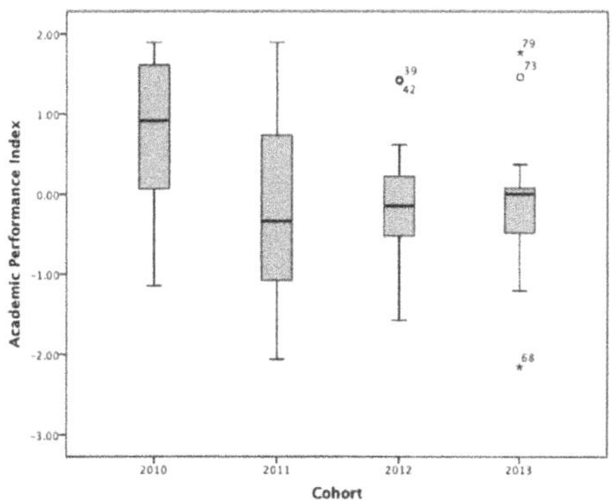

Figure 2. Academic performance index across cohorts

Some variation was evident across cohorts (Figure 2), but standardized residuals for the entire data set (2010 n = 12; 2011 n = 23; 2012 n = 22; 2013 n = 17) were normally distributed (Shapiro-Wilk stat. = 0.978, df = 74; p = 0.234; Skewness = 0.014 ± 0.279; Kurtosis = –0.459 ± 0.552).

Chemistry Data: A preliminary ANOVA indicated that achievement in Chemistry was not significantly different from achievement in other subjects (Figure 1). Variation within groups was not significantly different to variation between groups for English (F = 1.449; df = 71, 2; p = 0.495), Mathematics (F = 2.098; df = 72, 2;

p = 0.377), Other Science (F = 1.402 df = 61, 1; p = 0.598) or All Subjects (F = 3.821; df = 74, 2; p = 0.230) (Figure 4).

Performance in Chemistry varied across cohorts (Figure 2), but data were normally distributed when all cohorts were combined (S-W stat. = 0.981, df = 79, p = 0.271; skewness = 0.181 ± 0.271; kurtosis = –0.235 ± 0.535).

To capture maximum information regarding the impact of personalized assessment tasks, a range of Chemistry-specific achievement indices were generated, based on three mandated (syllabus) dimensions designated Knowledge and Conceptual Understanding (KCU), Investigative Processes (IP) and Evaluating and Concluding (EC).

According to the Senior Chemistry Syllabus (QSA, 2007):

- The KCU mark indicates the extent to which students are able to recall and interpret concepts, theories and principles; describe and explain processes and phenomena; and link and apply algorithms, concepts, theories and schema.

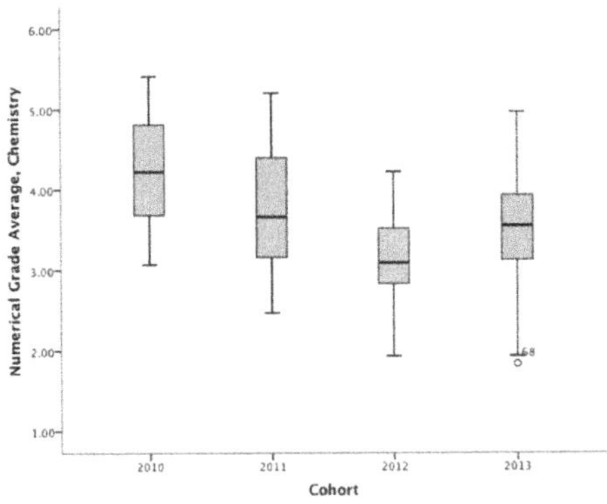

Figure 3. Chemistry grades across cohorts

- The IP mark indicates the extent to which students can conduct and appraise research tasks; operate chemical equipment and technology; and use primary and secondary data.
- The EC mark indicates the extent to which a student can determine, analyze and evaluate chemical interrelationships; predict outcomes and justify conclusions and recommendations; and communicate using a range of formats.

In addition to calculations of separate KCU, IP and EC averages, an overall level of achievement was generated by averaging all KCU, IP and EC grades.

To allow discrimination within levels of achievement, conversion of grades to numerical variables (A = 5, B = 4, C = 3, D = 2, E = 1) included plus and minus levels of achievement as decimal components. Plus grades were designated *X*.9

(e.g. $B^+ = 4.9$), mid-range grades were designated *X*.5 (e.g. C = 3.5) and minus grades were designated *X*.1 (e.g. $A^- = 5.1$).

Classification of Assessment Tasks: Categorization of assessment tasks was based on three instrument types recognized by the Queensland Chemistry syllabus (QSA, 2007):

- Supervised Assessment (SA): Instruments administered under supervised conditions to ensure authenticity of student work that may include short items, practical exercises, paragraph responses and responses to seen or unseen stimulus materials
- Extended Response Task (ERT): Instruments developed in response to a chemical question, circumstance or issues that are essentially non-experimental, but may draw on primary experimental data.
- Extended Experimental Investigation (EEI): Instruments developed to investigate a hypothesis or answer practical research questions through laboratory or field based self-directed experimentation and reporting.

Within each of these categories, the degree of personalization was assigned relative to the extent of student choice.

Tasks with low personalization (Non-Personalized) included six standard written exams (SA), where all students were required to generate responses to an identical set of multiple choice/short answer/medium length answer questions with a single opportunity to choose from one of two longer, complex questions at the end of the paper (Table 2).

Tasks with a medium level of personalization (Medium Personalization) included two EEIs and one SA (Table 2). These were classified as having a medium level of personalization because the overall problem to be solved required development/application of similar methods and techniques for all students and, although all individuals were required to produce individual outputs, a substantial degree of collaboration and overlap was possible in generation of solutions.

Tasks with a high degree of personalization (High Personalization) included one SA and two ERTs (Table 2). Like the EEIs, these tasks required all students to solve problems and produce outputs that were conceptually similar, but a key point of difference was that high-personalization meant the system to be studied was self-determined; although teacher assistance was provided and students who had difficulty choosing a system were given more substantial direction.

Using these groupings, eight Chemistry-specific achievement indices were generated by averaging grades (Overall, KCU, IP and EC) across personalized (P) and non-personalized (NP) tasks. These formed the central focus of the analysis, but additional sets of indices were later generated for medium and high personalization tasks.

Table 2. Summary of assessment tasks

Year Level	*Task No.*	*Type*	*Description of Task*		*Chemical Content*	*Product(s)*	*Personalization*
11	1a	SA	Practical Exam 3 x 70 minute lessons	Students provided with a set of reagents and relevant background information related to a central reaction in an area of self-nominated interest. Requires replication of the reaction in the laboratory and explanation of utility in both general and chemical terms. Focus is on linkage between experimental work and use/manipulation of chemical symbols and equations.	Molecular structure, ions, bonding, writing and balancing equations, mole concept, theoretical and empirical yield	Laboratory and research journal 10 minute presentation	High
	1b	SA	Theory Exam 90 minutes	Standard written examination: Mixture of multiple choice, short answer and extended response questions.	Atomic structure, electron configuration, ions, bonding, periodic trends, writing and balancing equations, mole concept	Written answers to examination questions	Low

Table 2. (Continued)

Year Level	*Task No.*	*Type*		*Description of Task*	*Chemical Content*	*Product(s)*	*Personalization*
	2a	ERT	Research Task 9 x 70 minute lessons	Students select a contemporary chemical topic (e.g. use of acid leaching in engineering/mining, drug identification in rainforest plants/fungi, identification of food trees in koala ecology, teeth whitening procedures in dentistry, nutritional content of junk foods) and conduct independent research into the underlying chemistry of the system, seeking information relating to practices designed to manipulate reaction rate and yield through manipulation of temperature, concentration, enzymes and catalysts.	Reaction rate, effect of temperature, concentration, enzymes and catalysts on reaction rate, rate laws	Research journal 1000 word written report	High
	2b	SA	Theory Exam 90 minutes	Standard written examination: Mixture of multiple choice, short answer and extended response questions	Endothermic and exothermic reactions, reaction rate, factors that affect reaction rate	Written answers to examination questions	Low
	3	SA	Theory Exam 90 minutes	Standard written examination: Mixture of multiple choice, short answer and extended response questions	Solubility, precipitation and acid-base chemistry	Written answers to examination questions	Low

(Continued)

Table 2. (Continued)

Year Level	*Task No.*	*Type*	*Description of Task*		*Chemical Content*	*Product(s)*	*Personalization*
	4a	EEI	Experimentation and Research Task 18 x 70 minute lessons	Students are presented with a forensic science sample kit and supporting story and asked to apply knowledge to design and conduct experiments that will determine chemical identity contents of 2 x liquid, 1 x metal and 4 x soil samples.	Analytical chemistry, quantitative chemical analysis (solubility, chemical identity)	Laboratory and research journal 2000 word written report	Medium
	4b	SA	Theory Exam 90 minutes	Standard written examination: Mixture of multiple choice, short answer and extended response questions	Quantitative analytical chemistry, acid-base chemistry	Written answers to examination questions	Low
12	5a	EEI	Experimentation and Research Task 18-24 x 70 minute lessons	Students adopt the role of resident chemist in establishing a self-sustaining research facility in a pristine landscape (each student assigned a unique bioregion). Task requires design and conduct of experiments that will allow them to identify, develop and refine sustainable methods of providing food, drinking water, fuel, refrigeration and electricity for the facility.	Organic chemistry, calorimetry, electrochemistry	Laboratory and research journal 10 minute presentation 2500 word written report	Medium

Table 2. (Continued)

Year Level	*Task No.*	*Type*	*Description of Task*		*Chemical Content*	*Product(s)*	*Personalization*
	5b	SA	Theory Exam 90 minutes	Standard written examination: Mixture of multiple choice, short answer and extended response questions		Written answers to examination questions	Low
	6a	ERT	Research Task 12 x 70 minute lessons	Students select an equilibrium system of commercial, social, biological, ecological or historical significance and conduct independent research into how kinetics of the system have been and can be manipulated to develop, maintain and improve chemical outcomes for the benefit of individuals, communities or industries.	Chemical equilibrium, equilibrium constants, reaction quotient, Le Chatelier's principle	Research journal 2500 word written report	High
	6b	SA	Theory Exam 90 minutes	Standard written examination: Mixture of multiple choice, short answer and extended response questions		Written answers to examination questions	Low

(Continued)

Table 2. (Continued)

Year Level	*Task No.*	*Type*	*Description of Task*		*Chemical Content*	*Product(s)*	*Personalization*
	7	SA	Practical Exam 3 x 70 minute lessons	Students are randomly allocated a specific clock reaction (maximum three students per reaction) and asked to develop and refine a strategy for achieving a permanent color change on, or close to, a designated target time (unique for each student). Students have the option of working individually, or with other students, during the experimentation and demonstration stages, but are required to submit individual laboratory/research notes.	Oxidation numbers, reduction-oxidation reactions and systems	Laboratory and research journal 10 minute practical demonstration	Medium
	8a	ERT	Experimentation and Research Task 10 x 70 minute lessons	Students select two from a folio of ten problems/questions and conduct independent research and experimentation to generate solutions/answers. Journal from Task 8a is taken into Task 8b examination.	Quantitative analytical chemistry, organic chemistry, polymer chemistry, reduction-oxidation reactions and systems, acid-base and buffer systems	Laboratory and research journal	Medium
	8b	SA	Theory Exam 90 minutes	Tailored written exam, with questions based on topics investigated during Task 8a.		Written answers to examination questions	Medium

Table 3. Correlation between achievement on personalized and non-personalized assessment tasks

	Pers. KCU	Pers. IP	Pers. EC	Pers. Overall	Non-Pers. KCU	Non-Pers. IP	Non-Pers. EC	Non-Pers. Overall
Pers. KCU	–							
Pers. IP	0.937	–						
Pers. EC	0.948	0.972	–					
Pers. Overall	0.962	0.989	0.991	–				
Non-Pers. KCU	0.933	0.805	0.817	0.824	–			
Non-Pers. IP	0.879	0.790	0.826	0.817	0.864	–		
Non-Pers. EC	0.936	0.847	0.852	0.861	0.955	0.887	–	
Non-Pers. Overall	0.948	0.844	0.862	0.865	0.974	0.945	0.981	–

All linear (Pearson's r) correlations between variables were significant at the $p < 0.001$ level for all pairwise comparisons.

Differences between Personalized and Non-Personalized Tasks: To determine whether variation in performance on personalized and non-personalized tasks was due to differences in general academic capacity required control for high levels of inter-correlation (Table 3).

An index of differential performance was generated through linear regression (dependent variable: Personalized Overall, independent variable: Non-Personalized Overall). The regression model (Figure 4) was highly significant ($R = 0.865$; $R^2 = 0.748$, $F = 222.496$; $p < 0.001$) and standardized residual scores were retained for use in regression (generalized linear model) against the Academic Performance Index.

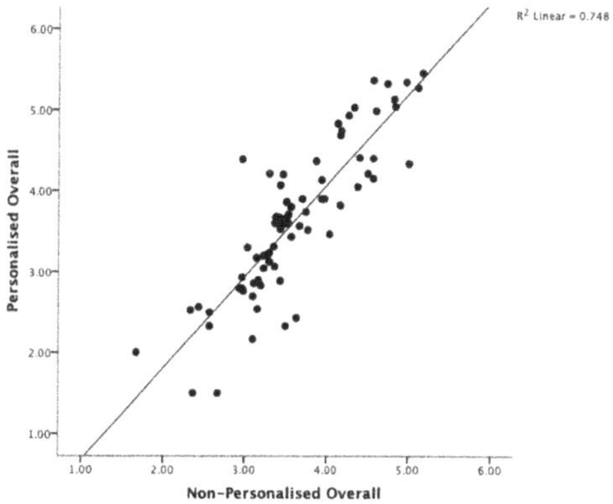

Figure 4. Linear relationship between performance on personalized and non-personalized assessment tasks

Table 4. Linear relationships between performance on personalized and non-personalized assessment tasks

Independent Variable	*Dependent Variable*	*R*	R^2	*df*	*F*	*P*
Non-personalized KCU	Personalized KCU	0.933	0.871	1, 76	514.505	<0.001
Non-personalized IP	Personalized IP	0.790	0.624	1, 75	124.408	<0.001
Non-personalized EC	Personalized EC	0.852	0.726	1, 75	199.007	<0.001
Non-personalized overall	Personalized overall	0.865	0.748	1, 75	222.496	<0.001

Similar regression-based transformations were performed for the personalized and non-personalized KCU, IP and EC grades (Table 4).

To determine whether KCU, IP and EC performance varied for personalized and non-personalized tasks, a series of comparative analyses were undertaken. The nature of the dataset meant that viability of parametric methods could not be confirmed through tests for homogeneity of variance and relatively low-power non-parametric tests were adopted as a conservative alternative.

To determine whether grades for personalized and non-personalized tasks were consistently similar or different, Kendall's co-efficient of agreement (W) was calculated for the Overall, KCU, IP and EC datasets.

Post-hoc pairwise comparisons between personalized and non-personalized KCU, IP and EC grades were then performed using Wilcoxon's signed ranks test, with a Bonferroni correction to significance levels ($\alpha = 0.05$/number of tests).

RESULTS

The academic performance index accounted for up to 57% of the variation in performance on personalized tasks (Figure 5a) and 82% of variation in performance on non-personalized (Figure 5b) tasks (Table 6). There was no significant linear relationship (Figure 5c) between API and differences in performance on personalized and non-personalized tasks (Table 6).

Table 5. Linear relationships between academic performance index and performance on chemistry assessment tasks

Independent Variable	*Dependent Variable*	*R*	R^2	*df*	*F*	*P*
Academic performance index	Personalized overall	0.752	0.566	1, 71	92.523	<0.001
	Non-personalized overall	0.817	0.663	1, 70	140.402	<0.001
	Personalized × non-personalized residual	0.086	0.007	1, 70	0.516	0.475

Table 6. Linear relationships between academic performance index and differences in performance on personalized and non-personalized assessment tasks

Dependent Variable	*R*	R^2	*df*	*F*	p_1
KCU Residual	0.102	0.010	1,70	0.730	0.396
IP Residual	0.197	0.039	1,70	2.815	0.098
EC Residual	0.101	0.010	1,70	0.726	0.397

When KCU, IP and EC residuals were analyzed separately, the Academic Performance Index explained no more than 10% of variation in KCU and EC (Table 6). The percentage of variance explained rose to 20% for the IP residual (Table 6), but none of the regression models were significant (Figure 6).

The Friedman test ($X^2 = 60.290$, $df = 5$; $p < 0.001$; n = 77) indicated that there were significant differences between personalized and non-personalized achievement (Figure 7). Pairwise Wilcoxon-rank tests indicate that the difference was due to a tendency for individuals to score higher for KCU on non-personalized tasks, and higher for IP on personalized tasks (Table 7). No significant differences in performance were detected for medium or high-level personalization (Table 7).

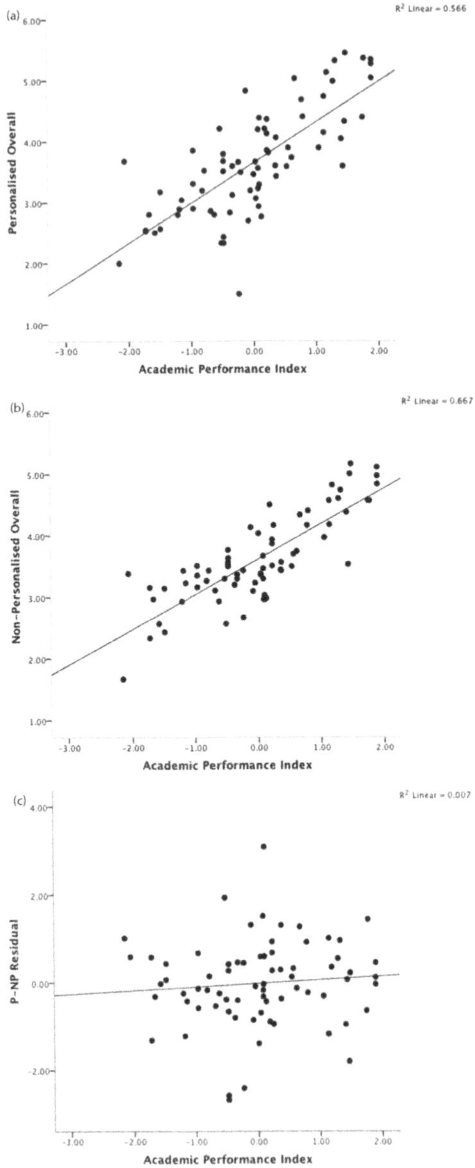

Figure 5. Achievement on chemistry assessment tasks as a function of academic performance index
The academic performance index showed a significant ($p < 0.001$) linear relationship with performance on both personalized (a) and non-personalized (b) tasks. Differences in performance on personalized and non-personalized tasks (c) were not significantly related to the academic performance index.

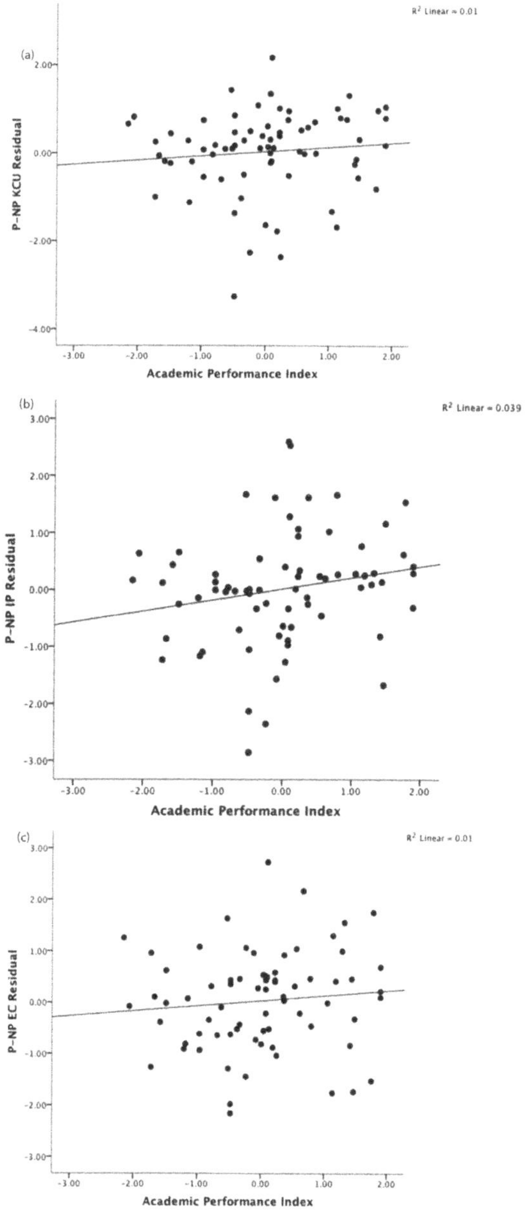

Figure 6. Differences in achievement on personalized and non-personalized tasks as a function of academic performance index

Table 7. Wilcoxon-Rank tests for differences in performance on personalized and non-personalized tasks

Dimension	*N*	*Z*	*p*	*Directionality*
KCU	78	–2.390	0.017	Non-personalized > Personalized
Med. vs. High		–1.143	0.253	No difference
IP	77	–3.503	<0.001	Non-personalized < Personalized
Med. vs. High		–0.034	0.973	No difference
EC	77	–0.107	0.915	No difference
Med. vs. High		–2.030	0.042	No difference

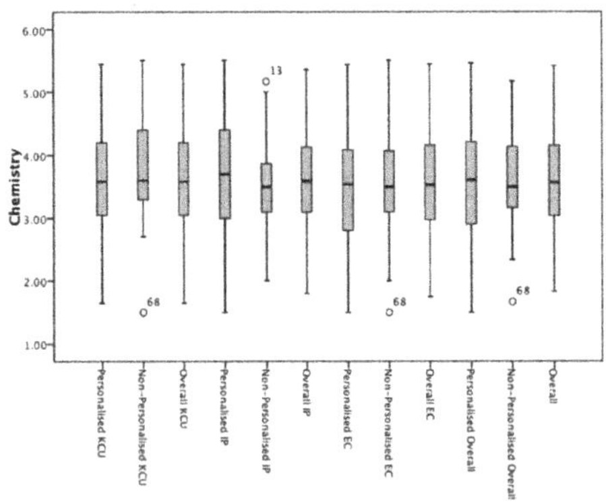

Figure 7. KCU, IP and EC achievement on personalized and non-personalized assessment tasks

DISCUSSION

Concern that numbers of academically competent students progressing to tertiary study of science are insufficient to meet the needs of new knowledge-based industries is catalyzing global reform of science education. Arguments for change often evoke a rhetoric of teacher deficiency, epitomized by statements about teachers that are "boring or lacking in subject knowledge" and the need to teach science "earlier and better" (p. 1; Universities Australia, 2012). As well as perpetuating a dysfunctional mythology that the ability to do, and therefore teach, science is a unique trait possessed by a relatively small number of elite individuals, such statements have limited utility within the classroom. This study demonstrates that there is a strong correlation between general academic ability and performance in senior chemistry, but as predicted by teaching and learning theory, a holistic, personalized approach to teaching and assessment has quantifiable potential to enrich the learning experiences of individuals and develop crucial awareness of the philosophies and practices of science. The decision to omit measurement of affective and social factors from this study was justified because any difference in performance on personalized and non-personalized tasks would be statistically significant only if it were able to transcend inter and intra personal factors. That tailoring assessment tasks to individual interests generates significant differences in engagement and/or investigative skills indicates that reform of science education should take care not to constrain flexibility.

Standardized Assessment

That general academic ability is a powerful predictor of grades is entirely consistent with expectations. Basic aptitude for learning is heritable (Vinkhuyzen, vanderSluis, Posthuma, & Boomsma, 2009) and individuals who do well in Mathematics would therefore be expected to perform well in subjects such as English and Science. It is also widely known that increases in core language, literacy and numeracy skills correlate with increased performance across all academic fields (Council for the Australian Federation, 2007; Hilton, 2006; Rubin, 2008; Sara, David, & Anthony, 2007; State of Queensland, 2002; Volante & Ben Jaafar, 2008).

The latter point has been used to justify standardization or nationalization of assessment in many countries, but critiques of national testing regimes suggest that one of their more insidious effects is establishment of merit-demerit cultures that reinforce disengagement of students who most need support, and encourage teachers to abandon creative, reflexive practices that foster higher-order thinking in favor of narrow, prescriptive methods designed solely to elevate test scores (Anagnostopoulos, 2006; Batagiannis, 2007; Creese, 2005; Hartley, 2008; Kyriakides, 2004; Leighton et al., 2010; Manzo, 2003; Schulte, Schulte, Slate, & Brooks, 2002).

Although criticism that national testing undermines the abilities of education professionals to diagnose the unique and situated instructional requirements of individual students has focused primarily on language, literacy and numeracy testing in lower grade levels (Kyriakides, 2004; Nagy, 2000), it is relevant in this context. Moon et al. (2003) have shown that classroom environments focused on external testing generate boredom and resentment in high ability primary students and emphasizing external measures of competitive attainment frustrates both performance and engagement even in high-achieving tertiary cohorts (Stallman, 2012).

The reality of teaching and learning practice is that reforms emphasizing external, nationalized tests of ability and aptitude reduce, rather than enhance, differentiated practice (Anderson, 2012; Moon et al., 2003). Tailoring tasks to meet the needs of different individuals and cohorts also requires adequate time for preparation, planning and reflection. This is acknowledged in some systems, where teachers are given a maximum of three classes (Gao, 2011), but it is also important to note that this is often due to expectations that primary and secondary teachers should be active in educational research and publication. This is not always realistic because it underestimates the value of time spent preparing individual learning plans for multiple classes, each of which may contain between twenty and thirty students. The tertiary sector is recognizing that there are reasons to separate the functions of teaching and research (M. Barrow & Grant, 2011; Bexley, Arkoudis, & James, 2012; Blackmore, 2009; Myer & Evans, 2005; Nair, Bennett, & Mertova, 2010; Ramoniene & Lanskoronskis, 2011), and the primary and secondary sectors must also acknowledge that imposition of research loads will constrain teaching.

Teacher Quality

By the time students reach the post-compulsory level, psychosocial factors become at least as, if not more, important than innate ability. In a study of students from over fifty countries, Montt (2011) found that opportunity to learn is crucial for student achievement, but links this to generic notions of teacher quality rather than any concrete recommendations for teaching practice.

Studies that do attempt to articulate a basis for quality teaching in science education often emphasize the importance of inquiry methods. Publications of this nature include countless theoretical expositions and applied examples of the inquiry method; the majority of which suggest, imply or demonstrate that inquiry is effective, while a handful focus on issues and problems with implementation in various settings. What this study adds to the body of literature is quantitative evidence that inquiry works because it goes beyond development of domain-specific knowledge and cultivates intrinsic motivation to learn.

In a secondary context, the problem is not that there has been no clear articulation of what constitutes an inquiry-based learning program, but that its manifestation can and should vary. Implementing inquiry requires educators who are able to diagnose, and respond to, the prior knowledge and metacognitive abilities of specific cohorts and individuals. This is one reason why teachers with similar sociocultural backgrounds to their students are often crucial (Kelly-Jackson & Jackson, 2011). It is important to note however, that this contradicts, rather than supports perceptions that those who would make good teachers can be identified prior to engagement with the profession.

Efficacy of education in a secondary context is heavily dependent on systems of shared belief. When teachers and students believe that they are working toward common goals, within a just and fair framework of attainment, the end result is an authenticity of self and society (Resh, 2009) that is reflected in, but not limited to, variations in power dynamics between students and teachers across different educational environments. In France for example, students expect, and therefore respond to, teachers who are distant and authoritative, but in the Netherlands, a more relaxed, informal approach delivers stronger interpersonal connection/validation (Hornikx, 2011).

The key point here is that attempts to articulate, ascertain or predict teacher quality are often counterproductive. A study of 368 education students show that specific personality types may be attracted to specific areas such as special education, or mathematics teaching (Rushton, Mariano, & Wallace, 2012) and there is no doubt that certain characteristics, such as suspension of judgment and flexibility, are essential when dealing with adolescents, but the exact mix of personal and academic characteristics required for success depends on complex, dynamic interactions that are underappreciated by those outside of the profession. Shortages of teachers with formal qualifications in pure science are a product of these interactions. Gaps between pedagogical and content knowledge may be filled through targeted programs such as content-specific Masters qualifications (Huntoon & Baltensperger, 2012), but content knowledge will not compensate for a disposition that is incompatible with teaching in general, or specific, contexts (Gawlik, Kearney, Addonizio, & LaPlante-Sosnowsky, 2010).

Personalized Education

The first indication of a difference in performance on personalized versus non-personalized tasks comes from the fact that general academic ability explains up to 87% of variance in performance on written exams, but only 53% of variance on ERTs and EEIs. To understand the full significance of this finding, it is necessary to consider the suite of skills and abilities that are tested in each type of task.

The structure of the QSA syllabus, and learning programs that are consistent with it, is such that the overall grade is derived from a combination of KCU, IP and EC. The KCU and EC strands map to classic conceptualizations of attributes that students should develop through exposure to secondary chemistry. In the case of KCU, this includes tasks such as reading and manipulating chemical formula and equations, and the quantitative information that pertains to, or arises from, them. In students of high ability, understanding of algorithms and procedures should be developed to such an extent that they can rearrange and reconfigure problem-solving schemata to fit unfamiliar scenarios. The EC strand focuses on articulating and conveying the meaning of chemical information and data in different contexts. These two strands have analogues in almost all areas of human endeavor, but IP is unique to science in that it focuses on philosophical frameworks based on generation and testing of hypotheses linked to the physical manipulation of scientific systems or models.

Written exams do not provide extensive opportunity for students to demonstrate IP skills because they are, by definition, generic question sets that are answered by all individuals in a given cohort or class: Responding with peripheral information detracts from, rather than adds to, the quality of the response. Written exams are important tools for allowing students to demonstrate KCU and EC, but IP is more effectively and appropriately assessed by other means. A disproportionate IP loading is therefore a diagnostic feature of any task other than a written exam because it requires the individual to explore what lies behind and beyond the model. ERTs and EEIs do, however, retain high loadings for KCU and EC. Task 1a, for example, is conceptually no different to a written exam in that the basic questions (mole/molarity and yield calculations) are identical for each student. What differs in this case is not the core content, (opportunity to demonstrate KCU), but the context (the system under investigation).

Despite differences in the number of assessment items included for each cohort, the fact that methods of generating indices of student performance did not capture information about exact traits and abilities associated with particular subjects (e.g. music versus mathematics) and a general trend for student performance to decline on transition to the senior years, the unique nature of the IP construct is supported by the results: KCU, IP and EC show differing degrees of dependence on general academic ability and, while KCU tends to be higher for written exams, IP reaches its maximum for all students when they are engaged in experimentation and research.

There is a degree of circularity in this. Performance against IP criteria is higher when tasks are personalized because any given example of a genuine inquiry task must be, to at least some extent, self-directed, but what this really means in terms of the impact of personalized research and experimentation (inquiry) tasks is that what is reflected in the IP grade is a combination of investigative ability *per se*, and the extent to which the individual engages with the process of investigation.

If we are serious about developing and maintaining a capacity for creativity within the field, science educators must not underestimate the significance of this point.

Personality Factors

A survey of 1,100 tertiary science students from the Netherlands shows that motivational factors do vary by discipline, with Law and Humanities students driven by generic conceptualizations of excellence, while physics students were motivated by the idea of learning itself (Scager et al., 2012). A commissioned study of Australian tertiary students enrolled in Science, Technology, Engineering and Mathematics subjects however, indicates that career and/or lifestyle aspirations have greater significance for science students than students of the humanities (Universities Australia, 2012). The Australian study also claims that science students are more likely to be identified as one of sixteen unique personality types (ISTJ - Introverted, Sensory, Thinking and Judgmental) on the Myer-Briggs personality scale.

There are two issues associated with this statement. The first is relatively minor in that the authors ignore the fact that this is one of three (from 16 in total) personality types that are also overrepresented in the general population. The second point however, is problematic because it reinforces perceptions that those who wish to succeed in science must be in possession of a set of pre-existing traits, characteristics and skills before they enter the science classroom. This is simply not consistent with what is known about how and why we learn.

The superhuman intellect of the uniquely creative 'mad scientist' is a myth: A study of 291 eminent individuals recognized as creative in their field of endeavor actually demonstrated that scientists are the least likely to display traits associated with, or predictive of, mental illness (Glazer, 2009) and creative output hinges upon essentially random, unpredictable interactions between personal, social and environmental characteristics (Simonton, 2003b). The likelihood of creative output however, can be increased by providing individuals with opportunities to develop high levels of domain-specific knowledge, become competent at applying it and find relevance in areas of personal interest (Schmidt, 2010, 2011a).

Highly variable, context-specific affective and interpersonal factors are, by nature, difficult to control and measure. A study of 579 British undergraduates for example, found no significant link between intelligence and learning style and only 25% of variance in learning was explained by the interplay between intelligence and personality (vonStumm & Furnham, 2012).

Societal Factors

Decreasing the variability in quality and quantity of learning experiences is an important step toward construction of a society where achievement in various fields of endeavor arises through talent, ambition and effort rather than the perpetuation of discriminatory policies and practices (Dewey, 1916; Montt, 2011). Widening participation in tertiary education is also an important mechanism of change because it is linked to emotional, mental and economic health (Cheung & Chan, 2009), but tertiary institutions represent one part of a far broader educational system.

Regardless of the field of endeavor, the current pace of social and technological change means what is taught or learned in education and training will be irrelevant to workplace practice within five years (Kilpatrick & Allen, 2001). Any reform of educational policy and practice will therefore be ineffective in the longer-term unless it is enacted in a manner which acknowledges that prescriptive approaches will only ever meet the needs of a relatively small number of individuals, for a limited period of time (Belanger, 1999). Rowlands (2011) places this in context by pointing out that the reality of scientific practice in the 21st century is multidisciplinary. In this environment, skills and knowledge are, and must be, acquired as required and meaningful creativity depends on intrinsic engagement.

Engagement with learning in any field is invariably personal. Triggering and sustaining student interest requires recognition that interest itself is unique because it consists of both cognitive and affective elements (Hidi, 2006). Kauffman et al. (2008) have previously cautioned against a tendency to misinterpret activities such as one-to-one instruction as genuine personalization and their point is supported by empirical evidence. A study of 123 undergraduates showed that neither learning style nor personality traits predict engagement with specific (ICT-supported) learning tasks (Nilsson et al., 2012), but students are less distracted and more engaged with learning when given materials that connect to areas of personal interest (Danzi, Reul, & Smith, 2008).

Vertical Alignment

Surveys of science teachers in middle and high school environments reveal deep awareness that calls for personalization and inquiry give rise to conflicting messages about good practice. Administration and government bodies emphasize a need to develop general academic skills, but tertiary science institutions and science education academics insist that rich, open-ended inquiry tasks are the only effective way to deliver quality science education (Aydeniz & Southerland, 2012).

Tensions between sectors are not unique to science education. All reform takes place in contested sociocultural space and delivers both positive engagement of teachers, and improved student outcomes, only when it is planned, designed and implemented through systems based on trust and mutual influence (Afdal, 2012). Recognition that quality education systems must allow flexibility in delivery of content is a hallmark of high-achievement: The high-performing Finnish system for example, is currently undergoing reform to restore teacher autonomy and increase recognition that progression to tertiary study is not, and should not be, the sole aim of the secondary system (Pyhalto, Soini, & Pietarinen, 2011).

This is a significant point. Despite widening participation in post-compulsory education, youth unemployment remains high even in OECD countries (Quintini & Martin, 2006) and over education creates as many problems as under education for individuals, communities and nations (Barone & Ortiz; Linsley, 2005; Messinis, 2007; Quinn & Rubb, 2011; Romanov, Tur-Sinae, & Eizman, 2008). This is particularly true in the sciences, where over graduation of PhD students has previously created an employment and training crises (Kendall, 2002; Gemme & Gingras, 2012; McCulloch & Thomas, 2012).

Declining enrolments in science subjects at secondary and tertiary level are potentially problematic, but tertiary science educators and practicing scientists are calling for reform of the secondary sector without any significant appreciation of the policies and practices that govern this domain. Numerous tertiary science educators for example, are operating in an environment of increased accountability for their own teaching and learning practices. As they encounter issues and problems associated with definitions and perceptions of inquiry, they assume that their own experiences are paralleled in the secondary sector. Buck et al. (2008) for example, point out that the call to inquiry in undergraduate education is ubiquitous, but claim that there has been little to no clarification of what inquiry means in terms of teaching and learning practice and Herron (2009) points out that this is particularly problematic in an environment where teaching staff are drawn from the ranks of graduate students, few of whom have any awareness of, or appreciation for, teaching and learning theory.

Gifted Stereotypes

In relation to the field of gifted and talented education, the most salient point to be drawn from the study is that skilled and competent use of personalized assessment tasks offers students with the highest levels of aptitude and interest an opportunity to achieve their academic potential without compromising their emotional and social needs.

Previous authors have noted that the perception of gifted and talented students as more likely to demonstrate personality difficulties, or issues with authority, is widespread, even among teachers (Geake & Gross, 2008). There is however, no evidence that this is the case. When age-appropriate personality tests are deployed in direct studies of gifted students, there is simply no evidence that gifted individuals are more likely to display personality disorders (Bain & Bell, 2004; Cross et al., 2008). There is, however, evidence that social and educational environments that impose labels and support negative stereotypes of gifted students can have a significant negative impact on the self-concept of gifted adolescents (Berlin, 2009; Eddles-Hirsch et al., 2010; Reis & Renzulli, 2009).

At a secondary level, gifted students, like their average peers, are generally grappling with issues of identity (Cross & Frazier, 2010) and this can manifest in a wide range of behaviors. Although gifted individuals often possess high levels of self-management and empathy (Chan, 2003), on the whole, the gifted population is no more homogenous than the general population (Reis & Renzulli, 2009). A study of gifted adolescents attending a residential high school for gifted students, for example, showed no correlation between social or academic success and IQ (Woitaszewski & Aalsha, 2004). Similarly, Skaar and Williams (2012) study of 15–24 year old students shows that those with high emotional intelligence and general IQ may be more likely to recognize risky and/or dysfunctional behavior, but this has little impact on their likelihood of engaging in these behaviors. Chan's (2003) study of 259 gifted adolescents also shows that measures of emotional intelligence have limited ability to predict which coping strategies any individual will adopt when faced with social and/or emotional challenge.

The capacity of gifted students to vary their coping strategies in response to different social environments was also noted by Cross and Frazier (2010), who suggested that differential treatment of gifted and talented students encourages these individuals to disguise their intellectual aptitude to avoid undesirable social consequences.

There is then, a clear case for deployment of teaching and learning strategies that reduce implicit reinforcement of negative stereotypes (in this case, the maladjusted, dysfunctional and socially inept gifted student). In fact, Peterson and Ray (2006) have suggested that integrated programs that emphasize the quality of education for all students are powerful mechanisms for reducing social problems such as bullying and other forms of violence.

Although there may be some benefit to educational programs that remove gifted and talented students from the mainstream classroom, skilful use of personalized learning and assessment tasks does offer an opportunity for gifted students to have their intellectual needs met within the socioculturally authentic context of the mixed ability classroom.

SUMMARY

This chapter examines the impact of teaching and learning strategies designed to foster personal engagement and creative thinking, without compromising foundation knowledge, in a Senior (Year 11 and 12) Chemistry program. Performance of students (four cohorts aged 15-18 years; n = 79) was assessed relative to a quantitative index of academic capacity, generated through factorial analysis of data from other subjects. Results indicate that the development of teaching, learning and assessment methods/instruments that challenge students to connect foundation knowledge to problems with personal relevance not only enhances general, affective factors but also supports realization of creative potential. The emphasis on differentiation within a mixed ability classroom may be of particular importance for gifted students as it provides an opportunity to explore and engage with more challenging material without inducing unnecessary emotional or social stress. In a broader sense, this study highlights a need for greater awareness and interaction across different sectors of science education. To evoke a culture of competitive attainment based on identification of individuals who possess innately superior ability in science, or science teaching, will do little to ensure quality outcomes without consideration of broader social factors. Science education is in danger of running aground in entrepreneurial and sociocultural terrain, when it is the developmental and psychometric discourses that hold the key to developing and implementing learning programs that activate and utilize students' intrinsic motivation to learn. Not only is this the only truly potent and effective stimulus for quality educational outcomes (McMeniman, 1989; Jacobs & Newstead, 2000; Nunan, 2000), it is the only pathway to genuine creativity (Schmidt, 2010, 2011a; Simonton, 2003b).

REFERENCES

Afdal, H. W. (2012). Policy making processes with respect to teacher education in Finland and Norway. *Higher Education Online First.* doi: 10.1007/s10734-012-9527-2

Agnoli, S., Mancini, G., Pozzoli, T., Baldaro, B., Russo, P. M., & Surcinelli, P. (2012). The interaction between emotional intelligence and cognitive ability in predicting scholastic performance in school-aged children. *Personality and Individual Differences, 53*(5), 660–665.

Anagnostopoulos, D. (2006). "Real students" and "true demotes": Ending social promotion and the moral ordering of urban high schools. *American Educational Research Journal, 43*(1), 5.

Anderson, K. J. B. (2012). Science education and test-based accountability: Reviewing their relationship and exploring implications for future policy. *Science Education, 96*, 104–129.

Areepattamannil, S., & Freeman, J. G. (2008). Academic achievement, academic self- concept, and academic motivation of immigrant adolescents in the greater Toronto area secondary schools. *Journal of Advanced Academics, 19*(4), 700–743.

Aslam, S., & Ahmad, I. (2010). Predictive ability of ability-based versus self-report EI measures for academic performance. *Pakistan Journal of Psychology, 41*(2), 91–104.

Aydeniz, M., & Southerland, S. A. (2012). A national survey of middle and high school science teachers' responses to standardized testing: Is Science being devalued in schools. *Journal of Science Teacher Education, 23*, 233–257.

Bain, S. K., & Bell, S. M. (2004). Social self-concept, social attribtutions, and peer relationships in fourth, fifth and sixth graders who are gifted compared to high achievers. *The Gifted Child Quarterley, 48*(3), 167–178.

Barone, C., & Ortiz, L. (2010). Overeducation among European university graduates: A comparative analysis of its incidence and the importance of higher education differentiation. *Higher Education, 61*(3), 325–337. doi: 10.1007/s10734-010-9380-0

Barrow, L. H. (2006). A brief history of inquiry: From Dewey to standards. *Journal of Science Teacher Education, 17*, 265–278.

Barrow, L. H. (2010). Encouraging creativity with scientific inquiry. *Creative Education, 1*, 1–6.

Barrow, M., & Grant, B. (2011). The 'truth' of academic development: How did it get to be about 'teaching and learning'? *Higher Education Research and Development, 31*(4), 465–477. doi:10.1080/07294360.2011.602393

Batagiannis, S. C. (2007). Leadership guided by courage: A challenge to instantaneous perfection. *The Journal of Educational Thought, 41*(2), 145–164.

Beres, I., Magyar, T. M., & Turcsanyi-Szabo, M. R. (2012). Towards a personalised, learning style based collaborative blended learning model with individual assessment. *Informatics in Education, 11*(1), 1–28.

Berlin, J. E. (2009) It's all a matter of perspective: Student perceptions on the impact of being labeled gifted and talented. *Roeper Review, 31*, 217–233.

Bexley, E., Arkoudis, S., & James, R. (2012). The motivations, values and future plans of Australian academics. *Higher Education* [Online First]. doi: 10.1007/s10734-012-9550-3

Blackmore, J. (2009). Academic pedagogies, quality logics and performative universities: Evaluating teaching and what students want. *Studies in Higher Education, 34*(8), 857–872.

Boden, M. (2001). Creativity and knowledge. In A. Craft, B. Jeffrey, & M. Leibling (Eds.), *Creativity in Education* (pp. 95–102). London, England: Continuum Publishing.

Buck, L. B., Bretz, S. L., & Towns, M. H. (2008). Characterizing the level of inquiry in the undergraduate laboratory. *Journal of College Science Teaching, 38*(1), 52–58.

Calhoun, C. (2009). Academic freedom: Public knowledge ann the structural transformation of the university. *Social Research, 75*(2), 561–600.

Chan, D. W. (2003) Dimensions of emotional intelligence and their relationships with social coping among gifted adolescents in Hong Kong. *Journal of Youth and Adolescence, 32*(6), 409–418.

Cheng, L., Fox, J., & Zheng, Y. (2007). Student accounts of the Ontario secondary school literacy test: A case for validation. *The Canadian Modern Language Review, 64*(1), 69.

Cheung, H. Y., & Chan, A. W. H. (2009). The effect of education on life satisfaction across countries. *Alberta Journal of Educational Research, 55*(1), 124–136.

Christine, C., & Glenn, E. S. (2007). General, artistic and scientific creativity attributes of engineering and music students. *Creativity Research Journal, 19*(2–3), 213.

Clegg, P. (2008). Creativity and critical thinking in the globalised university. *Innovations in Education and Teaching International, 45*(3), 219.

Coates, H., & Goedegebuure, L. (2012). High educ DOI recasting the academic workforce: Why the attractiveness of the academic profession needs to be increased and eight possible strategies for how to go about this from an Australian perspective. *Higher Education* [Online First]. doi: 10.1007/s10734-012-9534-3

Connor, C. M., Morrison, F. J., Fishman, B., Giuliani, S., Luck, M., Underwood, P. S., Schatschneider, C. (2010). Testing the impact of child characteristics × instruction interactions on third graders' reading comprehension by differentiating literacy instruction. *Reading Research Quarterly, 46*(3), 189–221. doi: dx.doi.org/10.1598/RRQ.46.3.1

Council_for_the_Australian_Federation. (2007). *Federalist paper 2 the future of schooling in Australia revised edition*. Department of Premier and Cabinet, Melbourne. Retrieved from http://deta.qld.gov.au/

Creese, A. (2005). Is this content-based language teaching? *Linguistics and Education: An International Research Journal, 16*(2), 188–204.

Cross, T. L., & Frazier, A. D. (2010) Guiding the psychosocial development of gifted studetns attending specialized residential STEM schools. *Roeper Review, 32*, 32–41.

Cross, T. L., Cassady, J. C., Dixon, F.A., & Adams, C. M. (2008) The psychology of gifted adolescents as measured by the MMPI-A. *The Gifted Child Quarterly, 52*(4), 326–339.

Danzi, J., Reul, K., & Smith, R. (2008). *Improving student motivation in mixed ability classrooms using differentiated instruction* (Master of Arts in Teaching and Leadership), Saint Xavier University, Chicago, IL.

Dewey, J. (1916). *Chapter 23: Vocational aspects of education: Democracy and education*. New York, NY: The Free Press.

DiFabio, A., & Palazzeschi, L. (2009). An in-depth look at scholastic success: Fluid intelligence, personality traits or emotional intelligence? *Personality and Individual Differences, 46*, 581–585. doi: 10.1016/j.paid.2008.12.012

Douglass, J. A., Thomson, G., & Zhao, C.-M. (2012). The learning outcomes race: The value of self-reported gains in large research universities. *Higher Education, 64*, 317–335. doi: 10.1007/s10734-011-9496-x

Eddles-Hirsch, K., Vialle, W., Rogers, K. B., McCormick, J. (2010). Just challenge those high-ability learners and they'll be all right! The impact of social context and challenging instruction on the affective development of high-ability studetns. *Journal of Advanced Academics, 22*(1), 106–128.

Fondacaro, M. R., Brank, E. M., Stuart, J., Villanueva-Abraham, S., Luescher, J., & McNatt, P. S. (2006). Identity orientation, voice, and judgments of procedural justice during late adolescence. *Journal of Youth and Adolescence, 35*(6), 987–997.

Frodeman, R. (2011). Interdisciplinary research and academic sustainability: Managing knowledge in an age of accountability. *Environmental Conservation, 38*(2), 105–112.

Gao, L. (2011). Eclecticism or principled eclecticism. *Creative Education, 2*(4), 363–369. doi: 10.4236/ce.2011.24051

Garbrecht, L. S. (2006). Schools' influence on identity formation in a time of change. *Educational Researcher, 35*(9), 42–48.

Gawlik, M. A., Kearney, C. P., Addonizio, M. F., & LaPlante-Sosnowsky, F. (2010). Teacher quality in michigan: A school-level analysis of the detroit metropolitan region. *Education and Urban Society, 44*, 412–448. doi: 10.1177/0013124510392568

Geake, J. G., & Gross, M. U. M. (2008). Teachers' negative affect toward academically gifted students. *The Gifted Child Quarterley, 52*(3), 217–231.

Gemme, B., & Gingras, Y. (2012). Academic careers for graduate students: A strong attractor in a changed environment. *Higher Education, 63*, 667–683. doi: 10.1007/s10734-011-9466-3

Glazer, E. (2009). Rephrasing the madness and creativity debate: What is the nature of the creativity construct. *Personality and Individual Differences, 46*, 755–764.

Griffin, T. M., Chavous, T., Cogburn, C., Branch, L., & Sellers, R. (2012). Dimensions of academic contingencies among African American college students. *Journal of Black Psychology, 38*, 201–229.

Grodsky, E., Warren, J. R., & Felts, E. (2008). Testing and social stratification in American education. *Annual Review of Sociology, 34*, 385–404.

Hargreaves, A., & Shirley, D. (2009). *The fourth way: The inspiring future for educational change*. San Francisco, CA: Corwin Press.

Harris, K.-L. (2012). *A background in science: What science means for Australian society* (p. 92). Australian Council of Deans of Science: Centre for Studies in Higher Education, The University of Melbourne.

Hartley, D. (2008). Education, markets and the pedagogy of personalisation. *British Journal of Educational Studies, 56*(4), 365–381.

Hayden, M., & Lam, Q. T. (2007). A 2020 vision for higher education in Vietnam. *International Educator, 16*(1), 14–17.

Herron, S. (2009). A curious thing happened on the way to constructivism. *Journal of College Science Teaching, 38*(6), 8.

Hidi, S. (2006). Interest: A unique motivational variable. *Educational Research Review, 1*, 69–82. doi: 10.1016/j.edurev.2006.09.001

Hilton, M. (2006). Measuring standards in primary English: Issues of validity and accountability with respect to PIRLS and national curriculum test scores. *British Educational Research Journal, 32*(6), 817–837.

Hogrebe, M., & Tate, W. (2010). School composition and context factors that moderate and predict 10th-grade science proficiency. *Teachers College Record, 112*(4), 1096.

Hornikx, J. (2011). Epistemic authority of professors and researchers: Differential perceptions by students from two cultural-educational systems. *Social Psychology and Education, 14*, 169–183. doi: 10.1007/s11218-010-9139-6

Huntoon, J., & Baltensperger, B. (2012). Increasing expertise in earth Science education through master's education. *Journal of Geoscience Education, 60*(2), 147–158.

Kauffman, J. M., Mock, D. R., Tankersley, M., & Landrum, T. J. (2008). Effective service delivery models. In R. J. Morris & N. Mather (Eds.), *Evidence-based interventions for students with learning and behavioral challenges* (pp. 359–378). Mahwah, NJ: Lawrence Erlbaum Associates.

Kelly-Jackson, C. P., & Jackson, T. O. (2011). Meeting their fullest potential: The beliefs and teaching of a culturally relevant science teacher. *Creative Education, 2*(4), 408–413. doi: 10.4236/ce.2011.24059

Kendall, G. (2002). The crisis in doctoral education: A sociological diagnosis. *Higher Education & Development, 21*(2), 131–141.

Kessels, U., Rau, M., & Hannover, B. (2006). What goes well with physics? Measuring and altering the image of science. *British Journal of Educational Psychology, 76*(4), 761–780.

Kitagawa, F., & Oba, J. (2010). Managing differentiation of higher education system in Japan: Connecting excellence and diversity. *Higher Education, 59*, 507–524. doi: 10.1007/s10734-009-9262-5

Klavir, R., & Gorodetsky, M. (2011). Features of creativity as expressed in the construction of new analogical problems by intellectually gifted students. *Creative Education, 2*(3), 164–173. doi: 10.4236/ce.2011.23023

Krause, K.-L. (2012). Addressing the wicked problem of quality in higher education: Theoretical approaches and implications. *Higher Education Research & Development, 31*(3), 285–297.

Kyriakides, L. (2004). Investigating validity from teachers' perspectives through their engagement in large-scale assessment: The emergent literacy baseline assessment project. *Assessment in Education: Principles, Policy & Practice, 11*(2), 143–165.

Lam, Y. Y. (2010). Impact of globalization on higher education: An empirical study of education policy & planning of design education in Hong Kong. *International Education Studies, 3*(4), 73–86.

Leighton, J. P., Gokiert, R. J., Cor, M. K., & Heffernan, C. (2010). Teacher beliefs about the cognitive diagnostic information of classroom versus large-scale tests: Implications for assessment literacy. *Assessment in Education: Principles, Policy & Practice, 17*(1), 7–21.

Linsley, I. (2005). *Causes of overeducation in the Australian labour market, department of economics* (Research paper 940). Melbourne, Australia: Department of Economics, The University of Melbourne.

Manzo, A. V. (2003). Literacy crisis or Cambrian period? Theory, practice, and public policy implications. *Journal of Adolescent and Adult Literacy, 46*(8), 654–661.

Martín, E., Martínez-Arias, R., Marches, A., & Pérez, E. M. (2008). Variables that predict academic achievement in the Spanish compulsory secondary educational system: A longitudinal, multi level analysis. *The Spanish Journal of Psychology, 11*(2), 400–413.

McBride, J. R., Ysseldyke, J., Milone, M., & Stickney, E. (2010). Technical adequacy and cost benefit of four measures of early literacy. *Canadian Journal of School Psychology, 252*, 189–204.

McCulloch, A., & Thomas, L. (2012). Widening participation to doctoral education and research degrees: A research agenda for an emerging policy issue. *Higher Education Research & Development, iFirst, 32*(2), 214–227. doi: 10.1080/07294360.2012.662630

McWilliam, E., Poronnik, P., & Taylor, P. G. (2008). Re-designing science pedagogy: Reversing the flight from science. *Journal of Science Education and Technology, 17*(3), 226–235.

Messinis, G. (2007). *Overeducation and overskilling in Australia: Second generation Greek-Australians and Italian-Australians CSES* (Working paper no. 37). Melbourne, Australia: Victoria University.

Miller, A. I. (2000). *Insights of genius: Imagery and creativity in science and art.* Cambridge, MA: Massachusetts Institute of Technology.

Milliband, D. (2004). *Personalised learning: Building a new relationship with schools.* Belfast, England: Department for Education and Skills, United Kingdom.

Montt, G. (2011). Cross-national differences in educational acheivement inequality. *Sociology of Education, 84*(1), 49–68.

Moon, T. R., Brighton, C. M., & Callahan, C. M. (2003). State standardized testing programs: Friend or foe of gifted education? *Roeper Review, 25*(2), 49–61.

Myer, L. H., & Evans, I. M. (2005). Supporting academic staff: Meeting new expectations in higher education without compromising traditional faculty values. *Higher Education Policy, 18*, 243–255. doi: 10.1057/palgrave.hep.8300086

Nadelson, L. S. (2009). How can true inquiry happen in K-16 science education? *Science Educator, 18*(1), 48.

Nagy, P. (2000). The three roles of assessment: Gatekeeping, accountability, and instructional diagnosis. *Canadian Journal of Education, 25*(4), 262–279.

Nair, C. S., Bennett, L., & Mertova, P. (2010). Responding to the student voice: A case study of a systematic improvement strategy. *The TQM Journal, 22*(5), 553–564. doi: 10.1108/17542731011072883

Nancy Butler, S., Hee-Sun, L., & Scott, M. (2003). Research towards an expanded understanding of inquiry science beyond one idealized standard. *Science Education, 87*(4), 490.

Newton, D. P., & Newton, L. D. (2009). Some student teachers' conceptions of creativity in school science. *Research in Science and Technological Education, 27*(1), 45.

Nilsson, M., Östergren, J., Fors, U., Rickenlund, A., Jorfeldt, L., Caidah, K., & Bolinder, G. (2012). Does individual learning styles influence the choice to use a web-based ECG learning programme in a blended learning setting. *BMC Medical Education, 12*(5), 1–9.

Nowak, J. A. (2007). The problem with using problem-based learning to teach middle school earth/space science in a high stakes testing society. *Journal of Geoscience Education, 55*(1), 62.

Obamba, M. O., & Mwema, J. K. (2009). Symmetry and asymmetry: New contours, paradigms, and politics in african academic partnerships. *Higher Education Policy, 22*, 349–371. doi: 10.1057/hep.2009.12

Oprescu, R. (2012). Bridging intellectual capital and the competitiveness of nations. *Managing & Marketing Challenges for the Knowledge Society, 7*(1), 73–88.

Peter, S. E., Bacon, E., & Dastbaz, M. (2010). *Personalised and adaptable learning: Designing the architecture.* London, England: Higher Education Academy.

Peterson, J. S., & Ray, K. E. (2006). Bullying and the gifted: Victims, perpetrators, prevalence and effects. *The Gifted Child Quarterly, 50*(2), 148–190.

Pyhalto, K., Soini, T., & Pietarinen, J. (2011). A systemic perspective on school reform: Principals' and chief education officers' perspectives on school development. *Journal of Educational Administration, 49*(1), 46–61. doi: 10.1108/09578231111102054

Queensland Studies Authority. (2007). *Senior syllabus chemistry 2007.* Brisbane, Queensland Australia.

Quinn, M. A., & Rubb, S. (2011). Spouse overeducation and family migration: Evidence form the US. *Journal of Family Economic Issues, 32*, 36–45.

Quintini, G., & Martin, S. (2006). *Starting well or losing their way? The position of youth in the labour market in OECD countries OECD social, employment and migration working papers.* Paris, France: Directorate for Employment, Labour and Social Affairs.

Ramoniene, L., & Lanskoronskis, M. (2011). Reflection of higher education aspects in the conception of national competitiveness. *Baltic Journal of Management, 6*(1), 124–139.

Reinheimer, D., & McKenzie, K. (2011). The impact of tutoring on the academic success of undeclared students. *Journal of College Reading and Learning, 41*(2), 22–36.

Reis, S. M., & Renzulli, J. S. (2009). Myth 1: The gifted and talented constitute one single homogeneous group and giftedness is a way of being that says in the person over time and experiences. *The Gifted Child Quarterly, 53*(4), 233–235.
Resh, N. (2009). Justice in grades allocation: Teachers' perspective social psychology and education. *Social Psychology of Education, 12*(3), 315–325. doi: 10.1007/s11218-008-9073-z
Rock, M. L., Gregg, M., Ellis, E., & Gable, R. A. (2008). REACH: A framework for differentiating classroom instruction. *Preventing School Failure, 52*(2), 31–58.
Romanov, D., Tur-Sinae, A., & Eizman, G. (2008). *Overeducation, job mobility, and earnings mobility among holders of first degrees in Israel.* Central Bureau of Statistics, Working paper.
Rowlands, S. (2011). Discussion article: Disciplinary boundaries for creativity. *Creative Education, 2*(1), 47–55. doi: 10.4236/ce.2011.21007
Rubin, J. (2008). Judging our students-keeping it just: The theorem of intellectual measure. *Journal of College Admission, 199*, 4–11.
Runco, M. A., & Chand, I. (1995). Cognition and creativity. *Educational Psychology Review, 7*(3), 243–267.
Runco, M. A., & Okuda, S. M. (1988). Problem discovery, divergent thinking, and the creative process. *Journal of Youth and Adolescence, 17*(3), 211–220.
Rushton, S., Mariano, J. M., & Wallace, T. L. (2012). Program selection among pre-service teachers: MBTI profiles within a college of education. *Creative Education, 3*(1), 16–23. doi: 10.4236/ce.2012.31003
Sampson, D., & Karagianidis, C. (2002). Personalised learning: Educational, technological and standardisation perspective. *Interactive Educational Multimedia, 4*, 24–39.
Sara, M., David, H., & Anthony, F. (2007). Improvement in national test reading scores at key stage 1; grade inflation or better achievement? *British Educational Research Journal, 33*(1), 47.
Scager, K., Akkerman, S. F., Keesen, F., Mainhard, M. T., Pilot, A., & Wubbels, T. (2012). Do honors students have more potential for excellence in their professional lives? *Higher Education, 64*, 19–39. doi: 10.1007/s10734-011-9478-z
Schmidt, A. L. (2010). The battle for creativity: Frontiers in science and science education. *BioEssays, 32*(12), 1016–1019.
Schmidt, A. L. (2011a). Creativity in science: Tensions between perception and practice. *Creative Education, 2*(5), 435–445. doi: 10.4236/ce.2011.25063
Schmidt, A. L. (2011b). Discourses of creativity. In E. G. Carayannis (Ed.), *Encylopedia of creativity, invention, innovation and entrepreneurship.* Freiburg, Germany: Springer International.
Schulte, D. P., Schulte, L., Slate, J. R., & Brooks, G. D. (2002). A test-driven accountability system in Texas: Principals' voices. *International Electronic Journal for Leadership in learning, 6*(7).
Settlage, J. (2007). Demythologizing science teacher education: Conquering the false ideal of open inquiry. *Journal of Science Teacher Education, 18*, 461–467.
Shay, S. (2012). Educational development as a field: Are we there yet? *Higher Education Research & Development, 31*(3), 311–323. doi: 10.1080/07294360.2011.631520
Simonton, D. K. (2003b). Scientific creativity as constrained stochastic behavior: The integration of product, person, and process perspectives. *Psychological Bulletin, 129*(4), 475–494.
Skaar, N. R., & Williams, J. E. (2012). Emotional intelligence as a predictor of adolescent risk behavior participation and perception. *Journal on Educational Psychology, 5*(4), 32–47.
Stallman, H. (2012). A qualitative evaluation of perceptions of the role of competition in the success and distress of law students. *Higher Education Research & Development, iFirst, 31*(6), 1–14. doi: 10.1080/07294360.2012.655241
State_of_Queensland. (2002). *Queensland the smart state: Education and training reforms for the future, A white paper.* Queensland Government. Retrieved from http://www.qld.gov.edu.au
Sweller, J. (2009). Cognitive basis of human creativity. *Educational Psychology Review, 21*, 11–19.
Taylor, A., Jones, M., Broadwell, B., & Oppewal, T. (2008). Creativity, inquiry, or accountability? Scientists' and teachers' perceptions of science education. *Science Education, 92*(6), 1058.
Tomlinson, C. A. (1999). Mapping a route toward differentiated instruction. *Educational Leadership, 59*, 12–16.

Tomlinson, C. A. (2000). Reconcilable differences: Standards-based teaching and differentiation. *Educational Leadership*, *58*, 6–11.
Tu, C.-H., Sujo-Montes, L., Yen, C.-J., Chan, J.-Y., & Blocher, M. (2012). The integration of personal learning environments and open network learning environments. *TechTrends*, *56*(3), 13–20.
Tzuriel, D., Bengio, E., & Kashy-Rosenbaum, G. (2011). Cognitive modifiability, emotional–motivational factors, and behavioral characteristics among gifted versus nongifted children. *Journal of Cognitive Education and Psychology*, *10*(3), 253–280.
Universities_Australia. (2012). *STEM and Non-STEM first year students*. Canberra, Australia: Australian Government Department of Indistry, innovation, Science, Research and Tertiary Education.
Verpoorten, D., Renson, J.-M., Westera, W., & Specht, M. (2009). *Personalised learning: A familiar concept to teachers? And which teachers?* Paper presented at the Making it personal!, Greenwich, England.
Vinkhuyzen, A. A. E., vanderSluis, S., Posthuma, D., & Boomsma, D. I. (2009). The heritability of aptitude and exceptional talent across different domains in adolescents and young adults. *Behavioural Genetics*, *39*, 380–392.
Volante, L., & Ben Jaafar, S. (2008). Educational assessment in Canada. *Assessment in Education: Principles, Policy and Practice*, *15*(2), 201–210.
vonStumm, S., & Furnham, A. F. (2012). Learning approaches: Associations with typical intellectual engagement, intelligence and the big five. *Personality and Individual Differences*, *53*, 720–723.
Whitchurch, C. (2012). Expanding the parameters of academia. *Higher Education*, *64*, 99–117.
William, A. S. (2005). Understanding students' practical epistemologies and their influence on learning through inquiry. *Science Education*, *89*(4), 634.
Windschitl, M., Thompson, J., & Braaten, M. (2008). Beyond the scientific method: Model-based inquiry as a new paradigm of preference for school science investigations. *Science Education*, *92*(5), 941.
Woitaszewski, S. A., & Aalsma, M. C. (2004) The contribution of emotional intelligence to the social and academic success of gifted adolescents as measured by the multifactor emotional intelligence scale – Adolescent version. *Roeper Review*, *27*(1), 25–30.
Wu, P.-L., & Chiou, W.-B. (2008). Postformal thinking and creativity among late adolescents: A post-Piagetian Approach. *Adolescence*, *43*(170), 237–251.

Adele L. Schmidt
School of Education and Professional Studies
Griffith University
Australia

ANDREA S. FOSTER

10. FOSTERING CREATIVITY IN SCIENCE CLASSROOMS

Lessons Learned from a Brigadier General

INTRODUCTION

This case study tells the story of fostering creativity in the science classroom through the lens of a gifted physician and now retired U.S. Army Brigadier General who was "unidentified" as gifted while a student and includes a series of informal interviews that illuminate the participant's formative elementary, middle, and high school and college experiences from 1936–1952. The central purpose of this chapter is to highlight these significant school experiences, the need for early identification of scientific talents, and the building of the necessary foundation for future scientific contributions. The chapter also promises a translation of wisdom into sound pedagogical practices – teaching science as inquiry, problem-based learning, etc… that guarantee the development of creative, gifted, and scientific thinkers, and includes recommendations regarding the specific role of creativity in science classrooms and in the nurturing of gifted children as well as key strategies that can make it all happen.

Elementary teachers shoulder the responsibility of teaching all subjects and provide the inspiration for our next generations to develop the necessary scientific habits of mind such as curiosity, informed scepticism, and openness to new ideas as well as playing a most significant role in establishing a science literate citizenry – Americans who have the capacity to solve critical world problems of today.

While observing and evaluating science teacher candidates in elementary and middle school classrooms in area school districts it has been demonstrated that very little time is left for children to express themselves creatively. More than one third of the school day is dedicated to test preparation which mostly involves completing worksheets and paper pencil benchmark testing. Teacher candidates struggle to negotiate the "ideal" inquiry-based classroom which they experience in the science methods classroom with the "reality" of the public school classroom where the emphasis is on worksheets and tests. They are often frustrated and disillusioned by the pedagogical disparities they face. The paradigm of our schools today must be shifted to make classrooms come alive, particularly at the elementary level, so that children are encouraged to think critically, solve problems, collaborate, and create.

M. K. Demetrikopoulos & J. L. Pecore (Eds.), Interplay of Creativity and Giftedness in Science, 187–200.

Although there is a current creativity crisis in America, previous eras included the wonderful world of Walt Disney and drawing in Anti-Coloring Books[1] where imagination and blank pages represented a world of endless possibility. Classrooms of today should be places where there are concerted efforts to nurture the creativity of all children, particularly in the sciences. The challenge is to design and model rigorous and relevant science experiences and inventive pedagogies for teaching science that come alive in the elementary and middle school classroom with the hope that prospective elementary and middle school teachers use these strategies with their students, particularly the gifted ones.

This chapter explores the interplay of creativity, science and giftedness by sharing a story of fostering creativity in the science classroom through the lens of a retired U.S. Army Brigadier General, Master American College of Physicians, Professor Emeritus in Internal Medicine from a medical center. I conducted a series of informal interviews about his elementary and middle school experiences from 1936–1945 where he attended public schools in Harlem and a High School specializing in Science in the Bronx. He graduated high school at age 16 and then, at age 20, graduated from Columbia University and entered medical school. His early formal school experiences provided a critical foundation for his future scientific contributions in medicine and offer insights into translating this story of science, creativity, and giftedness into sound pedagogical practices that provide meaningful strategies that promote creativity in science and scientific endeavours in order to develop creative, gifted, and scientific thinkers.

THE CREATIVITY CRISIS IN AMERICA

The July 2010 issue of *Newsweek* pointed out that we are currently in a creativity crisis – American creativity scores are falling. Kyung Hee Kim at the College of William & Mary made this discovery after analysing 300,000 Torrance[2] scores of children and adults. According to Kim, the decrease is very significant, with the most serious decline apparent in children from Kindergarten to sixth grade (Bronson & Merryman, 2010). The potential consequences are far-reaching. The necessity of human ingenuity is unquestionable. They report an IBM poll in which 1,500 CEO's identified creativity as the most essential leadership competency of the future, yet creativity is decreasing among Americans at time when it is most vital to the health of our future. "Creativity is necessary not simply to sustain our nation's economic growth, but also to help solve significant world problems like saving the Gulf of Mexico, bringing peace to Afghanistan, and delivering health care" (p. 45).

So what is to blame for our waning creativity among young school children? Likely culprits have been identified. Some claim it has to do with the number of hours kids spend in front of the television, playing videogames, or downloading music and Apps on their iPhones. Another is the lack of creativity development in our schools. Bronson and Merryman (2010) claim there is no concerted effort to nurture creativity in all children in schools today.

Sir Ken Robinson, an internationally recognized speaker and creativity in education expert, makes an entertaining (and profoundly moving) case for creating an education system that nurtures creativity rather than undermining it. Robinson points out the many ways our schools fail to recognize – much less cultivate—the talents of many brilliant people. "We are educating people out of their creativity," Robinson observes (Robinson, 2007). He points out many ways that our schools fail to recognize the talents of many brilliant people. Robinson claims that our schools are organized around an outdated factory model system where children are processed like automobiles or widgets. The number of children who are diagnosed with Attention Deficit Hyperactivity Disorder (ADHD) and are medicated is alarming. He claims that these children, who are living in the most stimulating and exciting time on Earth, are being, "anesthetized." We are using exactly the wrong approaches to educating these children. They should not sedated or 'anesthetized' but, stimulated through the arts or what he calls, "Aesthetic Education" or education in the arts. It is the arts that open up minds to creative thinking and problem solving.

STEM to STEAM

According to President Obama (2011), American 15-year olds rank 21st in science and 25th in math compared to their peers around the world. STEM education (Science, Technology, Engineering, and Math) may be fundamentally flawed. Richardson (2011) suggests STEM proponents should start focusing on creativity, originality, and design thinking. Here's why. The creativity crisis in our schools is not just one-dimensional. The European Union declared 2009 as the Year of Creativity, and Chinese faculty actually laughed when they found out the U.S. education trends were in "standardized curriculum, rote memorization, and nationalized testing."

NASA and Boeing are finding that recent graduates can technically render in two dimensions but can no longer think in three (Richardson, 2011). Also, STEM does not necessarily help create the "New Work" workers that are so highly valued in the evolving global community. In a report on "New Work," the Pew Charitable Trust wrote, "The creative jobs that drive the innovation are now the highest 'value added' jobs in the world – real creators of wealth. If states are going to stay competitive, they have to. develop a work force capable of doing creative work."

The Pew report acknowledges that creativity does not just come from artists. In fact, there are approximately 170 classifications that make up "New Work," which can be grouped into five major categories based on the types of knowledge, skills, and aptitudes needed. They are Creative, Education, Social, Technical, and Strategic. Based on these classifications, STEM appears to account for only one fifth of the training we will need to compete in coming decades.

Interestingly, in a recent study, creative jobs increased in Houston, Texas by 8 percent in the last ten years (11,268 new jobs) and is expected to grow by 7 percent by 2016. Creative businesses in Houston had an economic impact of more than $9.1 billion in 2011 (Glenzer, 2012). Ideal job candidates at these companies must now show they can "think with their hands" by having expertise or a second major in a musical instrument, auto repair, or sculpture. At Stanford, the rediscovery of hands-on learning arose partly from the frustration of engineering, architecture, and design professors who realized that their best students had never taken apart a bicycle or built a model airplane (Richardson, 2011).

STEM must keep up. According to Richardson (2011), "STEM's biggest flaw is that it continues to shine a bright light on engineering while relegating art and design to a dusty corner" (p. 2). The truth is that our biggest innovations come from both the arts and the sciences. John Maeda, president of the Rhode Island School of Design, hosted a workshop funded by the National Science Foundation to explore ways of turning STEM into STEAM (adding an A for "Arts). Students' brains need to be trained to think flexibily which can be achieved by engaging our creative potential.

The Lack of Creativity in Schools

The lack of creativity in schools, according to most teachers, stems from pressure to meet curriculum standards. Researchers (Bronson & Merryman, 2010) say creativity should be taken out of the art room and put into the homeroom. The argument that we can't teach creativity because kids already have too much to learn is a false trade off. Creativity is not about freedom from concrete facts. Rather, fact finding and deep research are vital stages in the creative process. With well-designed pedagogy, and project-based learning, curriculum standards can be met.

Creativity is not just about art projects, it is about the thinking process students use to solve problems in all fields. Bronson and Merryman (2010) suggest that students need problems that require them to first fact-find, and then move to problem finding, idea-finding and then solution-finding. This way, they are using divergent and convergent thinking to arrive at original solutions.

The good news is that students can learn techniques for uncovering and leveraging their creative potential. Schools are essential in helping students learn these techniques. Teachers are key to making this happen.

OLD SCHOOL – THE CASE STUDY

The General, a physician, was married to a Kindergarten teacher and both parents supported his children's early interest in art, science, and creativity and nurtured these creative interests. They limited their children's television watching to the moon landing, an occasional Brady Bunch episode, and Walt Disney on Sunday evenings and encouraged their children to explore outside interests and participate in outdoor and physical activities such as Blue Birds, Campfire Girls, swimming, diving, and dance. Their children were always engaged in artistic activities such as watercolour lessons, science projects, special colouring books with blank pages and inspirational prompts to spark creative thinking such as, "You had an amazing dream about the future last night, draw your dream."

The General had a strong work ethic and would leave early in the morning for the hospital where he took care of his patients and performed the administrative work of running the hospital. He would return home in time for dinner with his family. He and his wife devised a KP duty chart that indicated what their children's roles would be in the pre and post dinner preparation, i.e., setting the table, clearing the dishes, sweeping the floor, loading the dishwasher, etc. with each child given one night off per week. In 1969, he left for a tour in Vietnam leaving his wife with five children under the age of ten. Fortunately, his tour lasted only 10 months and he returned safely to a well-run, organized and happy home – thanks to his wife's excellent parenting and teaching skills. In time he became a Brigadiere General, Chief of Medicine at the age of 41 and Commanding General at an army medical center at the age of 48. Ultimately he became Master of the American College of Physicians and Professor Emeritus from a medical university where he contributed greatly to the profession.

His early school experiences shaped who he eventually would become. They provided a foundation for how he and his wife raised five children to become productive, successful adults. They always led by example. In addition, the stories of his schooling shed light on what works and what does not work when working with gifted children and inspiring gifted children to pursue careers in science. The following pages chronicle his early beginnings and educational journey.

The Kindergarten Clock Story & Early School Experiences – 1936–1945

He was born in 1931 to Italian-American parents in New York City. His father was a painting contractor and his mother a school teacher. His grandfather was an artist who immigrated to New York from Palermo, Sicily at the turn of the century. His ancestors came to America with a commitment to the promise of creating a better life for his future descendants.

During our interviews, I asked him to describe his early school experiences and to identify which teachers (K-College) motivated or inspired his creative thinking and problem solving capacities. I asked him to describe the strategies they used to keep him engaged. I also asked him to identify teachers that did *not* motivate his creativity thinking and problem solving capacities and to describe their approaches to learning. He attended Kindergarten in 1936 and the following is how he describes his first school experience that stifled his thinking. He remembers the following Kindergarten incident.

> My Kindergarten teacher, Mrs. Tweedy would say, 'Do what I tell you to do not what you can do.' I got an F for putting numbers on a clock face that she wanted to remain blank.

The very next day, his mother, a school teacher, stormed into the principal's office with the offending clock paper in hand and demanded that he be promoted to the first grade where, because of this new first grade teacher, he thrived.

> My first grade teacher gave me books to read (3rd and 4th grade readers). Many of these books, she bought and paid for herself so that I could be challenged. She often asked me to read to and explain the hard words to the other kids. She also let me help others in the class with their math problems. I felt empowered. She let me go where I wanted to go. I was given the freedom to explore. I was allowed to orchestrate my own experience. She would ask me what I wanted to do. I was pretty good at long division.

By 5th grade in a Manhattan elementary school, he experienced homogeneous grouping. His teacher divided the class into thirds based on ability level. He was okay with this because he was in the highest ability group and he did not mind helping others.

Rapid Development Classes and Specialized Schools – 1943

In 1943 he attended Junior High School in Harlem. His creativity was continuously being nurtured by his teachers. He was placed in a highly competitive, "Rapid Development Class." At the end of 7th grade students took entrance exams to determine where they would complete their high school experience. The idea was to honor the students' different interests and different capacities for learning. The students were tested to determine which specialized high school they would attend: School of Automotive Trades; School for Law; School of Economics; Brooklyn Technical and Engineering School; School of Performing Arts; George Washington General High school, and Bronx High School of Science. He recalls the following.

> In the 7th grade my teacher encouraged me to write to the Department of the Navy about my idea to put stretcher pods on an auto gyro (the grandfather of the helicopter) to evacuate wounded from the beaches in WW2. I got a letters saying thanks, but my idea was "impractical." (I wish I had saved the letter)

He remembered one Civics teacher who did not cultivate creativity during his Junior High School experience.

> On the other end of the spectrum was a 7th grade civics teacher who had a list of questions daily and required rote answers. There was no discussion. We never went into the creation of our constitution. We just memorized the results. She was mean to anyone who had a poor memory.

In spite of the rigid pedagogy, he tested high in mathematics and science and was admitted to the Bronx High School of Science at the age of thirteen.

Bronx High School of Science – 1945–1948

The Bronx High School of Science was founded in 1938, a few years before he became a student. Bronx Science started with 150 ninth grade students and 250 tenth grade students. In 1946, as a result of the efforts the principal, our faculty, and the Parents' Association, the school became co-ed. The achievements of the school have been many. Its graduates have gone on to success in almost every field, especially in science and mathematics. Many have become prominent in such fields as politics, atomic physics, and medicine, engineering, music and health careers.[3] The following is an excerpt from an electronic interview with him in which he considers the influences on his interest in science.

> For me math provides the basis for science to flourish. They seem inseparable. My 8th grade algebra teacher, brought math into astronomy measurements, engineering accomplishments, etc. He made algebra practical and real. Science really flourished in High School. I went to the Bronx High School of Science where all the pre-med students gravitated. We had the equivalent of 6 years of science curriculum including such visionary course as, "The historical development of modern science," which included biographies of great minds in science. I was immediately impressed that I must question, ask why? And think outside the box. It provided a basis as to how to think for the rest of my life I think the magnet school put me on track. The focus was on learning by application and by the excitement of participation. Memorizing alone is not learning even in math where memory up to a point is a necessity, somewhere along the line new equations need to emerge. Math exemplifies an exact science that ultimately needs creativity to advance. We need curricula to include the word, "Why." Why is the earth round? Not just tell "The earth is round" "Why is our blood pressure 120/70? "Why do we sleep?"

Columbia University and Internship and Residency at Bellevue – 1948–1957

After graduating from Bronx High School of Science at 16, he attended Columbia University and was accepted to medical school at New York University. He completed his internship and residency in Internal Medicine at Bellevue Hospital. At Columbia he expanded and broadened not only his understanding of science but also literature, the arts, philosophy, and economic theory. He became more focused in research science, medicine, and physics.

He recalled a favorite college English composition professor repeated the following to his students every day, "Write from your heart and experience: express yourself and release your creativity!" When asked how creativity should be cultivated among gifted children, he offered the following advice to educators.

> I believe we should encourage gifted children by just letting loose their capability. If a 3rd grader can do 8th grade math – that's what the child should be doing. Actually we need to start formal learning by age 3 and include a foreign language. They track the kids by age 5 or 6 to 3 or 4 tracks. The tracks should allow liberal shifting of students as they progress. By high school we should make available dozens of magnet schools: music and art,

> science, engineering, commerce, academic studies etc. Also vocational high schools such as, electrical, construction trades, medical careers, automotive trades, etc. All would have a basic curriculum embellished by their designated area of accomplishment. The non-gifted and slower students would have the opportunity to achieve on their track. No need to drop out or be left back. They now would have the opportunity to advance to a vocation. They would be looking at a successful future and a sense of accomplishment.

The following excerpt from the Bronson and Merryman (2010) *Newsweek* article on creativity about stability and hardship with regard to nurturing creativity in accomplished adults was shared with him and he was asked to respond to these findings.

> Having studied the childhoods of highly creative people for decades, Claremont Graduate University's Mihaly Csikszentmihalyi and University of Northern Iowa's Gary G. Gute found highly creative adults tended to grow up in families embodying opposites. Parents encouraged uniqueness, yet provided stability. They were highly responsive to kids' needs, yet challenged kids to develop skills. This resulted in a sort of adaptability: in times of anxiousness, clear rules could reduce chaos—yet when kids were bored, they could seek change, too. In the space between anxiety and boredom was where creativity flourished. It's also true that highly creative adults frequently grew up with hardship. Hardship by itself doesn't lead to creativity, but it does force kids to become more flexible—and flexibility helps with creativity.

The General offered the following response regarding the issue of hardship induced creativity and flexibility.

> Hardship can also lead to disaster. We don't know what kind of person rises above hardship or sinks with it. Some kids who are anxious and bored turn to crime, not scholastic creativity. I don't feel you can generalize. In my group in New York, growing up in the depression and WW2, we often said, "What are we going to do today?" Out of the group, we had 4 emeritus professors, one "walkie talkie burglar," a janitor and a CEO of a radio station. Who can predict?

STRATEGIES FOR FOSTERING CREATIVITY IN SCIENCE CLASSROOMS

His insights and experiences can be developed into a philosophy of science teaching that treats all students as potentially gifted. "Although creativity was long considered a gift of a select minority," according to Chrysikou (2012), "Psychologists have now revealed its seeds mental processes, such as decision making, language and memory, that all of us possess" (p. 26). The following section includes a sampling of teaching strategies and practical activities that work for teaching K-12 students about science and university students about science teaching. What is common among these examples is that they are problem driven, emphasize critical thinking, have hands-on experiences and are taught in the context of topics that students confront in their own lives. The following are examples of activities that boost creative problem solving in science classrooms.

It All Begins with a Droodle

When students came into the inner city 6th grade classroom, they would be confronted with a drawing on the chalkboard called a, "Droodle." *Droodles* are both a drawing and a riddle. They are simple, yet complex and can be quite humorous. The idea of a *droodle* is to kick start creative thinking, to warm up the brain to encourage out-of-the-box thinking – even though *droodles* are always constructed inside a box. Children enjoy solving these *droodles.* Often their ideas were far more creative than the answer given in the Roger Price's book of *droodles.*[4] This type of warm up activity provides the necessary mental practice for problem solving and decision making. The following is a sample *droodle.*

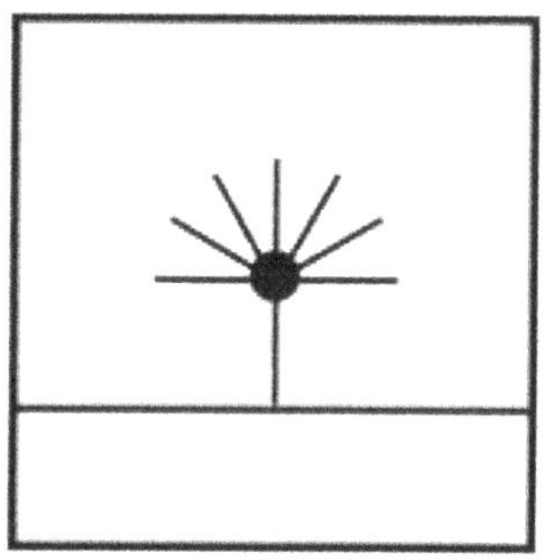

Figure 1. A sample Droodle – A spider performing a handstand

Modeling Inquiry and Teaching that Science Never Sucks with an Egg and a Bottle

An excellent way to inspire problem solving and introduce the inquiry process is to begin the school year with a discrepant event or a simple problem to solve like the classic egg-in-the-bottle demonstration. On the very first day of class, skip the typical syllabus review and dictating of the classroom rules and ask students how to get a hard-boiled egg into an old-fashioned milk bottle without breaking the egg or breaking the bottle. The bottle with the egg can be on top of the teacher's desk as the students come into the classroom. Most students will ask about the egg and the bottle and offer ways to solve the problem like pushing it in and using grease.

Eventually, with guided discussion, students might suggest lighting a match and placing it in the bottle with the egg on the top. When the match burns out, the air in the bottle escapes past the egg and creates a vacuum seal and a pressure differential. The egg appears to be "sucked" into the bottle which is exactly what the students will say. This provides a prime opportunity to help students understand that, "Science doesn't suck!" The egg was actually pushed and pulled by unbalanced force created by the increase of air pressure outside of the bottle. Invite students to then figure out how to get the egg out of the bottle – invert the bottle and blow air past the egg creating another pressure change.

Although the egg and bottle demonstration has been around for a many years, it still has tremendous impact on students' thinking about the science of everyday things. T science teachers have an arsenal of activities to draw from to liven up their classrooms including the work of Bill Nye the Science Guy®, Beakman from Beakman's World®, Sid the Science Kid®, Steve Wolf of Science in the Movies®, and even fictitious yet inspiring, Miss Frizzle®. The engaging works of all these science icons foster creativity in science classrooms.

Creating Science Eyes

Ask any elementary age child what their definition of science is and they will typically respond with something like this, "Science is the opposite of social studies." This response is far too common and the reason for the response, in my view, is tragic. Science, like social studies, in most elementary schools is still being taught opposite social studies. The emphasis on reading and mathematics pushes science and social studies to the end of the day. It is often the case that science is not taught all.

In order to prepare future elementary teachers to teach science, it is absolutely critical to develop ways to inspire teacher candidates to teach science in their classrooms. One approach to doing this is by inviting pre-service teachers to create their own pair of, "Science eyes."[5]

Figure 2 contains a photograph of teacher candidates wearing their newly created science eyes. These science eyes help the teacher candidates inspire their future elementary students to view the world as scientists—to see things in a different light.

Figure 2. Teacher candidates model their Science eyes

Science eyes (Foster, 1994) represent a physical model and a conceptual metaphor for teacher candidates to think through the lens of science and connect science with other subjects such as mathematics, social studies, and language arts as they plan lessons for their students. The act of constructing these creative science eyes helps teacher candidates shift their teaching paradigms to centre teaching and learning around science. Many teacher candidates report that they have their students construct science eyes and wear them during their science lessons to encourage, "Thinking like a scientist."

Project-Based Learning – 21st Century Skills

According to Jones (2012) science lends itself to teaching thinking skills; however, traditional text-based or "cookbook" forms of instruction do not foster scientific habits of mind. What is necessary is a classroom environment where learning strategies, inquiry, real-world, or authentic application, and exploration of relationships of major concepts are the norm.

Critical thinking, collaboration, and communication are three essential 21st century skills. One way to develop these skills is to engage students in Project-based learning (PBL). In Project Based Learning students go through an extended process of inquiry in response to a complex question, problem, or challenge (Pecore, 2015). Rigorous projects help students learn key academic content and practice 21st century skills.

Larmer and Mergendoller (2012) distinguish projects from project-based learning. Projects simply have students apply what they have learned from traditional instruction and; 'main course' project-based learning engages students to learn the material from completing the project. They identify the following attributes of PBL. A 'main course' project:

- Is intended to teach significant content
- Requires critical thinking and problem-solving, collaboration, and various forms of communication
- Requires inquiry as part of the process of learning and creating something new
- Is organized around an open-ended driving question
- Creates a need to know essential content and skills
- Allows some degree of student voice and choice
- Includes processes for revision and reflection
- Involves a public audience

Project-based learning has been at the center of effective science instruction for decades. Project-based learning is a highly effective way of engaging students in authentic problem-based experiences. For example, 7th graders could investigate evidence, from a staged murder in the classroom, to draw reasonable conclusions and learn how to properly use a microscope (Foster, 1995). Prepared evidence bags of hair fibers, an onion, ketchup for blood, and a ransom note were studied by "forensic teams" of students. Students communicated their findings orally and in a written forensic report. Argument and debate among teams were highly encouraged.

In another example, K-College students work in jigsaw (expert and home) groups on an Aquarium Problem in which they use their math skills, and knowledge of fish species, water chemistry, and aquatic plants, to stock a compatible tank of aquarium fish given a particular size tank and a pre-determined budget. Students write persuasive papers and present their solutions to their peers.

Similarly, elementary and middle school children learn about simple machines and Newton's Laws of Motion by exploring the physics of Bobble Head dogs and working in teams to create their own "Shaky Head Thing" using only recycled materials with a "bobability" factor of 5 seconds or more (Foster, 2003).

Classrooms have been transformed into a large human cell with a tarp (blown up with a fan) as a cell membrane and scale models of organelles inside. Students research, build and present their organelles from inside the class cell and then host a, "Cell-a-bration" of learning with parents. Honors Biology students were encouraged to build a museum quality biomes of the world and then invited first graders from a feeder school to tour the converted classroom and learn about biomes.

There really are no limits to creating problem-solving scenarios and driving questions in the classroom that engage students in developing their understanding of science and how they learn about their world. Ideas for projects can be found online, in current events, or they can be invented and created by you or your students. Inviting students to work together, not in homogeneous groups, but in what Fiero (2012) calls, "HeteroGenius," classrooms, is key to ensuring that all children, not just the gifted experience the challenge of learning in our world today and provide leadership opportunities for gifted learners.

C and the Box

The story of *C and the Box* by Frank A. Prince (1993) is a parable and reveals that people must break free of old assumptions and limitations if they want to grow and develop. By exploring outside of a familiar box, C, the leading character, becomes a role model for creativity and imagination, and shows that changing the old way of doing things is necessary for progress. The book is powerful and inspires individuals, particularly teachers, to feel that they can 1) overcome the constraints of conformity and bureaucracy, 2) find new ways to solve problems, 3) discover inner strengths, 4) be creative, and 5) motivate others by example.

LESSONS LEARNED FROM THE GENERAL

The General defines creativity as, "The ability to see things that others do not, particularly in an area that most see as *fait de accompli*,[6] we see it differently." He believes that creativity is intrinsic to the human condition. It comes in different doses and different flavours and we must find ways to unleash the potential for creativity and giftedness in young children. "Find it early and nurture it." His main concern is the lack of a process for identifying creativity and giftedness in

Pre-Kindergarten. Children's thinking must be stimulated early. So if this is the case, then it is imperative that we prepare future EC-6 teacher candidates to recognize the characteristics of gifted young in science and develop best practice "think outside of the box" methodologies to help them inspire our future citizenry to think critically, solve problems creatively and effectively. "Learning itself is a creative process involving thinking, analysing and doing. You don't learn much in a lecture unless you get involved in some way. A lecture may provide the "groundwork" but becomes useless without ultimate challenge, discussion and application. "The essential lessons concerning creativity, giftedness, and science include the following:

- Creativity is intrinsic to the human condition and it must be nurtured.
- Identify gifted young in science at early ages.
- Honour the individual talents and capacities of children.
- Promote Project-based learning as a strategy to inspire creative problem solving and critical thinking.
- Have children work together and let them explore.
- Convergent and divergent thinking are important to creativity and problem solving.
- Focus on effective preparation of future teachers of science, and
- If a child draws the numbers on a blank clock, recognize that there might be some creative potential, encourage it, and of course – do no harm.

Many people assume that creativity is an inborn talent that children either do or do not have; just as all children are not equally intelligent, all children are not equally creative. But, creativity has been demonstrated to be more skill than inborn talent and parents and teachers can help develop this skill. Creativity is essential to success of nearly everything we do and is a key component of health and happiness. Creativity allows people to be more flexible and be better problem solvers, which make them more able to adapt to technological advances and deal with change and take advantage of new opportunities. The key to changing the educational system of today and to shift the paradigm of the worksheet driven, test taking mentality, is to motivate future teachers to be creative, think differently, and be a change agent.

NOTES

1 Susan Striker's Anti-Coloring books for the Young at Art can be found at the following website http://www.susanstriker.com/

2 Torrance Tests of Creative Thinking are considered the 'Gold Standard' in creativity assessments (CQ) that indicates that people who are more creative as children grow up to be more successful than those who are less creative. The Torrrance tests were developed by E. Paul Torrance in the 1950's and 1960's.

3 Some of these individuals are, Harrison Goldin, New York City Comptroller; Oliver Koppel, New York State Assemblyman; Dr. Thomas Matthew, the first Black American neurosurgeon; Leon Cooper, Sheldon Glashow, Roy Glauber, Russell Hulse, David Politzer, Melvin Schwartz, and Steven Weinberg, Nobel Prize Winners in Physics; Harold Brown, Secretary of Defense; E.L. Doctorow and William Safire, authors; and Bobby Darin, a musician.

4 *Droodles* was a syndicated cartoon feature created by Roger Price and collected in his 1953 book *Droodles*. The trademarked name "Droodle" is a nonsense word suggesting "doodle", "drawing" and "riddle." Their general form is minimal: a square box containing a few abstract pictorial elements with a caption (or several) giving a humorous explanation of the picture's subject. For example, a Droodle depicting three concentric shapes—little circle, medium circle, big square—might have the caption "Aerial view of a cowboy in a Port-a-john." *Droodles* are (or were) purely a form of entertainment like any other nonsense cartoon and appeared in pretty much the same places (newspapers, paperback collections, bathroom walls) during their heyday in the 1950s and 1960s. The commercial success of Price's collections of *Droodles* led to the founding of the publishing house Price-Stern-Sloan, and also to the creation of a Droodles-themed game show.

5 Seeing things through science eyes: A case study of an exemplary elementary teacher *by Foster, Andrea Susan,* Ph.D., Texas A&M University, 1998, 229 pages; AAT 9903113
Science-eyed elementary teachers exhibit relentless passions for replacing traditional teaching with realistic, integrated, responsible instruction with science at its core. The purpose of this study was to explore an exemplary elementary teacher's thinking about science and how it serves as a vehicle for the learning that occurs in her primary classroom. Two research questions were investigated in this study. First, what does it mean for an exemplary elementary teacher to view all learning with science eyes? Second, in what ways does the science-oriented elementary teacher use her knowledge of science content, pedagogy, and practical experience to structure her students' learning and her classroom teaching?

6 Etymology. From French *fait accompli* ("an accomplished fact"), from fait ("a fact") accompli ("accomplished"). [edit] Pronunciation. IPA: /ˈfeɪt.ə.ˌkʌm.pli/.

REFERENCES

Bronson, P., & Merryman, A. (2012, July 19). The creativity crisis. *Newsweek, 20*, 45–50.

Chrysikou, E. (2012, July/August). Your creative brain at work. *Scientific American Mind, 23*(3), 24–31.

Fiero, A. (2012, Summer). HeteroGenius classes: Why inclusion and mixed grouping create a better science classroom. *Science Scope, 35*(9), 36–40.

Foster, A. S. (1995, February). Murder in the science lab? *Science Scope, 18*(2), 12–15.

Foster, A. S. (1998). *Seeing things through science eyes: A single case study of an exemplary elementary science teacher* [Dissertation abstracts international]. AAT 9903113.

Foster, A. S. (2003, April). Let the dogs out: The physics of bobble heads. *Science Scope, 26*(17), 16–19.

Glenzer, M. (2012, July 29). Creative juices flowing, growing. *Houston Chronicle*, p. 1.

Jones, R. A. (2012, March). What were they thinking? Instructional strategies that encourage critical thinking. *The Science Teacher, 70*(3), 66–70.

Larmer, J., & Mergendoller, J. (2012). The main course, not dessert: How are students reaching 21st century goals with 21st century project-based learning? *Project-based Learning 101*. Buck Institute for Education, Novato, CA.

Pecore, J. L. (2015). From Kilpatrick's project method to project-based learning. In J. Pecore (Ed.), *Past: Aims of progressive education: Section 1 of international handbook of progressive education* (Editors-in-chief: M. Y. Eryaman and B. C. Bruce). New York, NY: Peter Lang.

Prince, F. A. (1993). *C and the box: A paradigm parable*. San Francisco, CA: Josey-Bass/Pfeiffer, Inc.

Richardson, L. S. (2011, March 25). The creativity crisis: Why American schools need design. *The Atlantic*. Retrieved July, 25, 2012, from http://www.theatlantic.com/national/archive/2011/03/the-creativity-crisis-why-american-schools-need-design/73038

Robinson, K. [Sir Ken Robinson]. (2007, January 6). *Do schools kill creativity?* [TED.com]. Retrieved from http://www.youtube.com/watch?v=iG9CE55wbtY

Andrea S. Foster
Department of Curriculum & Instruction
Sam Houston State University
USA

SECTION 4

SCIENCE, CREATIVITY, AND GIFTEDNESS IN REAL WORLD CONTEXTS WITH DIVERSE LEARNERS

TANG WEE TEO, JIA QIAN WOO AND LAY KUAN LOH

11. AFFORDANCES IN SCHOOL SCIENCE RESEARCH

Narratives from Two Singapore Specialized Science School Students

INTRODUCTION

This chapter examines narratives written by two former students of a Singapore specialized science and mathematics school, Frontier Science High School, FSHS, (a pseudonym). In addition to the core subject-based curriculum, the school offers a complementary science and mathematics curriculum to develop students' research, innovation, and enterprising capacities over the six years of schooling. The courses offered in this program include research methodology and independent science research in which students learn to write research reports and carry out science research work mentored by school teachers with research experience, and/or scientists in external science research laboratories and universities. At the time when the former students were at the school, the complementary curriculum was implemented every Wednesday of the school week. The students were co-mentored by a chemistry teacher at the school and a doctoral level polytechnic (vocational tertiary institution) lecturer on an organic chemistry synthesis project. The lecturer conceptualized the project and came to the school every Wednesday to work with the students in the school chemistry laboratory and served as their research mentor, coach, and facilitator. The students learned organic synthesis research techniques and chemistry concepts undergirding the synthesis process that were typically taught to college level chemistry majors and graduate students. At the end of the project, the students presented a poster of the research findings at a national science symposium.

While student participation in science research was popular in FSHS, such opportunities were not prevalent in mainstream schools as prerequisite factors such as curriculum flexibility, availability of resources such as time, facilities, money, and expertise, and students with appropriate aptitude to do science research may be limited. Student participation in science research, however, has increasingly become a highlight of high-performing schools' branding in Singapore. This practice suggests that gifted students posses the ability and interest to acquire the knowledge, social, and disciplinary practice of the scientific enterprise. It also shows that this differentiated curriculum for the gifted and academically talented students provides an avenue for students' potential to be stretched in areas beyond academic learning. As recently featured in a local newspaper, students in some high-performing schools had reportedly coauthored peer-reviewed scientific publications with their

M. K. Demetrikopoulos & J. L. Pecore (Eds.), Interplay of Creativity and Giftedness in Science, 203–217.

collaborators including teacher mentors and scientists. Nonetheless, insights on the affordances present and emergent from the process of student participating in science research and effects they have on the lives of students is limited in the existing science education literature.

In this chapter, the theoretical construct "affordance" is used to examine two students' narratives of their past experiences participating in science research as middle school students. In particular, this chapter will unpack the affordances existent in making such opportunities available to them and emergent in the process of doing science research. In what follows, the theoretical concepts of "affordances" will be discussed and the ideas will be applied to analyze the narratives of their past experiences and how they shaped their current and future decisions. The affordances were complex and not always apparent to the students, or mentors at the time they worked collaboratively on scientific research. In addition, the findings will be discussed to illuminate the nested, interwoven, and sequential qualities of affordances.

AFFORDANCES

Affordance, a concept coined by perceptual psychologist James J. Gibson (1979) in his seminal book *The Ecological Approach to Visual Perception*, refers to an *action possibility* available to an individual in an environment. According to Gibson, there are three fundamental properties of an affordance. First, an affordance exists relative to the action capabilities of a particular actor. It is not a property of the experience of the actor. Second, the existence of an affordance is independent of the actor's ability to perceive it. Third, an affordance does not change as the needs and goals of the actor change. The second and third points suggest the invariance of affordance (McGrenere & Ho, 2000). In other words, the affordance exists whether or not the actor's experience and culture allows him or her to identify it. It is the ability to perceive the affordance that is experience and culture dependent. As such, the actor will need to exercise discretion in making judgment based upon the perceived information (McGrenere & Ho, 2000).

The dialectism in objectivity and subjectivity is played out as affordances are existent independent of its value, meaning, or interpretation and yet, an individual has to be present as an actor to make a direct perception and pick up the necessary information that specifies the affordance. In Gibson's view, being able to pick up the information is independent of the actor's experience, knowledge, culture, or ability to perceive something, but clearly, an individual is a frame of reference and is involved in the characterization of the existence of the affordance.

Donald Norman (1988), however, described 'affordance' differently in his book *The Psychology of Everyday Things*. The dialectism is played out in the relationship between the object and the actor acting on the object. In his definition, affordances are both actual and perceived properties. In his view, the perceived properties may not actually exist. Suggestions or clues on the use of properties can be present and affordances may be dependent on the experience, knowledge, or culture of the actor.

The most fundamental difference between Gibson's and Norman's ideas is that the former emphasized action possibility while the latter underscored that an action possibility is conveyed or made visible to the actor by having some perceptual information that specifies the affordance and allows it to be directly perceived (McGrenere & Ho, 2000). Another difference is that Norman implied the manipulability of the environment while Gibson did not. Norman also suggested that the existence of affordance is dichotomous; but rather, some grey areas exist due to different interpretations (McGrenere & Ho, 2000).

In extending Gibson's and Norman's ideas, Gaver (1991) distinguished affordances from the perceptual information that specifies affordances. In his framework, *false affordance* (Gibson's idea of 'misinformation') exists when the actor perceives the information about the affordance which does not exist. When the perceptual information is not present and there is no affordance, the actor makes a *correct rejection* that the affordance does not exist. *Hidden affordance* exists when the perceptual information is not there but the affordance is present. Last, when the perceptual information is present with the affordance, *perceptible affordance* exists. In Norman's term, the false affordance and perceptible affordance are "perceived/ apparent affordances". This type of affordance is perceived to exist whether or not the affordance exists or not.

The concept of affordance continues to evolve and efforts have been made to clarify the concept. Reed (1996) argued that affordances are resources in the environment or properties that might be exploitable. Chemro (2003) argued that affordances are relations between particular aspects of the actor and particular aspects of situations. He argued that affordances are not always properties but are about *placing features*—that is, seeing that the situation allows for certain activity—and they are not always in the environment but features of whole situations. Michaels (2003) provided six definitions of affordances summarized as follows: (a) Affordances are the actions that are goal-directed and encapsulates intention, (b) Affordances are multidimensional compounds of properties that are measurement-based, descriptive, or conceptual, (c) Affordances are not arbitrary actions, (d) Affordances exist independently of perception—they do not disappear when they are not perceived and taken advantage of, (e) Affordances are specified by the information and may be perceived as some action engaged in by the perceiver-actor and not others, and (f) Affordances entail an effectivity (Shaw, Turvey, & Mace, 1982)—properties that allow them to make use of affordances—for its actualization but not its existence (Turvey, 1992).

Other forms of affordances have also been discussed. McGrenere and Ho (2000) argued that Gibson had implied that affordance could be nested when an action possibility is composed of one or more action possibilities. Gibson had suggested an environment composed of nested objected and about nesting information that specifies information. McGrenere and Ho thus introduced the term *nested affordances* (p. 2). Gaver addressed the issue of complex affordances by coining the term *sequential affordances* to refer to situations in which the action on a perceptible affordance invokes other information that indicates new affordances that emerge over time.

While the concept of affordance is currently applied in the literature on human-computer interaction (HCI) to understand how designs may be improved to enhance the usability and usefulness of a product by creating affordances of possibilities for action, this concept is applied to a different context to analyze how gifted and academically talented students perceive information about their science research experiences. Through examining the complexity of the affordances, nested and sequential affordances are discussed. In addition, one more form of affordance is identified and interwoven from the analysis of the findings. While the idea of usefulness—meaning to contain the right functions for users to perform their work with efficiency and attain their goals—and usability—to mean clearly designing information to enhance the design—is used in product design in HCI community, the concept has been imported here to analyze the usefulness of doing science research and the usability of such affordances on their lives. The findings of this study will allude to the apparent and non-apparent benefits and limitations of science research so school administrators and teachers who want students to embark on science research may be informed about the possible effects it can have on students.

THE DATA

About the Mentees and Mentor

Grace (a pseudonym) was matriculated into FSHS in 2005 as a female 7th grader and Mary (a pseudonym) was matriculated into FSHS as a female 9th grader. At the time Grace and Mary embarked on their first science research project together, they were third and fourth year (Grades 9 and 10) students at the school. They were inspired by the strong research culture in the school and decided to take up the projects mostly offered to fifth year (Grade 11) students at the school.

At FSHS, Grace took courses in English, Chinese, Mathematics, Physics, Biology, Chemistry, Music, and Geography. At the end of Grade 10, she decided to leave for a mainstream high school and continued to study Physics and Chemistry in that school. Grace remained highly interested in science research and embarked on a new project at the high school under the supervision of a schoolteacher. Currently, she is a final year undergraduate at a local university and is majoring in Physics. Mary joined the specialized school as a 9th grader and then decided to continue her high school education at a junior college. She studied Mathematics, Physics, Biology, Chemistry, and other subjects at FSHS but disliked the Humanities. After the chemistry project, she embarked on another in bioengineering. Mary has recently graduated from a U.S. university and has found interest in computer science—a subject not offered at

FSHS and high schools. During one of the summers, she worked as a research intern at a university laboratory analyzing statistical data of volcano samples.

The mentor was a chemistry teacher at the school when she co-supervised Grace and Mary on the organic synthesis project with a polytechnic lecturer who has a doctorate degree in chemistry. During her chemistry honors year in the university, she had done research in organometallic synthesis and hence, was able to advise students on the research when the co-supervisor was not available. Currently, she is an Assistant Professor and conducts research in science education and teaches chemistry preservice teachers. Through analyzing her students' retrospective accounts of their science research experiences she had gained deeper insights into how such an informal curricular experience for her students had shaped their lives many years later.

The School Context

As compared to other Singapore mainstream schools, Frontier Science High School, FSHS, was unique in many ways. First, it was an independent school which did not follow the national curriculum. Second, as opposed to having a typical five-day teaching week, every Wednesday (at that time when Grace and Mary were students at the school) was devoted to informal curriculum programs including student independent research. On Wednesdays, Grace and Mary would meet their mentor in the school's chemistry laboratory and work for as long as six hours. This laboratory was atypical of most school laboratory as the infrastructure and layout mimics that of university chemistry research laboratories. There were three rows of at least five fume hoods each such that students were assigned a permanent working space, which allowed students to run overnight experiments. Nested within the laboratory was an instrument room equipped with ultraviolet spectrometers and other analytical instruments. Other apparatus not found in typical schools included the vacuum line, vacuum pump, and rotary evaporators used for filtration, purification, and extraction work. The mentor would demonstrate and coach Grace and Mary in setting up the apparatus for organic synthesis, purification, and analysis work. She also taught them how to keep records of scientific data. More importantly, she would teach them organic chemistry concepts and skills that were not taught in the formal curriculum as the students did not learn organic chemistry until their fifth year at the school. Grace and Mary would also do occasional follow up and preparation work on other weekdays if needed.

Narratives

According to Connelly and Clandinin (1988), experiences is the primary actor of education central to the understanding of what schools mean to those who spend a major part of their lives there. The narratives presented here display what the students went through, thought, and felt rather than what they did so that we could gain insights into what they lived through, construed their learning experience, and how those earlier experiences shaped their current higher education decisions. The narratives contained voices of their recollections and reflections on what they liked, disliked, learned, and did not learn in their science research work. Thus, the narratives were developmental and process-oriented, emphasizing the past, present, and future as a continuum and whole in (re)constructing meaning and understanding through story-telling about themselves (Connelly & Clandinin, 1988). Their education was a narrative of experience that developed the individual's capacity to handle their life matter in and outside the school. Narratives thus, helped to recover meanings of ideas people had in order to understand their impact on the curricular experience of these students beyond the school. The hidden curriculum—things which were taught but not intentionally planned for—may be illuminated.

Two separate narratives written by Grace and Mary are presented. Grace's narrative illuminated her informed understanding of science while Mary's narrative illuminated her additional understanding about her own niche and interests after participating in science research. Interestingly, both of them have now realized that that a career in science research involving synthesis work is not for them. When the first author contacted them to write the narratives on their reflections and recollections of the school and science research experience, and how the research experience had shaped their university major and career choice, they were not told how the narratives would be analyzed and what theoretical lens would be applied. The first author had decided to keep the narratives separate from the analysis so that Grace's and Mary's voices are clearly heard and uninterrupted by the analysis that follows.

Analysis

A mix of prescriptive and emergent codes was used to analyze the data. Using the concepts of affordances described earlier, prescriptive codes such as "perception of affordance", "manipulation of environment", "availability of resources, placing features", "multidimensionality of affordance", "nested affordance", "sequential affordance", and "relations between aspects of the actors and situations" were used. The two sets of narratives were coded separately using these codes. Nonetheless, other emergent codes were also identified in the process of reading the narratives. For example, "interwoven affordance" was derived as a code when Grace mentioned that she had learned from other mentors present in the laboratory. Grace's and Mary's realization that they would not want to pursue a career in science research was coded as "making more informed choices".

The process of coding was reiterative as the first narrative was reread and recoded from the beginning of the narrative as new codes were identified. The process was repeated with the second narrative. When new codes were identified in the second narrative, the first narrative was recoded again. The codes that were related were grouped together into categories. For example, the codes "nested affordance", "multidimensionality of affordance", and "interwoven affordance" were grouped together as a category to illustrate the multi-faceted nature of affordance which enriched the science research experience for students.

NARRATIVES FROM THE STUDENTS

Grace's Narrative

The school. It helped that Frontier Science High School is an independent school, meaning that the school is given autonomy in planning its own curriculum. I learned interesting content, which I would not be able to learn in a mainstream school. At that school, students could pick and choose electives for themselves. This way, we were given partial autonomy in deciding what we learn. Thus, in some areas, we got a chance to learn only what we want to learn and we will enjoy learning. Furthermore, the small class size compared to other schools promoted greater interaction and encouraged more discussion. More questions were raised and our quality of learning was raised to a higher level.

The research culture at FSHS was very strong. The school provided many platforms that promoted the exchange of research ideas such as the annual research congress. I found it really interesting to be given the chance to get to know more about research in other areas. It helped that the research facilities in FSHS were quite advanced, and this made it possible for us to do most parts of the research experiments in the school.

Science research experience. Research trained me to be creative and innovative when looking for alternatives. For example, there was once when my mentor modified a pipette so it could function as a spoon. I also learned to be more independent in thinking and doing my work. This was because in a laboratory, the mentors would not always be present and we had to make decisions on our own. In contrast, there was much more hand holding in the classroom, since the teacher was always present during lesson time.

It also helped that the teachers in charge of science research in FSHS and my teacher-in-charge for the research project were very helpful in taking us for Research Methodology Modules (RMMs). I had a better understanding of the research process as a result. It also helped that my mentor would give us work and checked our work regularly. This was important because I was new to the subject area and made conceptual mistakes very easily. By checking my work, any mistake made could be corrected early.

Whenever new procedures were introduced, my mentor was always there to supervise and correct any mistakes made while I was carrying out the experimental procedure. This way, I was more likely to master the procedure after a few tries. Besides, she was patient in explaining how different instruments worked. Furthermore, the mentor prompted me on what I could research on in my spare time and it had enriched my theoretical understanding.

It was good that the mentor and teacher-in-charge gave us extra theory lessons and lent us some resources that we need. This was because we did not have any prior knowledge in organic chemistry. They were also very patient in clarifying any doubts we have and gave us questions to think about. I benefitted more from the questions given by the mentors because they made me more aware of the areas I should be thinking about while doing the research project. However, I felt that it would have been better if, the mentor shared some of the challenges faced in her past research experiences at the beginning of the project and how she overcame them, so I that I would be mentally prepared about what I am in for.

While doing the project, there were a few other research groups in the same laboratory. I learnt quite a bit when the mentor from another group taught us relevant concepts once in a while. Thus, I realized that the laboratory was also a place for learning, and it was not just a place to carry out experiments.

I feel that I will learn more if I have a greater understanding of the different methods used to purify compounds obtained from different reactions. I remember various methods used, like recrystallization and column chromatography. At that time, I did not think about factors, which affect the purification method used. It also did not help that I did not think very critically at that time. Perhaps, the mentor could ask even more questions to push me to think at a higher level, and after that, have a discussion about the questions raised.

I saw the relevance of classroom learning and was more appreciative of the concepts learned in school. In the classroom, we just learned concepts, but hardly knew what the concepts were used for. By applying the concepts to the experiment, we saw how the knowledge learnt in school could be applied. For example, I appreciated how the varying polarities of different organic substances could affect the choice of eluent used in column chromatography. The experience had redefined the purpose of learning science for me. We did not learn science just because it was interesting. Rather, we learned it so that we would be able to have the necessary knowledge to carry out science research in future.

Also, the research experience allowed me to see how various fields of science came together. Even though the project was in the field of organic chemistry, I saw some applications of physical chemistry as well. This was unlike classroom learning where we isolated different fields of science and did not think about how the various fields could be integrated.

At the same time, the experience redefined my perception of science. In the classroom, I used to see myself as a person trying to absorb as much knowledge as I could. I did not appreciate science as a subject of inquiry. During the project experience, a number of questions were raised. These questions prompted me to think at a higher level. After the experience, I started to appreciate science as a subject of inquiry. Besides, the research project I undertook was not graded. Thus, I did not feel pressured during the experience and I saw the project as an opportunity to make mistakes and learn from them. Further, the research experience gave me a glimpse of what I could expect to do in my career if I chose a science discipline in the university.

If I could do the project again, I will check out books in the school library or in the university library. Even though I had access to the university library at that time, I did not think about making use of the resources there. My main source of information, whenever I needed help with theoretical knowledge, was the Internet. However, I did not find the Internet very useful, because many of the underlying theories were rather specific and information on the internet was rather general.

How the experience affected my university major and career choice. I felt that the research experience has provided many learning opportunities. Besides learning more technical knowledge, I also picked up life skills like perseverance and creativity. Thus, I will certainly want to do the honors year project in university, if my grades allow me to do so.

A few years after the project, I started thinking about my career and university options. I thought about my research experience, and felt that I will not want to do research involving chemical reactions all my life. I was sure about the reason behind my career decision. Thus, I would say that the experience allows me to make an informed decision about my career choice.

It also helped that the teacher-in-charge of the project shared about happenings in actual research laboratories, for example, possible conflict between different project groups. This has allowed me to make a more informed choice about going into research as a career.

Mary's Narrative

The school. FSHS was a young school, so there were many things to figure out for both the staff and the students. FSHS taught me to step out of my comfort zone. I had been a fairly good student in my previous secondary school. Some subjects came naturally to me, and hard work usually allowed me to do well at those which I was not naturally good at. That all changed in FSHS. I struggled a lot with Physics, and for two years it seemed that my best efforts could only yield disappointing results. Things got better after we studied Calculus. I eventually did figure Physics out, but I felt that I had never worked so hard at something.

Science research experience. My first experience with research was a project on organic chemistry. I worked with a fellow student. Since it was our first time doing research, we had a lot of guidance from the mentor who explained how to perform the steps first before we did them ourselves. I had not taken any organic chemistry classes at that point, so I felt that I did not do much critical thinking, but the experience was useful in preparing me for what to expect in future projects.

One of my more memorable research experiences was at an external research institute. I had a very dedicated supervisor who knew how to strike a balance between teaching and letting me figure things out myself. She also let me learn protocols which were not related to my project. I got to watch her dissect lab rats. It was more difficult that I thought, since the rats were able to sense when one was about to kill them and kept lashing out and biting.

A good supervisor is of course the most important thing to the research experience. I was lucky enough to get good supervisors most of the time, who took the time to explain things and think of projects which I could reasonably work on myself with a little guidance.

One thing I regret not knowing before I started research was computer programming. This is becoming a very crucial skill for anyone to know, especially those working in science. If I had known computer programming, there were some projects which I could have done differently, such as writing a physical model to simulate flowing through a micro fluidic device instead of running many actual experiments which were few and delayed due to problems with the equipment. After taking a computer science class in college, subsequent research supervisors have asked me to write programs for various reasons, and I felt that I did a lot more than with previous experiences.

How the experience affected my university major and career choice. After I graduated from high school, there was quite a wait before I entered college. I was not sure what to study yet. I knew it would be something to do with science and engineering. I thought about doing Physics, but I did not have spectacular experiences at the Physics labs in school. The theory made sense, but I strongly disliked the practicals. It took me another year of Physics in college to admit it, but I was not at all interested in experimental Physics and I was bad at it. Reluctantly, since I had invested a lot of time and energy into it, I dropped the idea of doing Physics major in college.

The previous summer at my university, I did an internship studying cryptography. It was a good experience. I had to learn computer programming language myself to write a program to attach a cipher. I also had to read up on cryptography and number theory, which I knew almost nothing about. Luckily, my supervisor was very patient and pointed out what I should know for the purpose of my internship. I decided that programming would be something I could see myself doing in the future, and decided to major in computer science.

I have always known that I wanted to do science, because I enjoyed the science classes in high school, but not so much the humanities ones. Some of my classmates from FSHS realized that math and science was not for them, and decided to pursue university degrees not related to these fields. I felt quite uncomfortable watching them, as I thought that it might be quite sad to have invested four years studying all the tough classes in math and science, and then do something completely unrelated to it. At the same time, that made me realized that I should be careful when choosing a career path as it was better to change directions than continue in a field I did not like. I was not sure which branch of science I would end up in. They were all quite interesting and useful, so I studied Math, Physics, Chemistry, and Biology until the end of Junior College. I decided in High school that I liked Physics and Chemistry best. I did a research project in Chemistry, then one in Bioengineering, and participated in Chemistry and Physics Olympiad. I thought they were quite fun and a good introduction to research within those fields, but I wasn't sure if I wanted to continue in the fields for the rest of my life.

After high school, the most natural thing was to try to get an internship related to Physics research to see if it was something I would want to do for a career. I knew from my experiences doing research in high school, that research in a field and studying it in high school were two very different things. Unfortunately, I found that I did not enjoy experimental Physics research, nor did I enjoy any of the experimental physics classes in college. I did another internship during the summer between by freshman and sophomore year in high school, this time in cryptography. I had just taken an introductory computer science class, and this internship made use of what I had learnt in math and computer science. I found that I enjoyed what I was doing, and decided to switch to math and computer science (CS) instead. I would most likely do something related to CS in my future career, since these are subjects, which are useful in many technical jobs, and something I enjoy.

FSHS has done a good job in preparing me for science classes in college, though I wish it could have been more emphatic in teaching how to write math proofs, and offering computer-programming classes. I did not know how to do either when I first started college, so it was quite a steep learning curve. However, all FSHS students are used to steep learning curves, so I was prepared for the amount of effort and work needed.

DISCUSSION

The affordances available to these two students were situated and not generic to all students in other Singapore schools. As Grace had mentioned, the school had partial autonomy in deciding what the students learn as opposed to mainstream schools, which follows more closely to the national curriculum. Situated affordances made available by having the science research program in the school, the necessary physical, human, and monetary resources including access to university libraries (due to the school's affiliation), and students' autonomy to decide their research topics and elective subjects. As Grace mentioned, the research facilities were advanced allowing most of the research experiments to be done in-school without having to travel to external research institutes. This gave them the flexibility to continue with their experiments even on non-designated research days. Being a boarding school, students who stayed in the hostels could also continue with their research later into the evenings.

The science research has provided opportunities for affordances to be acknowledged, generated, and engaged through several conduits. For example, affordances were provided by the human resource (mentors) to Grace and Mary as they learned to represent chemical molecules using symbols. This is a technique and language universally understood and used by chemists to communicate ideas about the reactions, reacting species, reaction sites, and reaction conditions. As such, affordances were provided by the introduction of conceptual entities and operations of discipline through the process of scientific discourses.

Affordances were made available to Grace and Mary by virtue of their participation in the social practice of science in a setting, which underscores the important status of science. The fact that science is one of the two core disciplines in the school, most students like Grace and Mary took up science related projects aligned to the school's niche. Grace recognized the availability of platforms such as the annual research congress where students showcased their work and presented their findings like research scientists. The resources including advanced research facilities were affordances, which Grace thought had allowed them to do most of their research in school and at ease. Nonetheless, the affordances existed even if she had not recognized them.

To use the affordance present, students need to be able to recognize its presence. The role of perceiving affordance is that such perception can set up action systems to act, direct attention to appropriate action-guiding information, and so on. Perceiving affordances is more than just perceiving relations. It involves making sense of the information and deliberating on issues in making independent decisions. As Grace noted, there was less handholding in the laboratory than classroom. What Grace and Mary had undergone was similar to that of what scientists do in real science laboratories. They had to make decisions without instructions or adequate information from existing literature as scientific research papers did not always provide all the details.

Affordances do not arise as a consequence of mental operations alone but from the action-related properties of the practice that may or may not be perceived. In the process of carrying out the experiments and making scientific records, they encountered and engaged gestures, facial expressions, and body language that provided information to each actor. The information, in turn, shaped their action, self-perception, knowledge-building, understanding of the practice of science—how theories may be applied and changed through the process of negotiation and gathering empirical data.

Affordances are action-permitting and the action is goal-directed; it entails intention and identification of information and a lawful relation between the information and the control of actions (Michaels, 2003). Mary commented that she had a supervisor who guided her and at the same time, allowed her to figure out things by herself. This afforded her the space and agency in decision-making. Grace talked about learning to be creative and innovative from this experience. She saw her mentor converted a pipette into a tool with spoon-like features. This was, to her, a valuable lesson in showing how one can be creative and innovative in the laboratory and address one's situated needs. As mentioned by Turvey (1992), the affordances are complemented by the effectivities of the actor to bring it to actualization. In this case, the mentor had perceived the possibility of using a pipette as a spoon and had effectively manipulated it to be used as one. Her mentor knew that a pipette was manipulatable through identifying its features that could be changed into a "spoon". In emplacing spoon-like features on the pipette and adapting it to her own needs, she had generated affordances for herself (to get the object out of a narrow neck vessel) and her students who learned how the object may be modified for other uses. The actor who adapts it for use thus subjected the perceived possibility of what a pipette may be used for to interpretations. Thus, science research could expose students to the idea that science and scientific practice is not rigid and fixed but adaptable to suit the situation.

Affordances are multi-dimensional. In doing research, Grace learned to apply concepts she learned in class and changed her perception of science. Drawing upon her knowledge of differences in bond polarities, she understood why various solvents were chosen as eluents in the separation of different organic compounds. Also, she saw how various science disciplines were less segregated than it seemed in the formal curriculum where science was taught as physics, biology, and chemistry. She had also learned from other research group mentors and hence, the learning took place across groups. This reflected the reality in scientific work where scientists and research groups collaborated on projects to support one another rather than work in isolation. This illuminates the interwoven affordances present as mentors in other groups may have relevant knowledge of various projects and may potentially be tapped upon as resources. Grace said that she learned that laboratory was a place for learning and not only for doing science research. This illuminated the nested affordances that different learning settings could offer to people. The university library, which students in this school could access due to its affiliation to the university, was nested within the school space but Grace had not recognized its importance earlier. Instead, she had turned to the Internet, which she eventually found to be limiting. As such, the recognition of affordances required the actors to perceive it; otherwise, its existence would be overlooked. In this case, the perception was aligned to Norman's (1988) idea that it was independent of the students' knowledge of the type of literature that could be found from different resources and experienced from doing literature search. Grace recognized that this experience from doing, listening, and learning had created longer term impact on her life in terms of her choice of career and major in the university. This illustrates that affordances are specified by information and perceived by the actor (Michaels, 2003). In making connections to what she had learned in class, Grace now understood how theoretical concepts could have practical applications. This understanding, she said, had redefined for her the purpose of learning science. Additionally, before her research experience, she learned science by trying to absorb as much information as possible rather than see science as a process of inquiry and that the knowledge is subjected to change. The questions asked during the research provoked her to think more. This was probably why she continued to embark on science projects in the junior college and university.

Interestingly, affordances do not only result in expected outcomes but also incite unexpected outcomes not aligned to the original goals. For example, from her research experience, Grace learned that she did not want to be doing research involving chemical reactions as a career. She was also adverse to the possible politics that could happen in laboratories. The affordance was thus one of action possibility in terms of her career choice and higher education. Similarly to Mary, her earlier exposure to science research made her realize that she disliked practical work, especially in physics. Her exposure to science and research made her rethink her future career and higher degree choices. Later, she found her interest in computer programming which was something she was not taught to do in the school. The unexpected outcomes of providing the science research experience to acquaint students to the practical and discourse of science had inevitably resulted in the students having a clearer idea of their interest and prepare them on what to expect when they do research work in future. This illustrates the sequential nature of affordances as her experience doing science incited her to look for other learning opportunities. Subsequently, she discovered her interest in an area which she had never been exposed to in her 10 years of schooling before college.

CONCLUSION AND IMPLICATIONS

In this chapter, the concept of affordance was applied to unpack the science research experiences of two students who attended a specialized science and mathematics school in Singapore. The findings show that affordances two students had were situated in the school they attended as the resources they had were not available to mainstream schools. The affordances that were available to them had to be perceived, acknowledged, and engaged by them even though they were made available to them through participation in the social practice of science. The affordances were abstract and real in that they arose in mental operations and from action-related properties of the practice of doing science. Finally, the affordances are multi-dimensional; they can be nested in contexts-within-contexts, interwoven, and sequential.

As a final note, let us return to the idea of usefulness and usability—two reasons why HCI communities engaged in the concept of affordance to theorize about the designs of products. The science research experiences were valuable to the students as they gained better understanding of what science was about, better ways to learn science, and about what they enjoyed or disliked doing. The usability—in terms of preparing them for a career in science—however, may be limiting as both students decided that they did not want to do organic chemistry, synthesis work, or to make science research a career. This suggests that while providing science research opportunities to students may provide students with some early exposure to the practices of the discipline, it may not guarantee that students would pursue a career or higher studies in this area. This may be an unexpected outcome of the science research program but it certainly, brings to light that affordances, or perceived possibilities, do not necessarily always lead to expected outcome of invoking more students' interest to become scientists. However, the affordances provided to students through participating in science research may enrich their schooling experience in more ways.

REFERENCES

Bourdieu, P. (1986). The forms of capital. In J. G. Richardson (Ed.), *Handbook of theory of research for the sociology of education* (pp. 241–258). New York, NY: Greenwood.

Chemero, A. (2003). Radical empiricism through the ages. *Contemporary Psychology, 48*, 18–20.

Connelly, M. F., & Clandinin, J. D. (1988). *Teachers as curriculum planners: Narratives of experience*. New York, NY: Teachers College.

Gaver, W. W. (1991). Technology affordances. In S. P. Robertson, G. M. Olson, & J. S. Olson (Eds.), *Proceedings of the ACM CHI 91 human factors in computing systems conference (April 28–June 25)*. New Orleans, LO.

Gibson, J. J. (1979). *The ecological approach o visual perception*. Boston, MA: Houghton Mifflin.

McGrenere, J., & Ho, W. (2000). *Affordances: Clarifying and evolving a concept*. Retrieved June 22, 2012, from http://teaching.polishedsolid.com/spring2006/iti/read/affordances.pdf

Michaels, C. F. (2003). Affordances: Four points of debate. *Ecological Psychology, 15*, 135–148.

Norman, D. A. (1988). *The design of everyday things*. New York, NY: Doubleday.

Reed, E. S. (1996). *Encountering the world*. New York, NY: Oxford University Press.

Shaw, R., Turvey, M., & Mace, W. (1982). Ecology psychology: The consequence of a commitment to realism. In W. Weimer & D. Palermo (Eds.), *Cognition and the symbolic processes* (pp. 159–226). Hillsdale, NJ: Lawrence Erlbaum.

Turvey, M. (1992). Affordances and prospective control: An outline of the ontology. *Ecological Psychology, 4*, 73–187.

Tang Wee Teo
National Institute of Education
Nanyang Technological University
Singapore

Jia Qian Woo
Independent Scholar
Singapore

Lay Kuan Loh
Independent Scholar
Singapore

JERRY EVERHART

12. THE GEOGRAPHY OF GIFTEDNESS

Growing Scientists in Rural Areas

INTRODUCTION

Rurality is best understood, not by census information, but by examination of the *space* between people. Distance between people and resources shape rural culture; and space can propagate independence, adaptability, lateral and critical thinking, and creativity. Sparse distribution of people impacts how residents approach problems and their exposure to new ideas; geography and population affect funding of essential services such as public education due to a marginal tax base. This chapter addresses three issues that impact transition of gifted rural children to university settings and, subsequently, into Science, Mathematics, Engineering, and Technology-oriented (STEM) careers: 1) an examination of space and an overview of rural communities and schools; 2) rural gifted students' readiness for university science studies including place-based knowledge, skills, and impediments, and 3) a possible plan-of-action leading to a science-oriented career trajectory linked to rurality. Although some of the following discussion focuses on culture and education in rural south-western United States, research indicates that space has a similar effect on children worldwide.

THE SPACE BETWEEN

Geographic Isolation

> Between Portales and Roswell, New Mexico lies one-hundred forty kilometres of mesquite, yucca, and sage. Well-kept dirt roads on either side of the highway indicate the presence of a ranch or oil field. Unlike major thoroughfares in urban areas, this stretch of four-lane highway contains no exit ramps, no advertisements, and cell service is intermittent at best; in fact, only one community is large enough to support a small convenience store and school (enrolment – 142 students pre-kindergarten to grade 12). Recognised for its high standards, the school operates on a 4-day week with sixteen teachers.

Geographic isolation impacts most aspects of daily life in rural America, and parallels exist with remote locales anywhere on earth. A study of space provides insights into physical, cultural, and economic factors that shape lives. As in the example above,

M. K. Demetrikopoulos & J. L. Pecore (Eds.), Interplay of Creativity and Giftedness in Science, 219–239.

somewhat innocuous aspects of where one lives mould, and sometimes dictate, a life path. An examination of rurality indicates that the environment is science-rich though relatively technology-poor, requires a high degree of independence and ability to problem-solve, but rarely meets the needs of gifted children in formal educational settings. Foundational to understanding the plight of children raised in predominantly agrarian communities is the fact that opportunities for some facets of growth and change may be restricted, but are not fixed. Since forty percent of American children attend rural schools, educators must devise place-conscious strategies to better help all children reach their potential (Kordosky, 2010).

Rural residents enjoy many benefits of living outside densely populated areas. Rurality is relatively free of crime, aesthetically pleasing, and quiet with less stress. Closely associated with the land, many country dwellers choose to maintain gardens and attend to livestock, freeze and preserve home-grown foods, and participate in environmental and conservation practices in varying degrees. Children in remote settings are often more in touch with nature and understand the relationships among space, wildlife, water, resources, and humans. Parents wishing to manage children's social and educational lives may have a distinct advantage by living without many distractions of urban areas.

Space governs what one can and cannot access. The absence of health services, home repair specialists, and markets in rural areas breed independence and flexibility – the phrase "rugged individualism" spawned by settlement of America's frontier is still applicable to many rural residents today. Electricians, plumbers, and carpenters are expensive luxuries and most tradesmen are unwilling to travel to remote sites for small jobs. Consequently, many rural residents become quite adept at do-it-yourself home repairs and routine maintenance. The independence bred by isolation is not necessarily recognized as understanding of scientific principles, ability to solve difficult problems, or indicators of giftedness by those living in rural areas.

Rurality also severely restricts employment opportunities and may limit children's options for future jobs regardless of their creative abilities or level of giftedness. Most rural residents work in agriculture or mining/energy extraction. Employment in these jobs tend to be inter-generational and often, children see few opportunities for work beyond the occupations of their parents and may feel bound to the open land. Community values and families' expectations have profound impacts on career and educational choices made by children (Hardre, 2009). Attempts to leave rural areas in pursuit of a better life may be viewed as "destabilising" to the community (Howley, 1998 cited by Lawrence, 2009). Rural families are among America's poor with average annual incomes of $36,920 (U.S. Census, 2010). Average wages for agriculture non-management workers in agriculture is $22,118 per year; miners and oil workers fare better at $42,240 annually (USDA, 2012; Technomine, 2012). These income data are easily misconstrued due to variability based on region, race, and annual versus seasonal employment; for example, 2010 data reveal that poverty in some rural population pockets in the United States approaches 50 per cent (Housing Assistance Council, 2012).

Geography's Effect on Schools and the Gifted

Like the environment in which they exist, rural schools are best characterised by extremes. In urban areas where political discussions include terms such as "competition" and "charter schools", rural schools are often the only viable option for children in remote areas. Consequently, some rural schools' heightened visibility and status within the community yield positive results. In a recent assessment of schools in New Mexico, outlying districts tended to receive higher marks overall that their urban counterparts (New Mexico Public Education Department, 2012); rural schools tended to have higher completion rates, higher tests scores, lower teacher-student ratios, and greater involvement of parents in all aspects of the educational process.

Although rural children enjoy benefits of small class size and more individualised attention, little research exists touting the advantages of being gifted in rural schools. Based on the body of current research, the state of rural education and the ability to serve gifted students seems quite bleak (Floyd, McGinnis, & Grantham, 2010). Decreases in budgets, non-compliance with state and national special education regulations, and lack of well-trained teachers does not bode well for creative children. According to Stambaugh (2010), rural communities' allocation of resources, classroom issues, teacher preparation and pay, educational attainment, and parental expectations are reasons for concern. Rural schools' inability to serve bright children results in boredom and general dislike for schooling (U.S. Dept. of Education, 1993).

Despite state and federal laws and regulations, few rural schools can sustain comprehensive special education programs and rural school administrators often view gifted and talented (GT) program services as optional. Using traditional measures of giftedness, the prevalence of gifted students at approximately two percent of the general population (intelligent quotients of 132 or above) underscores the problem (Gifted Development Centre, n.d.). Since most rural schools have enrolments of 400 students or less, separate GT programs comprised of eight to ten students are fiscally unsustainable. Many gifted educators assert that rural schools have no option (legally and morally) other than to serve the needs of bright, creative children. Kordosky (2010) states that forty per cent of children in the United States attend rural schools and assuming that giftedness is uniformly distributed across rural populations, gifted rural students are an untapped resource that demands attention and support.

Lawrence (2009) offers a practical rationale for why society should care about gifted students in rural areas – the obvious argument for enriched education for rural children is that they deserve every opportunity to reach their potential. No less important, and closely aligned with the concept of space, is that "rural communities cannot afford to lose the contributions gifted students can make to rural community, culture, and economy" (p. 462). Despite strong justifications for educational support of bright children, special education and gifted education services in rural schools tend to be uneven and student services are prioritised based on numbers and demand. Gentry et al. (2001) examined trends in rural schools associated with gifted education; they found that schools were less likely to provide appropriate placement and services because of program novelty (unproven or lack of empirical support), lack of financial support for programs, and low enrolment rates for gifted students. Other research suggests that most rural schools utilise differentiated instruction strategies to meet the needs of gifted children rather than allocation of additional personnel with specialised training (Stepanek, 1999). Kordosky (2010) dedicated a chapter in his book, *Rural Gifted Children: Victims of Public Education*, to extreme measures to force rural school districts to provide appropriate services to gifted children including litigation. However, Kordosky does not take in consideration rural culture and the implications of suing one's neighbour in a tightly knit, interdependent community.

A debate among gifted educators provides hope for rural GT programs. Since a link exists between resources available and numbers of students served, the modern gifted education movement advocates use of broader criteria for selection of gifted students. Traditionalists associate gifted abilities with intelligence quotients, yet IQ may not be the most reliable predictor of ability or performance (Matthews & Kelly, 2007). In addition to IQ, advocates urge use of criteria that include observations, interviews, performance data, and results from other standardised tests to determine whether students are gifted (Lawrence, 2009). A supporter of more flexible admission criteria for rural gifted programs, Stambaugh states, "Rural gifted students, especially those who are geographically or financially disadvantaged, show their talents in different ways and, therefore, may require multiple assessments to capitalise upon their unique needs." (p. 66) Brown, Renzulli, and others (2005) report that most special education administrators believe that objective measures as intelligence quotients tests must be used to assess gifted, but one-third of respondents in their research would like to include personality traits or performance-related criteria. These researchers concluded that how giftedness is identified is based on two questions or beliefs:

- Is giftedness an absolute or relative concept? That is, is a person gifted or not gifted (the absolute view) or can degrees of gifted behaviours be developed in certain people at certain times under certain circumstances (the relative view)? and
- Is giftedness a static concept (you have it or you do not have it) or is it a dynamic concept (it varies within the individual and learning/performance situations)? (p. 77)

If one believes that giftedness is relative, dynamic, and should be assessed using performance measures in addition to IQ tests, then numbers of children are likely to increase adding emphasis and legitimacy to rural gifted education programs.

Other challenges faced by rural schools are linked to personnel. Teachers and administrators in rural areas may be similarly place-bound as the students resulting in low attrition; when teachers do move or retire, schools may have difficulty recruiting replacements due to lack of housing and shopping or daunting commute times (Fowler et al., 2013). Some rural schools districts request emergency waivers from state education agencies to place non-certified or non-licensed personnel into classrooms. States with large rural populations are forced to offer alternative routes to licensure; these alternative licensure programs do infuse more teachers into the workforce, but attritions rates are high and self-efficacy to teach was low among those who participate in abbreviated preparation programs (Darling-Hammond, 2002). Alternative licensure programs may consist of as few as six university courses beyond an undergraduate degree. States employ a patchwork approach to alternative licensure in special education and an endorsement in gifted education can be secured after 12 hours of select university coursework (New Mexico Public Education Department, 2012). However, organisations such as "Teach for America" provide extensive professional support and other incentives for newly certified teachers seeking employment in rural areas.

The combination of low income and low taxes on agricultural properties results in difficult choices in funding services for rural populations. Monies that support schools are unevenly distributed across county and state boundaries. With local funding in decline, rural schools survive on small budgets; and since salaries and benefits make up nearly forty per cent of most school budget, rural schools operate with the least number of teachers and staff possible (Fiscal News, 2011). However, some innovative rural districts have entered into collaborative agreements in order to secure highly qualified personnel to work with students with special needs such that two or more small districts may share special education teachers in order to offset costs, comply with laws and regulations, and still attend to the needs of children with special needs. These shared teachers work in participating schools a few days per week and collaborate with regular classroom teachers in providing prescribed services. Other personnel issues affect rural education – teachers often instruct outside their area of expertise (Kordosky, 2010). For example, a rural school's mathematics teacher may serve as the science teacher and administrators may encourage physical education teachers to seek endorsements in special education. Offering advanced classes in mathematics, science, and foreign languages can be particularly difficult, in part, because rural districts are less able than others to attract teachers with specialised preparation and rural high schools are less likely offer a full complement of Advanced Placement (AP) courses (Lawrence, 2009). A school district's inability to provide advanced science and mathematics courses will impact students' attitudes toward science and their transition to entry level university courses (Everhart, 2012).

Little research exists that guide rural districts' efforts to improve the plight of gifted children. Arnold et al. (2005) state that funding for high quality educational research in rural settings is limited and that the few studies that have occurred are framed in a context used to preserve rural schools rather than to identify issues affecting quality of services. In his meta-analysis of reform research conducted in rural settings, Arnold found that researchers typically conducted only one or two formal substantive studies per year in gifted rural education in the United States.

Since the mid-1990s, the National Science Foundation through its Rural Systemic Initiative (RSI) has made strides to document exemplary programs, curricula and instruction reforms and improvements in policies and infrastructure that address needs of rural children while promoting science. RSI identified six "drivers" or tenets of change that include classroom, policy, resource, community, attainment, and equity (Systemic Research, 2005). Rural districts may use data collected in these programs to inform future decisions regarding funding, personnel, academic choices, and partnerships.

SPACE AND THE NATIVE SCIENTIST

In the previous section, issues relating to isolation and the effects on community and schools were explored. Although gifted students in rural schools do reap some benefits from the inherent structure of rural education including smaller class size, educational literature more often associates rurality with adversity. If gifted students' psychological needs are not being met by schools, and if they do not master requisite content knowledge and skills, then, are the rural gifted viable candidates for STEM-oriented careers – or are they, as VanTassel-Baska and Stambaugh described, "overlooked gems"? (2006).

Place-Based Science

Little primary research is available dealing with the interplay of "science careers, giftedness, and rural children." Some extrapolation is necessary to develop a complete picture of rural gifted students' potential as future scientists. With considerable variability among children identified as gifted (IQ, motivation, support from parents, economics, community values) and trends to use performance criteria when identifying gifted students, the issue of "space" provides a unifying concept among children raised in remote areas. Briggs, (2005) posited that the depth of the enculturation of local values and knowledge may facilitate positive change in rural areas. He stated that "local embeddedness of indigenous knowledge that imbues it with relevance, applicability and even power" (p. 21). Like their parents, space impacts young children's lives determining the types of play experienced, hobbies, and pastimes. Children play an integral role in the operations of the rural home. Farm life may dictate early assumption of chores and responsibilities for children (Extension, 2013); in eras before mechanisation, ranchers maintained large families that served not only as stabilising units for rural communities, but children contributed significantly to the workforce. During that time, adults passed skills and knowledge to children about animal husbandry, crop maintenance, food preparation, and considerable adroitness with hammers and saws – without regard to gender. Parents prepared children to be "generalists" or "jacks/jills-of-all-trades". Space and necessity of that period required children to assess issues as they arose, collect available information, consider and select appropriate solutions, and reflect on outcomes (Inwood, 2013).

The development of skills, acquisition and retention of useful knowledge and mastery of processes for making sense of the world are foundational to survival in rural areas, but these qualities are also fundamental to becoming successful scientists. Some data and anecdotal information suggest that native knowledge that is common to rural living might well serve as the precursors for careers in science (Avery, 2013). Coupled with traits of the gifted that include innate curiosity, persistence, ability to analyse and consider options, rural children offer a formidable repertoire of skills if they enter STEM-based jobs. In order for any knowledge and skills acquired in a rural context to be useful in transitioning to a university, two criteria must be met: 1) students and other rural constituencies must be able to connect their knowledge of the world to formal science and 2) children must see themselves as "native scientists".

Some research in science education shows that students are capable of linking their rural existence with scientific concepts. Avery and Kassam (2011) asked fifth and sixth graders to take photographs of elements of rural living that were representative of science and engineering concepts. The researchers conducted individual interviews with children to probe further the depth of student understanding. Based largely on Aristotle's supposition of "phronesis" or practical wisdom, the researchers attempted to determine if the children could make connections between their indigenous knowledge and science; the investigators also tried to locate sources from which native knowledge was derived. Avery and Kassam arrived at five key findings. They found that 1) students were able to identify science and engineering concepts embedded in rural living; 2) students learned about the scientific concepts by observing or doing (or both); 3) the primary source of information about native science was the family; 4) children used daily activities to understand science; and 5) students' understanding of place-based science was not fully formed and was dependent upon teachers and others to help students make relationships between practical wisdom and traditional classroom science. Although many of the students photographs were not exclusive rural existence (auto mobiles, gas grills, toys), this research suggest a strong link between rurality and native science.

In a similar study, Lloyd (2010) attempted to determine sources or "funds" of science knowledge from rural children. As in the Avery and Kassam research, some students' responses were not restricted to rural living. Lloyd developed criteria for separating general science concepts from those exclusive to living in rural areas. He found that students had place-specific understanding of water, farming, transportation, outdoor recreation, local geography, and knowledge associated to their parents' employment. Students in this study believed that certain conditions are necessary to glean value in their funds of knowledge. Students thought that any complementary science studies should mirror their way of understanding the world – activities should have local relevance, learning should occur in small teams, and science should involve hands-on activities and fieldwork. Participants expressed the importance of establishing community, possibly in the form of a club or student organisation, in order to help them learn science. Students liked the idea of working teams because they were closer to content, were able to manipulate equipment, and were not resigned to watching others complete tasks. Several of the student participants also expressed the importance of using authentic equipment, performing "real" science, and having active field experiences. The children indicated that a realistic context made the investigation seem like the type of science that scientists would do, not just learning things for a test. Many of the students' preferred learning styles reflected learning and work conditions common to farms and ranches.

Insights from the Ranch

In order to confirm some of the findings in the Avery and Kassam and Lloyd research, the author of this chapter sought addition information from resident experts in rurality, indigenous science, and gifted education (Everhart, 2012). Patrick and Robin Wallace (pseudonyms) are life-long residents of rural Colorado and New Mexico. They have two children and five grandchildren. Uniquely qualified to provide insights into rural living, Patrick and Robin are retired teachers; also, Patrick was a special education coordinator in a Colorado school district. The Wallace's are a mid-generational rural couple – *recipients* of information and skills from parents and community, as well as *transmitters* of rural culture to their own children and grandchildren. In a recent interview, Mr. and Ms. Wallace shared pertinent background about how they acquired science-oriented knowledge, the transference of their place-based knowledge to their children and grandchildren, and how skills, much of which learned sixty years ago, continue to serve them on their ranch in in rural, New Mexico.

Five themes emerged in their accounts of rural living: 1) parents and members of their community influence what and how rural children learn; 2) isolation affects the types of work and entertainment in which rural children engage; 3) families that exhibit and value inventiveness and creativity transition more easily to university settings and to STEM careers; 4) problem-solving on the ranch resembles some elements of the scientific method; and 5) rural children learn by doing. These themes underscore the convergence of creativity, science, and dictates of rural living.

Science is an integral part of growing up in a rural setting. Robin responded first to an open-ended question regarding an early experience that she believed had a basis in science. She shared that her mother frequently made pies for the family and since the family's orchard seldom produced much fruit due to lack of water; Robin's pie of choice was chocolate. Due to the delicate nature of a good pie crust and several missteps, Robin was relegated to the low-skill job of stirring the filling. As time passed, her mother taught the nuances of precise measurement and the responsibilities of pie making was eventually passed to Robin. She explained, "With that measurement you had to be real careful. How much water you put in and how much flour." The lesson continues to guide her cooking and the lesson of precise measurement has been passed on to her children.

Patrick recalled several stories of his youth; his accounts focused on getting information about the world from experts outside his immediate family. He shared that during summer months that he, his brother and their friends would pitch tents in the pasture. Before they slept, the group would spend time looking at the stars. Patrick continued, "Part of the thing being out there...you could see all the stars and you shared information and looked for different things and you learned about how to find directions...the North Star...and how the Big Dipper could help you find the North Star." Use of directions is a critical part of rural living in the southwestern United States. Directional information used regularly includes references to weather patterns, land usage, location of homes and real estate. Patrick's mastery of directions began sixty years ago under star-lit New Mexico skies.

Notable during the interview were several asides about "isolation" and other sources of information made by the couple. Patrick's rural sources of information were more global than Robin's due in large part to space. Patrick had access to other children because his family ran a business; Ms. Wallace had a more classic rural experience noting on a few occasions during the interview that no other children lived near her family's home.

During the last dozen years, the Wallace's have had a veterinarian visit their ranch only once. Both have served as sole health care providers to over five hundred cattle. When asked how they knew so much about animal care, Patrick stated:

> I worked on a ranch when I was 15. The fellow that I worked with had been the vet's helper. And so he did all kinds of stuff. He could deliver Caesarean. So I learned a lot about vet work with him. And also learned stuff from people like Mr. Russell Warner (a pseudonym) at the store in town. He (Mr. Russell) was the same way.not a vet but he worked for one a long time and he handled vet supplies (in his store). He learned what medicines on the shelves did what. So basically, he was a good resource for me. For example, you go to him (Mr. Russell) when you'd say, 'I got a cow with a snake bite.' And he'd tell you how to do it...and different things like that.

The discussion of science concepts utilised by rural residents led to another relevant issue – how "space" nurtured development of expertise in many science and engineering fields. Emphasising the notion that most rural people are forced to become their own veterinarian, he continued, "You can't get vets to come out for one or two animals. They just won't do it anymore. Now you have to haul the cows to the vet." Also, Patrick elaborated on his command of animal behaviour learned over many years. He described a furtive mare that would not birth a colt in his presence. He no longer expects to see livestock give birth noting that horses and cows will go to great lengths to deliver in private.

The interview revealed that use of native science concepts and skills can be generalised to new situations. This notion is particularly important if gifted high school students are expected to transition to university science classrooms and laboratories. Two years ago, Patrick decided to place a water source in a remote pasture for his cattle. Historically, few options were available to ranchers. They could haul water from another source (an expensive and time consuming option) or erect windmills to convert the abundant wind into mechanical energy lifting underground water to the surface. After much consideration, the Wallaces decided on a third, twenty-first century approach; they purchased two solar panels with complementary equipment and pump. After the water well was drilled and a few technical problems resolved, Patrick erected the panels, placed the pump, and completed all electrical circuits. The pump continues to work without problem despite the project being his first exposure to solar technology. When asked about how he was able to complete the complex bit of engineering, Patrick simply shrugged. He was able to confidently synthesise past experiences including knowledge of plumbing, electricity, hydrology, and even his command of ordinal directions to complete the construction of the solar pump. Since help from the outside was not an option, Patrick's life time of experience informed his decision to tackle and solve the problem of access to water using a high-tech solution.

Skills developed during youth are retained and used in new contexts. When asked if they remembered other stories relating to science and engineering, Robin told a story about how isolation affected how children entertained themselves. As a young girl with few local friends, she had fond memories of diagramming floor plans of imaginary houses. She stated that she still enjoys the process of laying out living spaces and visualising what household items would fit. Modestly, Robin shared that the interior of the home where this interview was conducted was a product of her imagination.

Much of the work that occurs on a working ranch is routine – feeding livestock, mending fences, and maintaining equipment. These routines are punctuated with intermittent problems that demand attention. Patrick and Robin accept machinery break-downs and animals becoming sick as part of their rural existence. Using personal stories, the Wallaces divided problems that they encountered into two categories – those that required immediate attention and responses to challenges that made rural existence easier. How rural residents approach problems offers insights into their role of native scientists. In response to these tests, Patrick and Robin follow similar processes of inquiry and problem solving commonly found in scientific laboratories. When confronted with a problem, the rural scientist assesses the situation and determines what is already known; the native scientist brings to bear all experiences and resources. They engage in extensive planning and gather tools and materials necessary to address the problem-at-hand. More than simple trial and error, the native scientist attempts to control a single variable at a time while implementing their plan. And finally, they assess outcomes and refine their process and product.

Creativity is always in play when confronting problems in a rural setting. The Wallaces used the term "inventive" often and interchangeably with creativity describing their work and the work of others. Patrick recounted a story about a family known in the community for approaching problems with a different perspective. He shared that the design of the neighbour's barn allowed more efficient loading and unloading materials, and that the children were "artistic" and were able to build their own toys to scale. Also, Patrick stated that the family modified common tools so that their work was less difficult. The family established a reputation among members of the tightly-knit community sincecreativity, like independence and persistence, are valued traits that are nurtured in rural homes. Robin shared an example that illustrated how children may be taught lessons dealing with personality traits. She recalled an instance when her neighbour told their son to remove the hydraulic system from a stalled tractor. When the boy replied, "I don't know how," the father told the boy, "You will figure it out" and drove away.

The Wallaces shared other stories about their own children (both attended college and became educators) and rural neighbours who made the transition to university settings. Many of the neighbouring children pursued careers in science, agriculture, and engineering. Patrick and Robin linked children's ability to succeed at the university with three inter-related conditions. They believe: 1) supportive parents and teachers are essential to children's success; 2) bright, "inventive" rural children are more likely to adapt to university life; and 3) strong connections exist between rural life and science content and pedagogies (inquiry, problem-solving, and research).

Adaptability, especially in a social context, is not considered a strength of rural children. The Wallaces cited university lecture-style classes as particularly problematic for rural children's transition to higher education. Many rural students have not been in classrooms with more than ten children. Teachers were attentive and addressed needs of individuals. The delivery of content in rural schools is contextualised and high interest. However, many university science courses are delivered in enormous lecture halls and university instructors rarely ask children what they know, but rather tell students what they must learn. Thus, the support systems found in rural schools are absent or not readily apparent in higher education. Therefore, Rural students migrate towards programs offering familiar systems that cultivate communities that resemble their rural support structures such as-agriculture. wildlife biology, and archaeology.

Despite the complexities and challenges associated with rural living, concepts associated with giftedness and native science do intersect. Nachtigal (1985) examined traits closely linked to rural life and his work supports assertions made in the Wallace interview. Nachtigal found that rurality was bound with tightly-linked personal relationships; broad understanding of one's environment; strong verbal communication skills; time measured by seasons of the year; entrepreneurship; responsive to the environment; and self-sufficiency. Although Nachtigal's rural qualities are generalisations, each represents individual and social tendencies shaped by isolation and each quality is easily transferable to a scientific setting and, ultimately, to a STEM career with the proper tutelage.

TRANSITION TO DIFFERENT SPACES

In order for students to be successful in their transition from high school to university settings (regardless of their career paths), they must believe that the challenges ahead will lead to a positive outcome and they must engage in the effort with a degree of self-efficacy or confidence. Dees (2006) likens the transition from rural school to a university campus to a shift in culture stating that,

> … when the dominant culture, represented in this study as the college classroom, presents ideas that conflict with the students' home culture, an added sense of stress is created in the students' lives. These student "immigrants" are forced to make very difficult choices: adopt the ideas of the dominant culture, deny these ideas, or negotiate some other form of cultural adjustment. (p. 2)

Rural gifted students encounter a high degree of dissonance associated with their home culture and that of any university campus. Survival and success depends largely upon students' ability to adapt, make friends and trustworthy contacts, and develop a new sense of place. Failure to make these difficult accommodations results in a perception of failure and possibly an early departure for more familiar environs. In an attempt to reduce early drop-outs, most universities have established "freshman seminars", early alert programs, or other types of orientation for incoming students (National Resource Center for First year Experience and Students in Transition, 2006). Institutions of higher education design these induction programs based on extensive research in the area of college readiness. Despite the care afforded to transition programs, little attention is paid to the backgrounds of incoming students (Gentry, 2001). Some institutions do acknowledge and track "first generation" college students, but the transition from high school to higher education is unidirectional. Universities' expectations are that gifted students change and adapt to the new academic setting and little attention is paid to students' rural legacy. Typically, admission offices' only required information are high school transcripts and standardised test scores. Higher education's failure to engage gifted rural students has contributed to low numbers of college graduates in rural areas. Only 17 per cent of rural adults age 25 and older had completed college in 2000, half the percentage of urban adults (Whitner & McGranahan, 2003).

A Place-Conscious Transition Model

Unequipped, unaware, or unwilling to capitalise on rural children's native scientific skills, public schools need assistance in identification of gifted students, affirming their abilities, providing enriching experiences, and transitioning to STEM-oriented career paths; higher education appears ill equipped to help rural gifted students transition to their campuses. Few models exist that focus on native science, innate creativity, and the process of nurturing gifted children living in rural areas. The following four-part theoretical framework edifies rural students' potential focusing on areas of independence, problem-solving, and persistence while offering prescriptions that maximise opportunities in science and mathematics-related careers. Subsumed under each area are activities that contribute to rural gifted students' movement toward STEM-oriented careers.

- *Affirmation* – acknowledgement of useful and valued abilities in native rural scientists;
- *Enrichment* – development of partnerships that expand experiences and help re-frame students' self images that include pathways to STEM careers;
- *Early Induction* – involvement in supervised, extended, hands-on experiences outside the rural environment; and
- *Mentorship* – participation in practical, long-term, career-specific work under tutelage of trained mentors in science.

Each part of the theoretical framework addresses a place-specific need linked with rural education. Based on a meta-analysis of results from multiple sources, students participating in programs that employ these four components will gain confidence and self-efficacy, develop new science-based knowledge and skills, transition more smoothly to higher education, and view STEM as a viable career option.

Affirmation

Affirmation is the intentional acknowledgement of other individual's or group's skills, knowledge, achievement or position. Marzano (2003) identifies affirmation as a critical ingredient for the smooth operation of healthy schools; Redding and Walberg (2012) believe that the interplay of affirmation and positive feedback, goal-setting, and persistence is a positivemix for rural learners. For gifted rural children, affirmation is particularly important for bridging one's history and belief system with future aspirations. "Affirmations are positive statements that fortify and strengthen us to achieve our goals. They are personally reinforcing, energising and self-motivating. Affirmations help correct negative concepts we have developed about ourselves, depending on our life experiences."(p. 8) (Mayland Community College). Theorists warn that transition from "less valued" or subordinate cultures to competitive settings can result in loss or distortion of identity (Lawrence, 2009; hooks, 2002). Rural children need confirmation that they have value and can succeed in an unfamiliar, academic environment. The process of affirming indigenous science should begin in public schools. Rural school culture must celebrate indigenous skills and information that rural children bring to class. This early affirmation is context for extending rural skills beyond local spaces. The outside culture's understanding that skills possessed by rural children are relevant and transferable to a broader context appear to be a barrier to higher education for rural children. Implications for gifted rural students entering science-intensive college programs are clear. In Hardre's extensive review of rural education, she observes, "Students with more positive motivational profiles in a particular subject area (high perceived ability, instrumentality, learning goals and success expectations) are more likely to take courses in that area and to choose related college majors and career paths" (p. 3).

Research regarding university affirmation of students with native science is limited. Some lessons about affirmation may be learned from members of the agricultural community. Agriculture programs thrive at community colleges, regional comprehensive universities, and primary research institutions positioned adjacent to rural areas. Students recognise course language and equipment in introductory courses and can relate to the credentials and backgrounds of agriculture professors. Agriculture instructors share stories about their own experiences that parallel those of the students in the class. Through familiarity and affinity for the content, students experience tacit affirmation in nearly every lecture. Like traditional STEM programs, agriculture programs are science intensive, yet few perceived barriers to success appear to exist for students with rural origins. In contrast, rural students may be unable to see connections between their life experiences, science content, and other science faculty.

In order to become an attractive alternative to agriculture, science faculty need to be aggressive in their attempts to recruit gifted students from rural areas. Dees (2006) stresses the need for higher education to be more responses to student needs. He posits:

> However, if we, as university educators, practice careful reflection of our own attitudes and values regarding rural/Appalachian students, we may identify strategies that will serve to reframe our own classroom practices. In this manner, we can help our students grow and adjust in meaningful ways. Additionally, we, ourselves, can learn from our students as we co-construct realities and practices that can help to reduce the acculturative stress that is created through the educational experience. (p. 11)

Curricula and instructional change with some emphasis on students' indigenous science knowledge can make a difference. Some institutions have developed complementary, interdisciplinary, science-intensive degrees that appeal to rural gifted students. Wildlife biology, informal science education, and food sciences have found niches at institutions serving rural communities. These science-intensive programs are high interest and affirm student prior knowledge.

Enrichment

In order for gifted students to reach their potential, educational enrichment must begin in elementary schools with increased emphasis at the middle and high schools. Public school educators must do more than simply provide additional content to gifted students; the gifted must have opportunities to explore concepts in greater depth and breadth. Although content is important, modern interpretations of what scientists and engineers do include communicating, collaborating, questioning, and thinking critically and creatively (Soto, 2009). Gifted rural education in science should mirror these values. Using Renzulli's Triad Model (1977), schools must design complementary science and mathematics curricula that challenge gifted students, bolster interest, and make clear connections with rural living.

Again, the literature base for integration of gifted education enrichment, STEM careers, and rural education is somewhat scarce and sometimes contradictory. The National Research Council (NRC) issued an extensive report detailing characteristics of successful STEM schools (2011). The Council outlined goals for STEM-based education and included a unifying theme for STEM schools in the United States. Criteria for STEM-based schools are rigorous curriculum that deepens STEM learning over time, more instructional time devoted to STEM, more resources available to teach STEM, and teachers who are more prepared to teach in the STEM disciplines (p. 7). The report did state that STEM schools target gifted students, but admits that no research indicates that investment into to this population results in STEM majors, related careers, or that graduates from STEM schools contribute to the fields of technology or science more than traditional schools. Using a case study format, the NRC document described select schools where some research is on-going and does attempt to better explain transitions from high school, university, and career. The select schools were all urban; the report did not contain any reference to rural communities, but cited educational inequities that are present across districts and states. It should be noted that most discussion of inequities dealt with structural factors such as facilities, supplies and credentialed or certified teachers. The report proposed five recommendations for improving STEM education. The recommendations reflected the Council's priorities citing content knowledge, qualified teachers, and school organisation with little emphasis placed on nature of students or their learning.

Thoughtful, research-based, student centred examples of enrichment do exist. In a recent grant-funded effort, the Department of Biological Sciences at San Jose State University engaged in a comprehensive reform to prepare incoming freshmen for the rigours of academic studies in biology while acculturating scientists-in-training to professional expectations (Soto, 2009). The biology faculty broach these issues using two strategies: 1) ongoing conversations regarding biology program expectations with high schools and community colleges; and 2) inclusion of enriched, innovative curricula and andralogically sound experiences for students in their introductory biology courses. The university faculty worked directly with their pre-university counterparts. Although much of the university-school interaction focused on meeting admission requirements and articulation agreements, information was shared regarding academics with the goals that incoming university students are better informed and better prepare for higher education science. With an emphasis on enrichment, San Jose State University has added a new component to their introductory biology course sequence that assists students in their transition to college-level science while building community among students. The biology department added an "activities" requirement to their two-course sequence for freshmen. This second laboratory incorporates hands-on activities, field experiences, simulations, problem-solving, and poster-style presentations – all while students work together in small groups. Early indicators show increases in student satisfaction and self-efficacy (Everhart, 2012). More than eighty per cent of students report that they feel that the introductory courses reinforced their goals of working in a biology-oriented career (medicine, research, veterinary science). Opportunities to work in small collaborative groups seem to counterbalance the more traditional impersonal lectures found on most college campuses. The formation of community has resulted in development of impromptu study groups and student-led use of social media to complement introductory biology courses. Although San Jose State University is an urban institution, development of community among entry-level students appears to help students acclimate to less familiar, more academically rigorous setting. Successes at San Jose State University can be adapted to better meet the needs of college-bound, gifted students in rural areas.

Early Induction

Induction is a process of re-orientation, typically moving from familiar to unfamiliar circumstances. In the case of gifted rural students, induction processes are particularly important; they may serve as invitation to succeed, permission to pursue long-term goals, and a means to build self-efficacy. Cook (2007) states that in-bound college students make up their minds quickly regarding whether they can or cannot adapt to new academic, social, and psychological challenges; historically, students withdraw from classes after the first six to eight weeks. Cook describes a well-constructed, early induction process that consists of three distinct parts:

- Separation (removal from former habits and habitat);
- Transition (learning behaviours appropriate to a new circumstance); and
- Incorporation (acceptance into a new society) (p. 7).

Since their commitment to place is strong, and rural gifted students may encounter a barrage of challenges upon during their first weeks on a university campus, induction processes should begin early. When possible, gifted high school students should physically visit college campuses. With the support of secondary school teachers, counsellors, and university personnel, students should build relationships with campuses of interest. Although campus activities such as science fairs and field experiences are memorable and viewed by college recruitment offices as effective tools, these activities are brief and may do little to get students to view themselves as scientists. Extended, structured experiences are necessary for lasting impressions and student ownership of science as a profession. As soon as middle school and high school students begin to view themselves as scientists, the more likely they will pursue careers in biology, physics, chemistry, and geology. Extended-stay opportunities as summer institutes and camps are more likely to meet Cook's induction criteria of "learning behaviours appropriate to a new circumstance" (p. 7). Campuses appear much friendlier places for visiting high school students during summer months. Issues associated with residential housing are eased after full-time students return home for summer. Funding, another possible impediment, may be less problematic because grants and external funding sources are supportive of many university summer programs for high school students. Other facilities such as laboratories and classrooms are less crowded and faculty's schedules are more flexible are accommodating.

Well-planned camps and summer institutes for pre-college students can positively impact cognitive and affective domains of learning. The curriculum for a summer camp should help establish connections with rural science, incorporate high interest science-oriented activities, and build student confidence that a university setting is less threatening and an environment that allows exploration.

Table 1. Sample activities

	Cognitive Domain "Think like a scientist"		*Affective Domain "Feel like a scientist"*	
Exploratory Cue/Activity	Contextualisation	Knowledge/ Skills Development	Behavior	Self-efficacy
In teams of three, build a structure made from simple materials that will support a five kilogram mass.	Rural storage buildings; barns; erection of farm equipment	Materials science; struts and beams, measurement; preparation of oral presentation; control variables;	Succeed as a team; collaborate; consider alternative approaches and opinions; improve structures at home.	Speak publicly about the research
Working in pairs, sort organisms by genotype and phenotype.	Rural fair animal competitions; purchasing and selling animals and crops for profit	Genetics; organismal classification; scientific systems; development of a poster	Share responsibilities; learn information related to animal purchases and sales.	Develop professional standards to share written information

Current efforts to initiate the induction process while students are in public schools are uneven and some activities may even dissuade gifted students from pursuit of careers in science. A recent trend in the United States is encouragement of bright, college-bound children to enrol in dual-enrolment programs. After students meet state high school requirements in language arts, mathematics, social studies, and science, they may take college courses that meet requirements for general education and for some majors. High school students either take these dual- enrolment courses on the college campus, using instructional television, or online. Since distances between rural high schools and universities prohibit face-to-face instruction, rural students' first introduction to a university science course may be relegated to sitting at a computer engaged in reading and low-level thinking. This form of learning reinforces the notions that science is a passive act and that university science is different from the native science experienced by rural children. Induction efforts must re-affirm student experiences and better represent a modern interpretation of careers in science.

Mentoring

Mentoring is a highly effective technique used to influence professional and personal attitudes, processes and techniques, and career paths of novices. Mentorship can be misunderstood and is often confused with internship, apprenticeship, and coaching. Mentorship is a time-intensive, comprehensive act dependent on mutual trust and respect between an expert and novice. Mentorship is an intentional act and the mentoring relationship is a venue for transmission of work habits, processes, skills, content knowledge, and professional ethics/standards. In a recent study, scientists serving as mentors to undergraduate researchers failed to agree on a definition of mentorship (Everhart, 2011). Only after two years in the program did scientists shift their primary mentorship emphases from content and laboratory techniques to modelling, networking, collaboration, and professional standards. Fine tuning mentorship skills happens over time and through extensive reflection and discussion.

Rural students need committed mentors after admittance to a university. More than academic advisors, science faculty may become "loco parentis" and serve as the rural students' point-of-contact and advocate similar to the adult figures that maintained a influential role in students' lives on the farm or ranch. The social support for incoming students can be as important to their success as the academic support (Martinez & Klopott, n.d.). Sambunjak, Straus, and Marusic (2010) provided an extensive review of qualitative research that deal with mentoring in the field of academic medicine. They found three dimensions of desirable characteristics among mentors – personal, relational, and professional. Subsumed in these dimensions are several characteristics particularly relevant for gifted mentees transitioning from rural environs to a university science department: altruistic, understanding, patient, responsive, nonjudgmental, motivator, accessible, able to identify potential strengths in mentees, and able to assist students in defining and reaching goals. These characteristics mimic familiar elements of rural students' home support system and fill in gaps that may occur for students on campus.

Rural Scientists

Although some rural students that acquire advanced degrees in science and math will strike out destinations more commensurate with their field of study, most students show preference for occupations that return them to rural locales (North Carolina State University, n.d.). Bonds with rural communities are strong. Failure to accept the desire to return to rural living could relegate forty per cent of gifted students in the United States incompatible with STEM-related jobs. Acceptance and integration of native science acquired during formative years by high school and university instructors may be the best recruitment tool for science programs. Universities must establish comprehensive transition programs for gifted rural students to meet their social, emotional, and psychological needs. Through affirmation, enrichment, early induction, and strong mentorship, students will reach their potential – unburdening them from any ill effects of space while capitalising upon its many benefits.

REFERENCES

ACT, Inc. (2011). *Affirming the goal: Is college and career readiness an international standard?* Retrieved July 17, 2012, from http://www.act.org/research/policymakers/pdf/AffirmingtheGoal.pdf

Arnold, M. L., Newman, J. H., Gaddy, B. B., & Dean, C. B. (2005, April 27). A look at the condition of rural education research: Setting a difference for future research. *Journal of Research in Rural Education, 20*(6), 1–25.

Avery, L. M. (2013). Rural science education: Valuing local knowledge. *Theory into Practice, 52*(1), 28–35.

Avery, L., & Kassam, K. (2011). Phronesis: Children's local rural knowledge of science and engineering. *Journal of Research in Rural Education, 26*(2), 1–18.

Briggs, J. (2005). The use of indigenous knowledge in development: Problems and challenges. *Progress in Development Studies, 5*(20), 99–114.

Brooke, R. (2003). *Rural voices: Place conscious education and the teaching of writing.* New York, NY: Teachers College Press.

Brown, S., Renzulli, J., Gubbins, E., Siegle, D., Zhang, W., & Chen, C. (2005, Winter). Assumptions underlying the identification of gifted and talented students. *Gifted Child Quarterly, 49*(1), 68–79. doi: 10.1177/001698620504900107

Complete College America. (2012). *Remediation: Higher education's bridge to nowhere.* Retrieved July 21, 2012, from http://www.completecollege.org/docs/CCA-Remediation-final.pdf

Cook, A., McIntoch, K., & Rushton, B. (2006). *Supporting students: Early induction.* University of Ulster, Coleraine, Northern Ireland. Retrieved July 17, 2012, from http://www.ulster.ac.uk/star/resources/(D)%20Supporting%20Students%20-%20Early%20Induction.pdf

Darling-Hammond, L., Chung, R., & Frelow, F. (2002). Variation in teacher preparation: How well do different pathways prepare teachers to teach? *Journal of Teacher Education, 53*(4), 286–302.

Dees, D. M. (2006, June). How do I deal with these new ideas?: The psychological acculturation of rural students. *Journal of Research in Rural Education, 21*(6). Retrieved September, 2006, from http://www.umaine.edu/jrre/21-6.pdf

Everhart, J. (2011). *Support for curriculum and research inquiry-based instruction – Annual external evaluation* [An unpublished report funded by Howard Hughes Medical Institute]. San Jose, CA: San Jose State University.

Everhart, J. (2012). *Support for curriculum and research inquiry-based instruction – Annual external evaluation* [An unpublished report funded by Howard Hughes Medical Institute] San Jose, CA: San Jose State University.

Extension, (2013, September). *Age appropriate tasks for children on farms and ranches.* Retrieved July 23, 2014, from http://www.extension.org/pages/63149/age-appropriate-tasks-for-children-on-farms-and-ranches#.U9kUbXLwvIU

Fiscal Times. (2011, June 6). *School budgets: The worst education money can buy.* Retrieved August 5, 2012, from http://www.thefiscaltimes.com/Articles/2011/06/06/School-Budgets-The-Worst-Education-Money-Can-Buy.aspx#page2

Floyd, E. F., McGinnis, J. L., & Grantham, T. C. (2010). Gifted education in rural environments. *Special populations in gifted education: Understanding our most able students from diverse backgrounds*, 27.

Fowler, J., Butler, J. S., Cowen, J. M., Streams, M. E., & Toma, E. F. (2013, January). Public employee quality in a geographic context: A study of rural teachers. *The American Review of Public Administration*, 1–19.

Gentry, M., & Ferriss, S. (1999, May/June). STATS: A model of collaboration to develop science talent among rural students. *Roeper Review, 21*, 316–320.

Gentry, M., Rizza, M., & Gable, R. (2001). Gifted students perceptions of the class activities: Differences among rural, urban and suburban student attitudes. *Gifted Child Quarterly, 45*, 115–129.

Hardre, P., Sullivan, D., & Crowson, H. (2009). Student motivation and characteristics in rural high schools. *Journal of Research in Rural Education, 24*(16).

hooks, b. (2002). Confronting class in the classroom. In A. Darder, R. Torres, & M. Baltodano (Eds.), *The critical padagogy reader.* New York, NY: Routledge-Falmer.

Housing Assistance Council. (2012). *Rural research note: Poverty in America- 2010.* Retrieved August 10, 2012, from http://www.slideshare.net/ruralhome/poverty-in-rural-america

Howley, C. B. (1998). *Distortions of rural student achievement in the era of globalization.* Charleston, WV: Appalachia Educational Laboratory.

Howley, C. B. (2009). Critique and fiction: Doing science right in rural education research. *Journal of Research in Rural Education, 24*(15). Retrieved August 1, 2012, from, http://jrre.psu.edu/articles/24-15.pdf

Inwood, S. (2013, 2nd Quarter). Social forces and cultural factors influencing farm transition. *Choices, 28*(2), 1–5.

Kordosky, D. (2010). *Rural gifted students: Victims of public education.* Indianapolis, IN: Dog Ear Publishing.

Lawrence, B. (2009). Rural gifted education: A comprehensive literature review. *Journal for the Education of the Gifted, 32*(4) ,461–494.

Lloyd, E. (2010). *Eliciting and utilizing rural students' funds of knowledge in the service of science learning: an action research study* [A unpublished dissertation from Warner School of education and human development]. New York, NY: University of Rochester.

Martinez, M., & Klopott, S. (n.d.). *How is school reform tied to college access and success for low-income and minority youth?* Adapted from Martinez, M., & Klopott, S. (2005). *The linkbetween high school reform and college access and success for low-income and minority youth.* Washington, DC: American Youth Policy Forum and Pathways to College Network.

Marzano, R. (2003). *What works in schools: Translating research into action.* Alexandria, VA: Association for Supervision and Curriculum Development.

Matthews, M., & Kelly, D. (2007). Grit: Perseverance and passion for long-term goals. *Journal of Personality and Social Psychology, 92*(6), 1087–1101.

Mayland Community College. (2004). *Setting goals for success: Workshop materials.* Retrieved August 4, 2012, from http://www.mayland.edu/aca111/GoalSetting.pdf

Nachtigal, P. (1982). *Rural education: In search for a better way.* Boulder, CO: Westview Press.

Nation Research Council. (2011). *Successful K-12 STEM education: Identifying effective approaches to science, technology, engineering, and mathematics.* Washington, DC: National Academies Press. Retrieved July 18, 2012, from http://www.stemreports.com/wp-content/uploads/2011/06/NRC_STEM_2.pdf

National Resource Center for First Year Experience and Students in Transition. (2006). *Preliminary summary of results from the 2006 national survey on freshman seminars.* Retrieved July 23, 2014, from http://sc.edu/fye/research/surveyfindings/surveys/survey06.html

Navajo Nation: Washington Office. (n.d.). *Navajo nation and Navajo tribal utility authority announce venture to deliver wireless phone and broadband services to Navajo nation.* Retrieved July 20, 2012, from http://nnwo.org/content/navajo-nation-and-navajo-tribal-utility-authority-announce-venture-deliver-wireless-phone-an

New Mexico Public Education Department. (2012). *Professional licensure bureau.* Retrieved August 2, 2012, from http://www.ped.state.nm.us/licensure/

North Carolina State University, Department of Agriculture and Life Sciences. (2014). *General facts about agriculture.* Retreived July 29, 2014, from http://www.cals.ncsu.edu/CollegeRelations/AGRICU.htm

Redding, S., & Walberg, H. (2012). *Promoting learning in rural schools.* Lincoln, IL: Center on Innovation and Improvement.

Renzulli, J. (1977). *The enriched triad model: A guide for developing defensible programs for gifted and talented.* Mansfield Center, CT: Creative Learning Press.

Ross, P. (1993). *National excellence: A case for developing America's talent.* Washington, DC: U.S. Office of Education.

Sambunjak, D., Straus, S., & Marusic, A. (2010). A systematic review of qualitative research on the meaning and characteristics of mentoring in academic medicine. *Journal of General Internal Medicine, 25*(1). Retrieved August 15, 2010, from http://www.ncbi.nlm.nih.gov/pmc/articles/PMC2811592

Silverman, L. (n.d.). *How to use new IQ tests in selecting gifted students.* Gifted Development Centre. Retrieved July, 19, 2012, from http://www.gifteddevelopment.com/About_GDC/newiqtests.htm

Solient Health. (2009). *Differences in rural and urban health.* Retrieved July 22, 2012, from http://blog.soliant.com/healthcare-news/rural-vs-urban-healthcare

Soto, J. (2009). *Support for curriculum and research inquiry-based instruction (SCRIBE).* A grant funded by Howard Hughes Medical Institute.

Source Water Assessment and Protection (SWAP). Retrieved July 27, 2012, from http://navajopublicwater.org/SWAP2.html

Stambaugh, T. (2010). The education of promising students in rural areas: What do we know and what can we do? In J. VanTassel-Baska (Ed.), *Patterns and profiles from promising learners in poverty* (pp. 59–83). Waco, TX: Prufrock Press, Inc.

Stepanek, J. (1999). *Meeting the needs of gifted students: Differentiated instruction for mathematics and science instruction.* Northwest Regional Education Laboratory. Retrieved August 8, 2012, from http://educationnorthwest.org/webfm_send/755

Systemic Research. (2005). *Bringing science and mathematics excellence to rural classroom: RSI program highlights and case stories from seven sites.* A Report for the National Science Foundation, Arlington, VA.

Technomine. (n.d.). *Mining technologies: Mining salaries.* Retrieved August 2, 2012, from http://technology.infomine.com/reviews/MiningSalaries/welcome.asp?view=full

VanTassel-Baska, J., & Stambaugh, T. (2007). *Overlooked gems: A national perspective on low-income promising learners.* Conference proceedings from the National Leadership Conference on Low-Income Promising Learners. Retrieved July 17, 2012, from http://www.nagc.org/uploadedFiles/Publications/Overlooked%20Gems%20(password%20protected%20-%20gifted).pdf

Wallace, F., & Wallace, L. (2012, July). *Understanding rural science: An interview with Patrick and Robin Wallace/Interviewer:* J. Everhart.

Whitener, L., & McGranahan, D. (2003). *Rural America: Rural opportunities and challenges.* Published originally by the United States Department of Agriculture. Retrieved August 10, 2012, from http://webarchives.cdlib.org/sw1vh5dg3r/http://ers.usda.gov/Amberwaves/Feb03/features/ruralamerica.htm

Jerry Everhart
College of Education and Technology
Eastern New Mexico University
USA

MARÍA G. ARREGUÍN-ANDERSON, J. JOY ESQUIERDO,
ADRIENNE GUILLEN AND LORENA VILLARREAL

13. IDENTIFYING GIFTED AND CREATIVE FUTURE SCIENTISTS WHO ARE LINGUISTICALLY AND CULTURALLY DIVERSE

INTRODUCTION

The underrepresentation of professionals of color across the overall working force in the science and engineering fields continues to be a matter of concern nationwide. In 2008, Hispanics accounted for only 4.9 percent, while Blacks constituted only 3.9 percent of all employed scientists and engineers (National Science Foundation, 2012). For the linguistically and culturally diverse learner, this meek presence among science, technology, engineering, and mathematics (STEM) professionals does not occur in the vacuum. It is also mirrored in educational structures, practices, and programs that marginalize and stigmatize diverse learners early in their schooling. In 2006, only 4.2% of Hispanics and 3.6% of African American were identified as gifted and talented in the public school system. In general, white students are 3.8 times more likely to be accepted into the gifted and talented group than minority learners (McBee, 2010; Yoon Yoon & Gentry, 2009).

SCIENCE FOR ALL: A CRITICAL PERSPECTIVE ON GIFTEDNESS AND CREATIVITY

We contend that a more liberal definition of giftedness is essential as educators seek to materialize the 'science for all' ideals. In this chapter, we use a critical race theoretical framework to argue that in most scenarios, issues of power, race, and ideology influence identification of gifted and creative individuals. In general, "the closer a student operates in relation to dominant power, the more likely she/he is likely to be labeled intelligent" (Kincheloe, 2004, p. 121) and creative. We concur with a critical complex cognition that "takes into account a wide variety of social, cultural, political, cognitive, and pedagogical discourses", specifically embracing the idea that intelligence is learnable (Kincheloe, 2004, p. 123) and creativity is to a great extent a socially determined quality (Boden, 1991; Nickerson, 1999). Often, research on "characteristics of creative individuals reports on characteristics of European or European-American men" (Starko, 2002, p. 109), thereby the need to adopt approaches that are more inclusive of "those who in the past have largely been

M. K. Demetrikopoulos & J. L. Pecore (Eds.), Interplay of Creativity and Giftedness in Science, 241–256.

bypassed in science and mathematics education: ethnic, and language minorities and girls" (American Association for the Advancement of Science, 1990, p. xviii). This standpoint contradicts pre-established or canned definitions of the intellectual and creative process bringing environmental and social factors to the forefront. Starko (2001) indicates that "the more we come to understand the complexities underlying the creative process, the more difficult it is to give a one-size-fits-all description of a creative individual" (p. 108).

OVERVIEW OF CRITICAL RACE THEORY

We adhere to critical race theory (CRT) as a framework that assists us in asserting that identification of scientifically inclined students is deeply influenced by racial factors. Key in this assertion is the notion that racism is "normal, not aberrant, in American society" (Delgado, 1995, p. xiv). Race is considered to be a social and historical system that includes the social, cultural, economic and political areas of an individual in order to establish power (DeCuir-Gunby, 2006, p. 93) and is determinant of how much power the individual will possess. The critical analysis of race is explored in the CRT framework, which can be used to challenge and question how race affects social structures. CRT questions long standing thoughts and ideologies that have been accepted and unquestioned by society (Yosso, 2005; Zamudio, Russell, Rios, & Bridgeman, 2010). In brief, CRT speaks out against racism; challenges traditional society ideologies; is dedicated to social justice; honors experiential knowledge; and maintains an interdisciplinary perspective (Solórzano & Yosso, 2001; Zamudio et al., 2010).

Stemming from an initial focus on the slow pace of racial justice following the Civil Rights movement and its subsequent legislation, CRT began appearing as a theoretical framework that, in education, serves to analyze and challenge traditional multicultural paradigms (Ladson-Billings & Tate, 1995) and racism in curricular structures, processes, and discourses (Yosso, 2002). This is particularly relevant to science education. Currently, "the relationship between science and the children from diverse cultures and languages is problematic because school science typically reflects middle-class experiences and excludes the lives of students most on the margins of school science" (Barton & Osborne, 2001, p. 20). In this chapter, the interplay of science, creativity, and giftedness related to culturally and linguistically diverse individuals is discussed in terms of narrow definitions of giftedness and educators' conceptions of creativity that disregard expressions that fall outside mainstream definitions of novelty and appropriateness (Starko, 2001). We begin with a review of current views of giftedness and propose that a more liberal definition of giftedness is more likely to embrace potential scientists from diverse backgrounds (Solórzano & Delgado Bernal, 2001). We discuss the role of teachers in identifying and valuing students' experiential knowledge and their historically accumulated social and cultural capital. In our conclusion, we offer recommendations for practice and challenge the dominant deficit ideology with which minority students are perceived.

FROM NARROW TO LIBERAL DEFINITIONS OF GIFTEDNESS

Educators, policy makers, and researchers continue to debate the overall definition of giftedness (D. Y. Ford & Grantham, 2003; Lara-Alecio & Irby, 2006; Valdes, 2003). Giftedness has been assigned an assortment of definitions throughout the years that has affected the identification of linguistically and culturally diverse (LCD) gifted students (Ford & Trotman Scott, 2010). The reason for this phenomenon is that the majority of the definitions of giftedness tend to center on the middle-class, Anglo-American population. For example, many school districts execute a definition of a "typical," mainstream gifted child as a student that has a supportive home environment that offers verbal enhancement opportunities that nourish his/her natural abilities in ways that allow him/her to be highly successful on standardized tests (Castellano & Frazier, 2010). These types of definitions of giftedness do not apply to all children enrolled today in American public schools. In fact, these types of definitions apply to a small portion of the student population. With any one of those factors missing (i.e. college-educated parents, middle class, etc.), it is very difficult for schools and society in general to understand the concept of giftedness in other groups. The method by which schools identify potentially gifted students is founded on the definition of giftedness adopted by that educational institution. Each school district adopts a particular and specific definition of giftedness that will be implemented in the process of identification, instruction, and advisement. For example, if the definition is based on a conservative ideology, then the identification process will weigh more on IQ testing and other standardized assessments. The instruction may not recognize and/or include opportunities that foster creativity, leadership, and other talents that involve the arts. Under the conservative definition, the school district may advise teachers and parents on how to identify gifted children based exclusively on academic performance or standardized assessments.

A more liberal definition of giftedness is needed at a time when the global market strongly relies on the upsurge of a diverse workforce in the fields of science. To satisfy the demands of the evolving technological and scientific progress, it is essential that an interest in these fields be fostered. Society in general inclines to view professions in the field of science as difficult compared to the humanities (Taber, 2010). LCD students tend to find the field of science challenging and engaging. Due to the complexity of science and technology, LCD gifted students continue their interest in their learning and are motivated to continue studying in a science-related field (Irby & Lara-Alecio, 1996). The identification of, and service to, linguistically and culturally diverse (LCD) gifted students is a critical necessity considering the increasing need of preparing students for STEM fields (Hubbard & Stage, 2010) and global scientific progress.

Main Ideologies of Giftedness

In the educational field of giftedness, there are two main ideological groups: the conservatives and the liberals (Robinson, 1998; Valdes, 2003). Conservatives equate giftedness with intelligence as measured by IQ testing. Therefore, students that score in the top 5% on IQ tests tend to be considered gifted. This notion of giftedness dates back to the beginning of the 20th century, when psychologists used IQ testing to confirm the superiority of some racial groups over others (McClellan, 1985) narrowing down the identification possibilities for other diverse groups (Ford & Grantham, 2003). One noted conservative scholar was Lewis Terman, who promoted the idea that intelligence could be expressed in a single numerical ratio (Renzulli, 1999). Terman's views, highly permeated by deficit thinking, "meant that people of color would be virtually locked out of the upward flow of educational and social mobility" (Valencia & Suzuki, 2001, p. 8). In addition to claiming that solely IQ can measure giftedness, conservatives support the notion that giftedness is genetically inherent and cannot be enhanced (Renzulli, 1977). The conservative view of giftedness influenced the educational field for decades (McClellan, 1985). Scholars on the liberal side of the debate such as Renzulli, Tannerbaum, and Sternberg argue that giftedness is a complex, problematic, and evolving construct that cannot be easily measured with IQ tests. Renzulli (1999) defined gifted children as those that exhibit, or have the potential to exhibit, three distinguished traits: above average ability, task commitment, and creativity. The inclusion of creativity within Renzulli's definition of giftedness is significant to science educators of culturally and linguistically diverse learners. Although creativity is often defined in terms of novelty and appropriateness, under a humanist psychological perspective, creativity is not for the few, or the elite, but for everyday people. Creativity, Maslow (1968) contends, can be described as "a fundamental characteristic, inherent in human nature, a potentiality given to all or most human beings at birth, which is most often lost as the person gets enculturated" (p. 143). However, although all individuals can potentially produce novel products or ideas, the environment and its components "have a profound impact on creative expression" (Lubart, 1999 p. 339). Therefore science teachers who operate from the assumption that all students are potentially creative, include knowledge that is relevant to students' lives, provide supportive environments, and develop a range of teaching approaches based on the recognition that not every student learns the same way (Giroux & Schmidt, 2004). Congruent with this view, Renzulli (1977) supported the conception that schools need to provide academic environments that are conducive to the development of students' talents and present opportunities for them to utilize these talents (Renzulli, 1999). Such opportunities, from a CRT perspective, must be congruent with a situated pedagogy that is rooted on who the students are. Ultimately, cognitive ability cannot be separated from the social, cultural, ideological, and political context or specific learning conditions in which it takes place (Vygotsky, 1978). Critical educators "know that despite the power of generations of cognitive determinists operating under the flag of IQ, human beings can learn to become more intelligent" (Kincheloe, 2004, p. 134).

Towards Definitions that Incorporate a Multicultural Perspective

The underrepresentation of minority groups, Hispanic, African-American and Native American, in gifted education has been as high as 70% (Esquierdo, Irby, & Lara-Alecio, 2008). In an attempt to respond to this crisis, several definitions have been crafted at the federal level. For example, The *Jacob K. Javits Gifted and Talented Act of 1988*, established a program that seeks to serve traditionally underrepresented gifted students. Drawing from the Elementary and Education Act, the Javits program defines gifted children as "Students, children, or youth who give evidence of high achievement capability in areas such as intellectual, creative, artistic, or leadership capacity, or in specific academic fields" (Stephens & Karnes, 2000, p. 222). Subsequently, and progressing toward a notion that giftedness can be found in children of all ethnicities and social economic groups, the U. S. Department of Education (1993), defined gifted and talented students in National Excellence: The Case for Developing America's Talent as:

> children and youth with outstanding talent perform or show the potential for performing at remarkably high levels of accomplishment when compared with others of their age, experience, or environment. These children and youth exhibit high performance capability in intellectual, creative, and/or artistic areas, possess an unusual leadership capacity, or excel in specific academic fields. Outstanding talents are present in children and youth from all cultural groups, across all economic strata, and in all areas of human endeavor. (p. 26)

In contrast with previous definitions that heavily emphasized the role of IQ scores in the perception of giftedness, the above description of gifted children incorporates a new element of giftedness: the identification and development of potential talent. This flexible perspective recognizes the complex nature of the cognitive act. Specifically, it recognizes intelligence as a dynamic construct continuously shaped by the social context, and gives new meaning to a pedagogy that is inclusive of students of color whose talents have historically gone unrecognized or undervalued in educational settings. Most importantly, it steers away from a deficit perspective that regards culturally and linguistically diverse learners as 'in need of fixing'. Deficit ideologies, CRT scholars contend, must be openly discussed, not only as an initial step in transforming general practice, but also as a way to empower victims of oppression to find their voice.

Advancing from deficit to dynamic views on giftedness is a crucial step in addressing issues of underrepresentation. In doing so, educators must recognize that gifted students exist at all levels of society, regardless of gender, race, socioeconomic, or ethnic origin (Ford & Grantham, 2003). In recent years, a variety of studies have been conducted to determine the unique characteristics of gifted minority children. These students' exceptional intellectual capabilities, academic aptitudes, and/or creativity must be used to identify their giftedness. There exists a need to steer away from stereotypes and focus on strengths minority students bring to school in order to adequately identify their giftedness. Lara-Alecio and Irby (2006) and Vanderslice (1998), noted four major problems in the identification of linguistically and culturally diverse (LCD) gifted students:

- vague definition of giftedness;
- educational equity;
- misuse of identification instruments; and
- testing during inappropriate stages of the identification process.

The definition of giftedness varies within the realm of researchers: the liberals and the conservatives. Therefore, when schools begin the identification process, their first challenge is selecting a definition to use as a guide. The linguistically and culturally diverse gifted need a unique definition that is specific to the population's characteristics. For example, Lara-Alecio and Irby (2000) defined gifted Hispanic students as those who possess above average intelligence, task commitment, and creativity, considering the socio-linguistic-cultural context. They referenced Renzulli on this portion of their definition, remarking that this broader definition was more inclusive for Hispanic bilingual gifted, but they also needed additional consideration more specific to their realities. Additionally, eleven characteristics for LCD students have been described by Irby and Lara-Alecio (1996): motivation for learning, keen social and academic language (both in English and Spanish), cultural sensitivity, strong familial connections, use of collaboration, ability to be imaginative, high academic achievement, creative performance, utilizes environmental support, ability to problem solve, and internal locus of control.

Moreover, gifted students in the sciences exhibit other set of unique traits. These particular students accomplish to extraordinarily high levels of achievement in all or some facets of the standard curriculum requirements in school science (Taber, 2007). More specifically, researchers have noted that gifted science students' characteristics can be categorized into four groups: cognitive skills, curiosity, metacognitive sophistication, and group-work skills (Gilbert, 2002; Stepanek, 1999). Additionally, Taber (2010) explained that with conventional academic ability and performance, gifted science students demonstrate a deeper understanding of complex science concepts, propose innovative ideas, and utilize advanced scientific vocabulary. He also suggested that gifted science students can identify real world connections with the science content, academically work with highly abstract and theoretical concepts, and construct inferences.

Examining the two lists of characteristics of gifted LCD and science students, one can note commonalities. For example, a gifted LCD student can exhibit giftedness through strong verbal abilities in both languages and a science-gifted student can use advanced technical vocabulary. Another similarity between these two types of gifted characteristics is the ability to effectively learn in cooperative groups. Therefore, teachers that instruct LCD gifted students in the sciences must consider these characteristics in order to differentiate the curriculum to best fit their academic needs. In the next section we elaborate on the intersections between giftedness, scientific endeavors, and LCD students' cultural capital.

Educators' Role in Validation and Identification of LCD Students' Cultural Capital

Teachers occupy a critical position as cultural agents and gatekeepers. As initial evaluators of children's products, teachers are in a position to determine, not only who will be labeled smart; but also who is creative. Because we operate within a specific cultural context, it is no surprise that "teachers reward those children who were socialized into a view of intelligence that happens to correspond to their own" (Sternberg, 2007, p. 18). Generally, teachers operate in reference their own socialization and racial and cultural background. This becomes an issue in a nation where students of color make up more than 40 percent of the student population while teachers of color make up only 17 percent of the teaching force (Boser, 2011). Access to certain domains, such as science and mathematics, is not automatic as "sometimes rules and knowledge become the monopoly of a protective class or caste, and others are not admitted to it" (Csikszentmihalyi, 1999, p. 320).

Reports released from government institutions, media agencies, and academic researchers denote that public school teachers are not identifying LCD gifted, especially those from low socio-economic status (Bernal, 2002; Castellano, 1998; Lara-Alecio & Irby, 2000). State and district level administrators responsible for recommending and monitoring the identification procedures need to be cognizant of the possible reasons for the under identification of LCD gifted. In order to effectively identify LCD gifted, educators need to acknowledge the traditional perceptions of giftedness and the biases that can exist in the nominations. CRT scholars propose that educators adopt pedagogies of possibility that empower students to capitalize on the cultural wealth they posses and that materialize in at least six "forms of capital such as aspirational, navigational, social, linguistic, familial, and resistant capital" (Yosso, 2005). It is essential that educators understand traits that often contradict the characteristics of mainstream giftedness.

Familial capital. The attribute of having a strong family relationship and respect for authority figures is not typically considered a "gifted" trait. This is because most "gifted" checklists include non-conformity as characteristics of giftedness (Lara-Alecio, Irby, & Walker, 1997). Therefore, most teachers who observe LCD students with strong family ties or respect for authority do not consider them gifted since it goes against the "norm" of giftedness (Esquierdo et al., 2008, Spring). Familial Capital involves knowledge and relationships that are nurtured, transmitted, and continuously emerge within the context of a family. The "pedagogies of the home" (Delgado Bernal, 2001) and the curriculum of the home (Wong Fillmore, 2000) comprise the lived experiences, traditions, language, and diverse ways of knowing generated and used with their family and community. It is in the context of their familiar environment, that children absorb a wealth of "every day" concepts that serve as a foundation for a more systematic acquisition of scientific concepts (Vygotsky, 1986). Informal science learning that occurs within the dynamics of familial connections is vast and sophisticated. In students' households and communities, the "historically accumulated and culturally developed bodies of knowledge and skills" also known as funds of knowledge may include, for example, agriculture and mining, material and scientific knowledge, medicine, etc. (Gonzalez, Moll, & Amanti, 2005; Vélez-Ibáñez & Greenberg, 1992). Additionally, Shepard (1998) suggests encounters with nature are key in facilitating human language and thought. This cognitive and linguistic relationship between children and nature can potentially be "used to facilitate human intellectual development" (Lawrence, 1993). In African American communities, communal bonds infuse a strong sense of cultural identity, and belonging. In the same way, tribal communities have played a key role in Native American students' cultural and linguistic maintenance (Yosso, 2005). Bea Medicine, a Native American anthropologist, warned about the common practice of overlooking students' familial capital and strongly criticized the "ways in which schools function to disenfranchise these students and their families" (Deyhle & McCarty, 2007).

Linguistic and social capital. According to CRT scholars, LCD students posses linguistic capital, a trait that encompasses the cognitive, "intellectual and social skills attained through communication experiences in more than one language and/or style" (Yosso, 2005, p. 78). The multiple language and communication skills that students of color exhibit are particularly valuable in science, a domain that calls for "the ability to communicate ideas and share information with fidelity and clarity, and to read and listen with understanding" (American Association for the Advancement of Science, 1990, p. 192). Students that demonstrate gifted traits in the sciences show the propensity to lead student groups in discussions and explorations of scientific topics (Taber, 2010). Additionally, LCD students posses social capital, which relates to the variety of contacts, outside of the family circle that and individual can access to accomplish a variety of goals. Historically, Yosso (2005) suggests, people of color "have utilized their social capital to attain education, legal justice, employment, and health care" (p. 80). Because historically, ethnic and linguistic minorities have utilized their social networks for a variety of purposes, it can be argued that opening access to resources in the scientific domain is likely to draw the attention of creative individuals whose careers in science might have been impossible, unlikely, or not accomplished to its fullest potential. Social capital, in many instances can be tapped into and expanded through existing community organizations such as science clubs, after-school programs, etc. New, useful, ideas, are likely to emerge in contexts where people from diverse backgrounds converge, interact, and encourage each other in the quest for educational achievement. This type of gifted trait counters the "typical" characteristics of giftedness that describes gifted students as working more effectively independently. Both, their advanced vocabulary and social capital allow science-gifted students to employ and discuss derivations of specialized terms and ideas/theories. Their motivation for inquiry work keeps them task committed. Because language shapes thought (Whorf, 1956 in Sternberg, p. 344), it has been suggested that scientifically inclined bilingual learners exhibit the capacity to creatively conceive of questions and solutions. This is due to the mental flexibility they possess; the multiplicity of associations they generate in connection with the same concept; and their tolerance for ambiguity (Lubart, 1999). Decades of research in bilingual education have demonstrated that when the school system approaches students' linguistic diversity as an advantage and a resource and capitalize on this advantage through the implementation of enrichment bilingual programs, they enhance students' long-term school success (Ramirez, 1991; Thomas & Collier, 2002) thereby increasing the possibility of minority students' involvement in science related fields.

Aspirational and resistant capital. Educators, according to Paulo Freire, have the responsibility to "reveal situations of oppression", but just as importantly, educators must create the context for a "pedagogy of desire' (P. Freire, 2007). That is, teachers must recognize that students of color often enter their classrooms with a clear understanding of their reality and with projections of a better future. This ability to maintain hope in spite of economic or social obstacles represents a form of aspirational and resistant capital modeled and instilled through familial interactions. In describing his journey towards becoming a NASA scientist, José Hernández narrated how, through his parents advice (consejos) he began to dream of a future and realities that were distant from the fields he and his family picked as migrant workers. Aspirational capital, one might argue, is a valuable trait for minority students interested in science. After all, "the most eminent and creative scientists tend to be more driven, ambitious, and achievement oriented than their less eminent peers" (Feist, 1999, p. 280). This ability to perceive reality as mutable and transformable and to aim one's efforts towards goals that are attainable in one's minds, yet unknown (P. Freire, 2007), represent habits of mind that are valuable, not only in the quest for social mobility, but in the search of scientific advancement as well.

Along with their aspirational capital, communities of color have historically developed attitudes of resistance that confront racism and inequality (Yosso, 2005). In practical terms, resistant capital often translates into what teachers perceive as 'confrontational' attitudes, when in reality, "parents of color are consciously instructing their children to engage in behaviors and maintain attitudes that challenge the status quo" (Yosso, 2005, p. 81). In many ways, this form of capital operates under the assumption of an instilled or self-developed consciousness that empowers individuals to transform conditions that perpetuate oppression. In the domain of science, educators can capitalize on this form of cultural capital by encouraging their students to challenge an environment impregnated with images of white male scientists and to question, 'Why not me?' Congruent with this view, educators of students of color can systematically provide opportunities to discuss the role of Latinos, African American, Native Americans and other minorities in the science fields.

Navigational capital Students of color use a set of psychological skills that continuously develop as a means to survive and successfully overcome obstacles faced when encountering institutional mismatches in terms of who they are culturally and linguistically. Navigational capital often mirrors the type of habits of mind that are highly sought after in the scientific enterprise in which significant contributions are not produced in the first attempt and where the scientific community is not always welcoming of breakthroughs that contradict the established accepted paradigms. It is under these circumstances, that LCD students' navigational skills can assist them not only in maneuvering through institutions, but in maintaining commitment to task until novel and appropriate results, as perceived by experts in the field, are produced in their scientific endeavors.

TEACHERS' UNDERLYING IDEOLOGIES OF ASSIMILATION

Many LCD students are not nominated for gifted programs because the cultural capital they possess upon entering the school system is not recognized and valued. Although LCD students' cultural wealth is often part of the discussion in teacher preparation programs, very few teachers receive in-depth academic preparation to work with LCD gifted students, and even fewer to work with LCD science-gifted students (Ford & Trotman Scott, 2010). Esquierdo, Irby, and Lara-Alecio (2008) stated that most teachers who are certified in gifted education are English–only speakers who are not trained to work with the LCD science-gifted. These circumstances place LCD gifted in a triple disadvantage in the school setting (cognitive, linguistic, and content-specific). Therefore, it is crucial that teachers become aware of the characteristics of LCD science-gifted so that they can successfully serve them in the classroom. It is vital that educators understand these attributes since they generally begin the identification process for most gifted programs often resorting to an assimilationist lens.

Masten and Plata (2000) found that teachers are most likely to rate high-acculturated Hispanics higher on a gifted checklist compared to low acculturated Hispanics. Therefore, if educators do not comprehend the characteristics of LCD gifted, they are likely to regard them as less gifted. This strengthens the argument that trained and informed educators need to advocate for the LCD gifted in the identification, instruction, and advisement process. Unfortunately, science education is "nestled in the politics of assimilation and meritocracy" (Barton & Osborne, 2001, p. 12). The traditional American view of education being the great equalizer and that all can be attained by individuals who strive forward and persevere is a great deception. This concept of equality and fairness is defined within the term and myth of meritocracy

(Zamudio et al., 2010). It is believed that those who fail in society have no one to blame but themselves ignoring the existence of any educational inequalities. Until a student linguistically assimilates, and masters the official language, his/her linguistic limitations will keep him from accessing information, resources, and opportunities to be recognized as creative in science. From the beginning of their educational experience, children learn that languages other than English, and 'ways of knowing' other than those officially endorsed in their textbooks, are tacitly deemed as temporary and inferior. This subtractive approach is alarming at a time when close to 4 million English language learners are enrolled in the US public system and generally lose their language in the process of learning English. The illusion of equality could clearly define what is currently believed about the current education system, forgetting that it is centered on specific mainstream students. CRT looks to question the concept of meritocracy and the "even playing field" theory (Zamudio et al., 2010). The playing field ignores the forms of cultural and linguistic capital students already posses and that make them different. These differences play a part in academic success, since the playing field was not created with these distinctions in mind.

The Challenge to the Dominant Ideology

One overall goal, under a CRT perspective, is to challenge deficit perceptions commonly held in education regarding students of color. Deficit views permeate institutional structures, discourse, and practice, often disguised as excellence in education movements. "Science for all" ideals, for example, operate under three assumptions: a) that schools are meritocratic in nature i.e. they are color-blind, and students' success is contingent on their achievement (b) science reform movements adhere to a deficit model under which certain groups are either culturally or linguistically deprived, and (c) there is an assumption that students will choose to adapt to middle-class cultural values when their own are shown to be inferior (Barton & Osborne, 2001). CRT openly challenges these assumptions and argues that claims of race neutrality and equal opportunity perpetuate the power of dominant groups in U.S society (Solórzano & Delgado Bernal, 2001). Therefore, when national science reform initiatives invite educators to join in "science for all' projects, the "all" often translates into a concept that implies homogeneity, as well as cultural and linguistic assimilation. Eventually, cultural deprivation theories and IQ testing served as key instruments to justify preferential treatment of Anglo-Saxons (Menchaca & Valencia, 1990) consequently translating into 'deficit ideologies' that severely limit minority students' access to gifted and talented (GT) programs. Deficit thinking, Ford and Grantham (2003) state, is the driving force behind underrepresentation of culturally diverse gifted students and is reflected in limited definitions of intelligence; standard testing and assessment; policies and practices; questionable teacher preparation in multicultural education, gifted education, testing and assessment; a lack of communication/relationships with diverse families and communities; and students perceptions about gifted education.

CRT scholars propose a renewed appreciation of students of color as holders and creators of knowledge (Delgado Bernal, 2002). Creative individuals who are linguistically and culturally diverse often encounter a home-school mismatch, specifically in terms of the type of science that is valued in formal academic settings. In asking the fundamental question "Whose science, and whose knowledge? Are valued in society, Harding (1991) exposed not only the need to examine what is portrayed as valuable knowledge in science, but the privileged position of certain groups in dictating the paths of science education and the scientific enterprise in general. The privileged position of certain groups extends its influence in all sectors including education. This dominant school of thought allows the perceived act of racism as acceptable and inevitably contributes to the white hegemonic larger picture in society (Taylor, 1998). Freire (1994) further addresses the issue of the dominant group by acknowledging that the educational system has already established a set social order, which encourages students to accept the existing perceptions and values of the dominant group as true. According to Zeus (2009), the concept of a dominant group refers to the idea that white or Anglo American individuals enjoy extra benefits due to being born white. Therefore, in accordance with this concept, education should have a greater benefit for those within groups of privilege. In addition, Zeus (2009) associates domination to power and highlights the dominant group's power to control education, which is exemplified in how, and whose history is recorded and studied in school. This same notion extends to science.

RECOMMENDATIONS

This chapter centered on the notion that the creative scientists who are culturally and linguistically diverse are not likely to be formally identified as gifted unless there is radical change in the educational structures, practices, and discourses that perceive LCD students from a deficit perspective. With this in mind, we outline two key recommendations described below:

First, we suggest that in gifted science education initiatives, programs, and processes, schools must address the mismatch between home and school. Without systematic efforts that attend to the "distinction between home and school language and culture, educational endeavors aimed at these distinct students are likely to fail" (Garcia, 1993, p. 54). To students of color, the home-school mismatch is multidimensional and is evident in the omission of their native languages in the science curriculum and materials. This mismatch is also reflected in practices that lack cultural responsiveness despite officially adopted definitions of giftedness that call for inclusiveness and initiatives that promote excellence. Currently, these practices favor and facilitate access to the scientific domain to students that have assimilated. In short, the gifted and talented curriculum must be revised to include a multicultural perspective (Bernal, 2002; D. Y. Ford & Whiting, 2008).

Second, given the key role that teachers play in the identification of gifted or potentially gifted future scientists, it is essential that schools make concerted efforts to increase multicultural sensitivity awareness (Bernal, 2002; Ford & Whiting, 2008). Specifically, teachers must discuss and explore the cultural wealth that students of color posses and that is evident in the aspirational, social, familial, linguistic, resistant, and navigational capital acquired and constantly developed in the context of their families and their communities (Yosso, 2005). Furthermore, multicultural discussion and education will not only lead to the improvement of teachers' cultural sensitivity but would greatly enhance teachers' expectations of LCD students (Ford, 2012).

CONCLUSION

Opening spaces to the linguistically and culturally diverse gifted learners in the science fields requires changes in practice, but most importantly it requires changes in ideology. In the field of gifted education, this change translates into a shift from deficit ideology to dynamic views of linguistically and culturally diverse learners.

The racist nature of deficit perspectives that subject minority learners to remediation practices aimed at correcting what is perceived as a 'cultural deprivation' has had a profound impact not only on students' of color self-perception, but on their academic inclination to pursue careers in science. The ramifications of "racism and racial stigmatization harm not only the victim and the perpetrator of individual racist acts but also society as a whole" (Delgado, 1995, p. 161) depriving the science field of potentially gifted and creative scientists of color. Validating students' creativity and intellectual production in science, implies adopting a pedagogy that functions in terms of students' potential. In Frereian terms, a pedagogy of the "unfinished" embraces students' cultural capital and overcomes fatalistic views of their future highlighting their ability to change reality.

REFERENCES

American Association for the Advancement of Science. (1990). *Science for all Americans*. New York, NY: Oxford University Press.

Barton, A. C., & Osborne, M. D. (2001). Marginalized discourses and pedagogies: Constructively confronting science for all in classroom practice. In A. C. Barton & M. D. Osborne (Eds.), *Teaching science in diverse settings: Marginalized discourses and classroom practice*. New York, NY: Peter Lang Publishing.

Bernal, E. (2002). Three ways to achieve a more equitable representation of culturally and linguistically diverse students in GT programs. *Roeper Review*, *24*(2), 82–88.

Boden, M. A. (1991). *The creative mind: Myths and mechanisms*. New York, NY: Basic.

Boser, U. (2011). *Teacher diversity matters: A state-by-state analysis of teachers of color*. Washington, DC: Center for American Progress.

Castellano, J. (1998). *Identifying and assessing gifted and talented bilingual hispanic students*. Charleston, WV: Clearinghouse on Rural Education and Small Schools.

Castellano, J. A., & Frazier, A. D. (2010). *Special populations in gifted education: Understanding our most able students from diverse backgrounds*. Waco, TX: Prufrock Press.

Csikszentmihalyi, M. (1999). Implications of the systems perspective for the study of creativity. In R. Sternberg (Ed.), *Handbook of creativity*. New York, NY: Cambridge University Press.
DeCuir-Gunby, J. T. (2006). Providing your skin is White, you can have everything: Race, racial identity, and property rights in whteness in the supreme court case of Josephine Decuir. In A. D. Dixson (Ed.), *Critical race theory in education: All god's children got a song* (pp. 89–111). New York, NY: Routledge.
Delgado Bernal, D. (2001). Living and learning pedagogies of the home: The mestiza consciousness of Chicana students. *International Journal of Qualitative Studies in Education, 14*(5), 623–639.
Delgado Bernal, D. (2002). Critical race theory, Latino critical theory, and critical raced-gendered epistemologies: Recognizing students of color as holders and creators of knowledge. *Qualitative Inquiry, 8*(1), 105–126.
Delgado, R. (1995). *Critical race theory: The cutting edge*. Philadelphia, PA: Temple University Press.
Deyhle, D., & McCarty, T. L. (2007). Beatrice Medicine and the anthropology of education: Legacy and vision for critical race/critical language research and praxis. *Anthropology & Education, 38*(3), 209–220.
Esquierdo, J. J., Irby, B., & Lara-Alecio, R. (2008, Spring). Initial screening for gifted and talented programs: Increasing participation of Hispanic English language learners. *TEMPO, 28*(2), 25–29.
Feist, G. J. (1999). The influence of personality on artistic and scientific creativity. In R. J. Sternberg (Ed.), *Handbook of creativity*. New York, NY: Cambridge University Press.
Ford, D. Y., & Grantham, T. C. (2003). Providing access to culurally diverse gifted students: From deficit to dynamic thinking. *Theory into Practice, 42*(3), 217–225.
Ford, D., & Trotman Scott, M. (2010). Under-representation of African American students in gifted education: Nine theories and frameworks for information, understanding, and change. *Gifted Education Press Quarterly, 24*(3), 2–6.
Ford, D. Y., & Whiting, G. W. (2008). Cultural competence: Preparing gifted students for a diverse society. *Roeper Review, 30*(2), 104–110.
Freire, P. (1994). *Pedagogy of the oppressed*. New York, NY: Continuum.
Freire, P. (2007). *Daring to dream: Toward a pedaogy of the unfinished*. Boulder, CO: Paradigm Publishers.
Garcia, E. E. (1993). Language, culture, and education. *Review of Research in Education, 19*, 51–98.
Gilbert, J. K. (2002). *Characteristics of gifted and talented pupils in science*. Retrieved from http//www.educ.cam.ac.uk/apecs/
Giroux, H. A., & Schmidt, M. (2004). Closing the achievement gap: A metaphor for children left behind. *Journal of Educational Change, 5*, 213–228.
Gonzalez, N., Moll, L. C., & Amanti, C. (2005). *Funds of knowledge*. Mahwah, NJ: Lawrence Erlbaum.
Harding, S. (1991). *Whose science? Whose knowledge?* New York, NY: Cornell.
Hubbard, S. M., & Stage, F. K. (2010). Identifying comprehensive public institutions that develop minority scientists. *New Directions for Institutional Research*, (148), 53–62.
Irby, B., & Lara-Alecio, R. (1996). Attributes of hispanic gifted bilingual students as perceived by bilingual educators in Texas. *SABE Journal, 11*, 120–140.
Kincheloe, J. L. (2004). *Critical pedagogy*. New York, NY: Peter Lang.
Ladson-Billings, G., & Tate, W. F. (1995). Toward a critical race theory of education. *Teachers College Record, 97*, 47–68.
Lara-Alecio, R., & Irby, B. (2000). The culturally and linguistically diverse gifted. In C. Reynolds (Ed.), *Encyclopedia of special education*. New York, NY: John Wiley & Son.
Lara-Alecio, R., & Irby, B. J. (2006). The culturally and linguistically diverse gifted. In C. Reynolds & E. Fletcher-Janzen (Eds.), *Encyclopedia of special education* (3rd ed., Vol. 1, pp. 506–510). New York, NY: John Wiley & Son.
Lara-Alecio, R., Irby, B., & Walker, M. (1997). Identification of hispanic, bilingual, gifted students. *Tempo: Texas Association for the Gifted and Talented, 17*(2), 20–25.
Lawrence, E. A. (1993). The sacred bee, the filthy pig, and the bat out of hell: Animal symbolism as cognitive biophilia. In S. R. Kellert & E. O. Willson (Eds.), *The biophilia hypothesis* (pp. 301–341). Washington, DC: Island Press.

Lubart, T. I. (1999). Creativity across cultures. In R. J. Sternberg (Ed.), *Handbook of creativity* (pp. 339–350). New York, NY: Cambridge University Press.

Maslow, A. H. (1968). *Toward a psychology of being* (2nd ed.). Princeton, NJ: Van Nostrand.

Masten, W., & Plata, M. (2000). Acculturation and teacher ratings of Hispanic and Anglo-American students. *Roeper Review, 23*(1), 45–46.

McBee, M. (2010). Examining the probability of identification for gifted programs for students in Georgia elementary schools: A multilevel path analysis study. *Gifted Child Quarterly, 54*(4), 283–297.

McClellan, E. (1985). *Defining giftedness.* Reston, VA: ERIC clearinghouse on handicapped and gifted children.

Menchaca, M., & Valencia, R. (1990). Anglo-Saxon ideologies in the 1920s–1930s: The impact on the segregation of Mexican students in California. *Anthropology & Education Quarterly, 21*(3), 222–249.

National Science Foundation. (2012). *Science and engineering indicators.* Arlingron, VA: National Science Foundation.

Nickerson, R. S. (1999). Enhancing creativity. In R. J. Sternberg (Ed.), *Handbook of creativity* (pp. 392–430). New York, NY: Cambridge University Press.

Ramirez, J., Pasta, D., Yuen, S., Ramey, D., & Billings, D. (1991). *Final report: Longitudinal study of English immersion strategy, early-exit, and late-exit bilingual education programs for minority children.* San Mateo, CA: Aguirre International.

Renzulli, J. S. (1977). *The enrichment triad model: A guide for developing defensible programs for the gifted and talented.* Wethersfield, CT: Creative Learning Press.

Renzulli, J. S. (1999). What is this thing called giftedness, and how do we develop it? A twenty-five year perspective. *Journal for the Education of the Gifted, 23,* 3–54.

Robinson, A. (1998). Giftedness: An exceptionality examined. *Review of Psychology, 49,* 117–139.

Shepard, P. (1998). *Thinking animals: Animals and the development of human intelligence.* Athens, GA: The University of Georgia Press.

Solórzano, D. G., & Delgado Bernal, D. (2001). Examining transformational resistance through a critical race and LatCrit theory framework. *Urban Education, 36*(3), 308–342.

Solórzano, D. G., & Yosso, T. J. (2001). From racial stereotyping and deficit discourse toward a critical race theory in teacher education. *Multicultural Education, 9*(1), 2–8.

Starko, A. J. (2001). *Creativity in the classroom: Schools of curious delight.* Mahwah, NJ: Lawrence Erlbaum Associates.

Stepanek, J. (1999). *Meeting the needs of gifted students: Differentiating mathematics and science instruction.* Portland, OR: Northwest Regional Educational Laboratory.

Stephens, K. R., & Karnes, F. A. (2000). State definitions for the giftd and talented revisited. *Exceptional Children, 66*(2), 219–238.

Sternberg, R. J. (2007). Culture, instruction, and assessment. *Comparative Education, 43*(1), 5–22.

Taber, K. S. (2007). Science education for gifted learners? In K. S. Taber (Ed.), *Science education for gifted learners* (pp. 1–14). London, England: Routledge.

Taber, K. S. (2010). Challenging gifted learners: General principles for science educators, and exemplification in the context of teaching chemistry. *Science Education International, 21*(1), 5–30.

Taylor, E. (1998). A primer on critical race theory: Who are the critical race theorists and what are they saying? *The Journal of Blacks in Higher Education, 19,* 122–124.

Thomas, W. P., & Collier, V. P. (2002). *A national study of school effectivenss for language minority students' long-term academic achievement.* Santa Cruz, CA: Center for Research on Education, Diversity, and Excellence, University of California-Santa Cruz.

U.S. Department of Education. (1993). *National excellence: A case for developing America's talented.* Washington, DC: Author.

Valdes, G. (2003). *Expanding definitions of giftedness, the case of young interpreters in immigrant communities.* Mahwah, NJ: Lawrence Erlbaum Associates.

Valencia, R. R., & Suzuki, L. A. (2001). *Intelligence testing and minority students: Foundations, performance factors, and assessment issues.* Thousand Oaks, CA: Sage Publications.

Vanderslice, R. (1998). Hispanic children and giftedness: Why the difficulty in identification. *The Delta Kappa Gamma Bulletin, 64*(3), 18–23.

Vélez-Ibáñez, C. G., & Greenberg, J. B. (1992). Formation and transformation of funds of knowledge among U.S.-Mexican households. *Anthropology & Education Quarterly, 23*(4), 313–335.

Vygotsky, L. (1978). *Mind in society*. Cambridge, England: Harvard University Press.

Vygotsky, L. (1986). *Thought and language*. Boston, MA: MIT Press.

Wong Fillmore, L. (2000). Loss of family languages: Should educators be concerned? *Theory into Practice, 39*, 203–210.

Yoon, S., & Gentry, M. (2009). Racial and ethnic representation in gifted programs: Current status of and implications for gifted Asian American students. *Gifted Child Quarterly, 53*(2), 121–136.

Yosso, T. J. (2002). Toward a critical race curriculum. *Equity & Excellence in Education, 35*(2), 93–107.

Yosso, T. J. (2005). Whose culture has capital? A critical race theory discussion of community cultural wealth. *Race Ethnicity and Education, 8*(1), 69–91.

Zamudio, M., Russell, C., Rios, F., & Bridgeman, J. L. (2010). *Critical race theory matters: Education and ideology*. New York, NY: Routledge.

Zeus, L. (2009). The color of supremacy: Beyond the discourse of 'white privilege'. In L. Zeus (Ed.), *Critical pedagogy* (pp. 127–139). Malden, MA: Blackwell Publishing Ltd.

María G. Arreguín-Anderson
Department of Interdisciplinary Learning and Teaching
University of Texas at San Antonio
USA

J. Joy Esquierdo
Department of Curriculum and Instruction
University of Texas-Pan American
USA

Adrienne Guillen
Department of Interdisciplinary Learning and Teaching
University of Texas at San Antonio
USA

Lorena Villarreal
Department of Interdisciplinary Learning and Teaching
University of Texas at San Antonio
USA

CORIN BARSILY GOODWIN AND MIKA GUSTAVSON

14. SCIENCE, CREATIVITY AND THE REAL WORLD

Lessons Learned from the U.S. Homeschool Community

INTRODUCTION

There is a great deal of concern regarding the lack of quality in science education in the United States. A simple Google search on "science education in America" brings up link after link on the horrific failure of standards and the many additional problems inherent in the current system of education. Moreover, the need for appropriate education for gifted students is also under attack – as it has been for many years. The combination of these problems is damaging to society as a whole due to the wasted potential and overall lack of scientific literacy, and it is also harmful to the gifted individuals whose needs are not being met. Giftedness often comes hand-in-hand with a great deal of creativity, which is also the foundation of scientific discovery (Van Tessel-Baska, 2004). The U.S. school system is designed in a way that often limits critical thinking and exploration (Weill, 2012), thereby constricting learning by all students and by gifted students of science in particular.

A 2012 study by the Fordham Institute (Eberhardt, 2012) identifies four main factors for the failure of science standards to produce a flock of achievers: an undermining of evolutionary theory, vague goals, not enough guidance for teachers on how to integrate the history of science and the concept of scientific inquiry into their lessons, and not enough math instruction. While a greater quantity of guidance for teachers might be helpful, we propose that a focus on increased quality and opportunities for creativity through experimentation, exploration and failure might be a better approach.

This chapter will first examine the needs of the gifted learner, with a particular emphasis on creativity, and how those needs are expressed in the context of science learning. Next, it will discuss the limitations of the rote method of scientific education for these learners and consider some alternative options gleaned from homeschoolers and others who take a more flexible approach to education. The fact is that a gifted scientist needs room to think, to ponder, to consider outside-the-box possibilities, in short, to be creative. Students need to learn from their mistakes, and society is failing our students when learning is restricted to memorizing only what other people already think they know. We need to learn from our mistakes.

M. K. Demetrikopoulos & J. L. Pecore (Eds.), Interplay of Creativity and Giftedness in Science, 257–266.

> Creativity is any act, idea, or product that changes an existing domain, or that transforms an existing domain into a new one. ~ Mihaly Csikszentmihalyi

So what exactly does it mean to be gifted, and how does that enter the equation of science education and creativity? The term "gifted" can have a multitude of meanings, depending on context and desired outcome. Considerations such as IQ, achievement and aptitude measures are only a few pieces of the puzzle. Typically, an IQ score falling in the range of 130 or higher on the most recently normed testing tools is used to identify gifted students (Kottmeyer, 2014). However, there are many other traits and qualities that frequently go along with the more quantifiable measures of giftedness, and those are the ones that come into play here: abstract thinking skills, rapid non-obvious (to others) connections, keen observation skills, a need for novelty, a dislike of rote repetition, a need to do work that "matters" in the world, concern for fairness and the well-being of others and an intolerance for boredom (Duke Talent Identification Program, 2014). Similarly, according to most standard definitions, creativity requires the ability to view things in new ways or from a different perspective, with motivation for this coming from the need for novel, varied, and complex stimulation as well as the need to solve problems (Franken, 2006). If this list sounds familiar, it should. Looking at it, it is easy to see how a traditional "drill and kill" approach to science would leave a gifted student cold and uninterested.

Another trait of many gifted learners that does not always work well within a traditional classroom is their goal-driven motivation of whole-to-part learning. The usual sequential, standardized approach to teaching tends to take a part-to-whole view, building upwards from smaller units of information into larger ideas. Gifted students may be better served by considering the goals and then working backward to create the needed path. Not unlike a child who learns to read by recognizing whole words and then figuring out the individual letters that make it up, a gifted student may decide they want to understand an idea or cure a disease or invent a way to do something new and will then figure out how to get there. Gifted students are often frustrated by a slow, sequential pace where they are taught to take Step A and then wait to take Step B with the rest of the class without ever being told where they are going. Oftentimes, if they knew the intended endpoint, these children would happily figure out a different, more creative – and possibly more effective – approach on their own.

This more whole-to-part manner of learning about science is particularly suited to gifted students who may think in pictures rather than words, as some do, and who can "see" ideas long before they are able to translate them into language (Goodwin & Gustavson, 2011). A gifted child who sits at a desk, day after day, using workbooks at a set pace or performing only experiments with known results will quickly lose interest in the topic at hand (Willis, 2012). A good teacher not only imparts information, but answers questions, inspires curiosity and injects applicability into the process. A good teacher will find a way for these children to work at a pace appropriate for them.

> Education is not the filling of a pail, but the lighting of a fire.
>
> ~William Butler Yeats

In the typical classroom, a science teacher has a specific group of facts which they are expected to feed into the minds of their students. Little leeway is available in how that information in imparted, as the teacher has limited time to teach the lessons and move along in order to prepare for the next standardized test. Students are therefore restricted in their ability to process and assimilate the information. Rather, they must memorize names and formulae, and any hands-on opportunities are guided by step-by-step directions for Getting the Right Answer (and thus a good grade). Data goes into their working memory, sticks around long enough to be regurgitated on the test, and then it's gone. There's little opportunity for a meaningful consideration, there is no 'hook' to hang it on in their brains to give it context, and there is no motivation to think about any additional implications.

Equally important is that the student who does attempt to think creatively often gets graded down for not following directions, an approach that may not allow the student to demonstrate their mastery of the subject. Moreover, it eliminates any possible chance for a student to let their creativity flourish. It is understandable that school curricula are geared toward mastery of basic, broad scientific content – not every student will have the desire to pursue further study in the sciences, and there are many other topics about which to learn – but there need to be opportunities for those students who *would* be fired up by the sciences. Further, when the curiosity of gifted children is tamped down, they lose not only the spark of interest but may actively avoid following their ideas for fear of unpleasant consequences. Since gifted children are frequently highly sensitive and likely to take feedback personally, there is also the potential for serious negative impact on their emerging self-concept as a result (Webb, 1994). These children need to be urged along in an environment where creativity and mistakes are valued, and where opportunities to explore are better supported. Homeschoolers often have such chances while they are learning on their own or in smaller groups, unhampered by the need to pass a standardized test and unlimited by structured curriculum.

SCIENCE AND HOMESCHOOLING

Despite the misperception that homeschoolers are anti-science, the reality is that a growing U.S. homeschool demographic is very interested in cutting-edge science, technology and education. Many of these families have gifted children whose needs were not being met in school. Some of these families have chosen to homeschool specifically because they live in areas where science is not a valued part of the local school curriculum; and some simply appreciate the freedom to be creative and learn "outside of the box". These families – and the teachers who sometimes work with these families – incorporate science into life, and vice versa. This population embraces new technologies and views creative thinking as a feature, not a problem to be overcome. There is enormous benefit to allowing gifted students to follow the paths that beckon to them, jumping from question to question, and considering new possibilities whether someone else has already invented them or not (Duke TIP, 2014).

Science is essentially the story of life. As Dr. Elizabeth Murray has written, "In general, human beings are curious and we're also pretty good problem solvers. Every day – in restaurants, grocery stores and airports – we make observations and invent explanations for them" (Murray, n.d.). This kind of exploration is, in fact, what we have always done. When our cave-dwelling forebears encountered a new animal or plant, they would pick up a long stick and give it a poke, testing their hypothesis about whether or not this was a dangerous thing. To this day, humans solve problems using the same scientific method based on logical patterns of inquiry, whether it is named the scientific method or not. Why should adults not simply provide students with a running narrative, supplemented by books, videos and – especially – first-hand interactions? When science is compartmentalized as an isolated subject, as it is in traditional school settings, students often end up feeling intimidated by the scientific pursuits which are really only another way of understanding what they see around them on a daily basis (Joyner, 2011).

Some of the ways that homeschoolers learn about science come from boxed curricula or classes at local science centers. Many families get together with small groups or co-ops (including, but not limited to, scouting and 4-H programs) to try experiments and discuss new concepts. Most importantly, however, many of these families incorporate science into life. For example, at the ice skating rink, a young gifted child might ask why there are all of those droplets on the ceiling – offering the perfect opening to introduce condensation and beginning chemistry. A hike in the woods offers ample opportunities for lessons on botany, geology and ecosystems, as well as more interdisciplinary topics such as local history, anthropology, art, and politics. The child who does not want to brush his teeth may be regaled with stories illustrating germ theory, while the adolescent girl who is fascinated with cars may be able to get under the hood with a parent or mentor and learn about physics and engineering firsthand.

The number of books and videos available for a range of ages – both fiction and non-fiction – with science tie-ins is increasing at a tremendous rate, allowing wonderful possibilities for self-teaching. State and national parks have Junior Ranger (and Junior Paleontology) programs for kids to participate in at the parks and online. The array of quality virtual resources is astounding, and many are free or very low cost. Some homeschoolers enroll in courses at their local community colleges as a supplement to their homeschool activities, and many find mentors in areas which especially interest them. These options are available, as well, to students in classrooms where the teachers and administrators are willing to think creatively about education.

Gifted students often thrive with a mentor who shares their enthusiasm for a specialized or unusual interest, particularly when they are willing and able to provide a depth of exploration which is unlikely to be covered in standard curriculae (Goodwin & Gustavson, 2015). The opportunity to find a mentor who will share knowledge and toss ideas around with gifted young people is extremely valuable. The flexibility inherent in homeschooling has allowed many families of gifted children to take advantage of human capital in ways that a public school schedule can only accommodate with the cooperation of the teachers and administrators. Some homeschool families have found mentors for their children in adult friends and neighbors; others are located more haphazardly. A family might allow their children to spend time at the local reptile shop to soak up herpetology, while another might stop by a professor's office hours for a discussion of string theory or climate science. A public school schedule can be adjusted to allow for time to work with mentors by bringing the mentors into the classroom or making time and space available on- or off-campus. The relationship between mentor and student encourages a tailored learning experience as well as the sense that real people are scientists and the child can become one, too (Hood, 2005). The student can ask questions as they arise with no worry that they will be brushed aside due to lack of time or asking the question out of order.

A GOOD APPROACH FOR FUTURE SCIENTISTS

It's important to note, of course, *why* it matters that future scientists are not discouraged from exercising their imagination. Throughout history, creative leaps have been taken to arrive at new and unexpected conclusions (Kean, 2010), just as problems have been solved through application of creativity rather than the following of predetermined instructions. We, as a society, are so accustomed to living with such seemingly simple technologies as rubber bands and post-it notes, that we forget that someone first had to come up with the idea (or determine the use for an accidental invention – such as penicillin). It's a safe bet they weren't prompted to discover these ideas from reading a textbook. A prime example of the creativity needed to solve complex problems in today's world is the landing of the rover Curiosity on Mars in August 2012 (Mars Science Laboratory, NASA, 2012). The Mars Science Laboratory (MSL) team faced a situation that had no precedent. They were given a problem to solve – landing the rover in a precise spot in a specific crater – and had to begin by brainstorming ideas and identifying obstacles. This is how science works outside of the classroom – either an individual experiments with their own ideas, or teams of people share ideas and collaborate to create something that never before existed. Either way, neither the group nor the individual is working with a blueprint initially; they are advancing ideas, exploring concepts and solving problems through experimentation. Testing failures become opportunities for further innovation. This skill set is not, unfortunately, being encouraged in the classroom. Science educators need to acknowledge that cultivating this mindset is far more useful and constructive than rote learning. It simply would not have been possible for the MSL team to have begun by saying, "Hey, let's build an all terrain vehicle and figure out how to ship it and land it and drive it from another planet," if the scientists in question had sought answers solely by looking backward at what had already been done.

MOVING HOMESCHOOL-STYLE INNOVATION INTO THE CLASSROOM SETTING

The innovative practices that emerge from the highly individualized, small-scale educational settings which make up the homeschool world could also effectively be viewed as an idea incubator for the public school system. Pilot projects within most school districts are often impractical due to size and administrative or regulatory limitations. Whereas formal partnerships such as independent study programs already exist in some areas allowing educators and families to collaborate, informal opportunities are less common and entirely dependent on the flexibility and willingness of the individuals involved. Parents and educators of gifted and twice-exceptional children frequently find themselves needing to work together to develop creative solutions for the appropriate education of these children. It makes sense to incorporate lessons learned under the umbrella of homeschooling options into the bigger picture of our system for learning in the U.S.

One brilliant instance of how creativity can be cultivated in the classroom and translated into a real life experience might be that of Kenneth Boehr, an elementary school science teacher in Kansas City, MO, who allowed his students time for creativity – and took their ideas seriously. When his student Clara Lazen, age 10, modeled an unusual-looking chemical compound, he photographed it and sent it to Robert Zoellner, a friend and chemistry professor at Humboldt State University (Huffington Post, 2012). Zoellner (2012) realized it had never been seen before, and published a paper on it in Computational and Theoretical Chemistry, listing Boehr and Lazen as co-authors.

This is a good example of an effective classroom strategy. Creativity is at play on several levels of this story – Clara had a good grasp of the big picture of how chemistry works, the teacher allowed her the time and space to experiment with the ideas that she was being taught, he recognized and respected innovative thought, and he was able and willing to extend himself beyond the confines of the typical educational environment to ask interesting questions and invite collaboration. Each of these steps illustrates a key point in the differences between business-as-usual test-based science education and a creative approach that is more akin to work in the real world. It also shows that what works well for many homeschoolers can be brought into the schoolroom, with dramatic and highly successful outcomes.

Another way to apply homeschool experience to the classroom could be the implementation of a mentorship program. There are many ways to locate an appropriate mentor (Tan, 2004). Many corporations have programs to promote volunteerism or to host interns in exactly this manner. They recognize the business sense of essentially training future employees. One benefit to students is the chance to see how scientific concepts apply in the adult world, as well as the chance to be guided through aspects of a professional or business environment. Retirees may also be excited about sharing their expertise with students. College students are often willing or even required to spend time in a mentorship role. For students living in rural areas or who cannot find a local mentor, online possibilities are everywhere. Skype, Google Plus, Facebook and other web-based services make such communication free and easy (Bierema, 2002). A variety of non-profit and for-profit clearinghouses also exist to play matchmaker.

One other pioneering idea that comes from the business sector is the concept of the 20% project. Where a homeschooler might be free to pursue the interests of their own choosing a majority of the time, a school setting – like a typical office environment – rarely leaves time for such flexibility. The corporate sector has come up with an interesting take on this problem: giving employees one day per week (20% of their time) to work on their own side projects. Companies have realized huge benefits from such a policy – GMail is one of many products that were developed as result of Google's implementation of this idea.

Innovative teachers have come up with ways to apply this to classroom time, with intriguing results (Petty, 2013). Interestingly, while the concept arose in a science and technology-focused sector of the business world, the educational world seems to have adopted the idea primarily in the "softer" subjects, notably English and Social Studies. It would seem there is a great opportunity here for science educators that is largely untapped.

One English teacher, Kevin Brookhouser, who teaches at York School in Monterey, California, sent a letter to his students and their families about the 20% project in his classroom. In it, he stated,

> Before I get into the details of the project, I want to explain why we're asking students to participate in this activity. For over 20 years a trend in education has been gaining momentum that suggests the role of the teacher ought to shift away from an industrial model where the teacher stands in the front of the classroom to dispense knowledge through lectures, and the students sit to consume the information. Rather than being the "sage on the stage" as some pedagogical experts maintain, teachers increasingly ought to play the role of the "guide on the side." In this role, the students play a much more active role

> in how the content and knowledge is acquired. In this model, teachers provide resources, ask questions, and suggest projects for students to explore their content. While I will play the "sage on the stage" role in much of this English class, the 20% project is one place where I will be the "guide on the side." Put simply, this is a student-centered project rather than a teacher-centered project. (Brookhouser, 2012)

This is another excellent example of how creative teaching can lead to creative learning opportunities.

One final classroom idea that educators have been putting to use with exciting results is modeling class projects on real-world problem-solving challenges. The Apollo 13 near-disaster-turned-engineering-triumph is a popular example. In that situation, engineers on the ground had to solve the problem of CO2 buildup in the spacecraft using a motley collection of bits and pieces that happened to be available to the astronauts, and they had to do so within a rigid time frame (Dumoulin, 2001). They also had the challenge of communicating how do to this from a great distance, with astronauts who were not at their peak of functioning.

It's interesting to note that this historic example of science in action has spawned a multitude of lesson plans – some of which limit themselves to a study of how the engineers solved the problem, along with a list of materials and steps to duplicate the process. Other educators have seen the potential for modeling the spirit of the situation, and have created analog classroom projects which focus on the problem-solving, individually or in small groups, with a limited and specific inventory of materials, using a host of different problems and materials. (Fleetham, 2012) These educators are moving beyond the content to teach the essential processes that underlie the work of "doing science"

CONCLUSION

Faced with a decline in science education, it can be tempting to solve the problem by a frantic attempt to cram more information into young minds, hoping that a "more is better" approach will result in desired improvements. In a crisis, however, it can be far more effective, as it was with the Apollo 13 challenge, to pool our knowledge and materials in order to create an innovative solution using the existing resources at hand. Homeschooling experience is one of the many existing resources available, but is frequently ignored or dismissed as being irrelevant. If we are facing our own Apollo 13 in science education – and it appears that we are – does it make sense to leave any potential solutions out of the problem-solving process? It's time that homeschooling, as both an educational option and a breeding ground for educational innovation, is given its rightful place at the table. The risks are few, but the potential for positive outcomes – for educators, for gifted students, for the development of creativity and for the benefit of the world – seem enormous.

REFERENCES

Bierema, L. L. (2002). E-mentoring: Using computer mediated coummunication to enhance the mentoring process. *Innovative Higher Education, 26*(3), 211–227.

Brookhouser, K. (2012, August 4). *A letter to my students and parents about the 20% project.* Retrieved from I Teach, I Think website: http://www.iteachithink.com/2012/08/a-letter-to-my-students-and-parents.html

Duke Talent Identification Program. (2014). *Gifted characterisitics.* Retrieved from Duke TIP website: http://tip.duke.edu/node/99#

Dumoulin, J. (2001, June 29). *Apollo 13.* Retrieved from NASA website: http://science.ksc.nasa.gov/history/apollo/apollo-13/apollo-13.html

Eberhardt, T. (2012, January 31). *The state of state science standards.* Retrieved from Thomas B. Fordham Instititute website: http://edexcellence.net/publications/the-state-of-state-science-standards-2012.html

Fleetham, M. (2012, August 11). *Use the famous Apollo 13 mission to boost children's thinking and learning skills.* Retrieved from Thinking Classroom website: http://www.thinkingclassroom.co.uk/LinkClick.aspx?fileticket=5kPP5yP5brk%3D&tabid=92

Franken, R. E. (2006). *Human motivation* (5th ed.). Boston, MA: Wadsworth Publishing.

Goodwin, C. B., & Gustavson, M. (2011). *Making the choice: When typical school doesn't fit your atypical child.* California, CA: GHF Press.

Goodwin, C. B., & Gustavson, M. (2015). *Writing your own script: The role of the parent in the gifted child's social development.* California, CA: GHF Press.

Hood, S. (2005). *The value of a mentor.* Retrieved from Gifted Homeschoolers Forum website: http://giftedhomeschoolers.org/resources/parent-and-professional-resources/articles/the-value-of-a-mentor/

Huffington Post. (2012, February 3). *Clara Lazen, 10-year-old fifth grader, Discovers new molecule.* Retrieved from Huffington Post website: http://www.huffingtonpost.com/2012/02/02/ten-year-old-discovers-chemistry-molecule-published_n_1250825.html

Joyner, J. (2011, November 9). *Why more Americans don't major in the math and sciences.* Retrieved from Outside the Beltway website: http://www.outsidethebeltway.com/why-more-americans-dont-major-in-the-math-and-science/

Kean, S. (2010). *The disappearing spoon: And other true tales of madness, love and the history of the world from the periodic table of elements.* London, UK: Little, Brown and Co.

Kottmeyer, C. (2014, September 10). *What is highly gifted? Exceptionally gifted? Profoundly gifted? and what does it mean?* Retrieved from Hoagies Gifted website: http://www.hoagiesgifted.org/highly_profoundly.htm

Mars Science Laboratory, NASA. (2012, August 11). *Mission overview.* Retrieved from Mars Science Laboratory website: http://mars.jpl.nasa.gov/msl/mission/overview/

Murray, E. A. (n.d.). Trails of evidence: How forensic science works (Video Course Handbook). *The great courses: Course 1190.* Virginia, VA: The Teaching Company, LLC.

Petty, K. (2013, March 20). *Reflextions on the 20% project.* Retrieved from The Tech Classroom website: http://www.thetechclassroom.com/20-project/reflections-on-the-20-project

Tan, T. (2004, November). *How do I homeschool my gifted child?* Retrieved from Gifted Homeschoolers Forum website: http://giftedhomeschoolers.org/how-do-i-homeschool-my-gifted-child/

Van Tessel-Baska, J. (2004). *Creativity as an elusive factor in giftedness.* Retrieved from Davidson Institute for Talent Development website: http://www.davidsongifted.org/db/Articles_id_10333.aspx

Webb, J. T. (1994). *What are the social and emotional needs of gifted children.* Retrieved from Gifted Kids website: http://www.giftedkids.ie/emotionalsens.html

Weill, D. (2012, July 7). *Texas GOP declares: No more teaching of 'critical thinking skills' in Texas public schools.* Retrieved from Truth Out website: http://truth-out.org/news/item/10144-texas-gop-declares-no-more-teaching-of-critical-thinking-skills-in-texas-public-schools

Willis, J. (2012). *Boredom is a brain turn-off.* Retrieved from Great Potential Press website: http://www.greatpotentialpress.com/authors/author-articles/boredom-is-a-brain-turn-off

Zoellner, R. W. (2012). A computational study of novel nitratoxycarbon, nitritocarbonyl, and nitrate compounds and their potential as high energy materials. *Computational and Theoretical Chemistry, 979*, 33–37.

Corin Barsily Goodwin
Gifted Homeschoolers Forum
USA

Mika Gustavson
Gifted Matters
USA

LAUREN MADDEN AND KRISTIN DELL'ARMO

15. SCIENTIFIC CREATIVITY WITHIN THE RULES

Suggestions for Teaching Science to Gifted Children with Autism

INTRODUCTION

Many of the world's most notable, gifted, and creative scientists, such as Newton, Einstein, and Tesla, are suspected of having had autism, or at the very least, to fall somewhere "on the spectrum" of autism disorders. Recently, Buchen (2011) presented an article in *Nature,* suggesting that the reason so many notable scientists fall on this spectrum is because individuals with autism are drawn to the rules and formulas associated with scientific thinking. Buchen reports on the work of renowned autism expert Simon Baron-Cohen, suggesting that, "the parents of autistic children, and the children themselves, have an aptitude for understanding and analysing predictable rule-based systems—think machines, mathematics, or computer programs." (p. 25) The author noted that many scientists and other science, technology, engineering, and mathematics (STEM) professionals exhibit milder forms of these traits, and when these scientists have children, their children are more likely to be autistic. Other recent reports in popular media (e.g. Coghlan, 2011; Tate, 2012) support Buchen's findings that more children with autism spectrum disorders (ASD) have parents who are scientists. Coghlan and Tate both reported that geographic areas with high numbers of scientific and high-tech companies, such as Silicon Valley, California, and Eindhoven, Holland, have significantly higher than average incidences of children with autism. Correlation does not imply causation, but does leave one to wonder whether traits associated with being good scientists—such as an aptitude for rules and formulas—are passed down from parent to child, perhaps with some other traits associated with ASD, such as difficulty in social situations.[1]

However, rules and structure are not the only defining characteristics of science, and certainly not the only characteristics of scientific genius. Scientific careers also involve creativity, innovation, and exploration—some things that could be seen as breaking the rules. In order to be a successful scientist, one must be willing to think outside the box and challenge what is already known. Though an affinity for rules and order might be what leads autistic scientists to their chosen professions, it leaves science educators with a challenge: how can we best structure learning experiences for children with autism to foster creativity within these rules?

M. K. Demetrikopoulos & J. L. Pecore (Eds.), Interplay of Creativity and Giftedness in Science, 267–280.

TEACHING SCIENCE TO CHILDREN WITH ASD

Researchers and teachers from across the education disciplines—including science education and special education—recommend the use of many common practices. For example, constructivism, or the act of learners creating meaning for themselves based on their own interactions with the world around them and prior knowledge, is generally seen as a best practice for planning instructional activities (Piaget, 1964). In a science classroom, use of constructivism involves allowing student exploration of scientific phenomena before presenting content or vocabulary. In a special education classroom, a teacher might encourage students to use preferred strategies for solving problems before introducing a new method in an effort to build on prior knowledge. Social constructivism, based on the work of Lev Vygotsky, is the idea that learners create meaning for themselves as a result of interactions with others (Wertsch, 1985). This also tends to influence instruction in many educational contexts. The use and exchange of tools, expertise, and language among peers from a variety of ability levels can be structured in such a way that students make meaning of the world around them through social interactions. However, despite the prevalence of shared practices among education disciplines, there are best practices specific to individual disciplines that are important to consider when planning instruction, especially instruction that fosters creativity in students with ASD. Below, we delineate guiding principles of teaching science to children with ASD.

SCIENCE TEACHING PRACTICES

After the Cold War, the US hoped to modify science education programs in a way that fostered the development of creative and genius scientists. Since the 1950s, the United States has put forth numerous efforts aimed at reforming science education (deBoer, 1991). These reforms have included multiple goals, most of which center on inquiry and problem solving. Most notably, the National Science Education Standards (NSES) emphasized a shift in science teaching to include less emphasis on laboratory investigations for verification and "activity for activity's sake," and more emphasis on investigations that promote further questions, understanding scientific ideas that cut across multiple content areas, scientific communication, and use of evidence, argumentation and explanation (NRC, 1996). The notion of using one "scientific method," and memorizing scientific facts has been replaced with a push for teaching students to think creatively about science. The evidence as to whether science education reform efforts—stemming from research funded at large universities and disseminated to teachers through professional development—have been adopted by teachers across the United States is mixed. Many teachers across the US (and perhaps globally) rely on more "traditional" teaching methods such as reading from a text and use of "cookbook" laboratories (Fulp, 2002). Thus, much of what children know as science as learned in typical school settings may be centered on rules rather than creativity.

However, reform efforts continue to move forward. A new framework for K-12 science education (NRC, 2012) elaborates and expands upon the goals of the NSES by incorporating goals for engineering education. The authors described several purposes and goals of scientific endeavors as such: "many scientific studies, such as the search for the planets orbiting distant stars, are driven by curiosity and undertaken with the aim of answering a question about the world or understanding an observed pattern" (NRC, 2012, p. 47). This statement about science as an enterprise illustrates the importance of understanding patterns and rules alongside curiosity and creativity. This new framework puts forth strategies for developing new standards that mesh these two seemingly disparate goals in an effort to better prepare our students for the scientific and engineering challenges of the future.

Scientific inquiry is often seen as central to creative science teaching. However, the term *scientific inquiry* can often be seen as vague and open for interpretation. In the most general sense, inquiry-based science instruction includes all science instruction that starts with a question—either generated by the student, teacher, or text. A model for a continuum of types of inquiry-based instruction can be used to help categorize inquiry-based learning experiences (Martin-Hansen, 2002; Banchi & Bell, 2008). On one end of this continuum is structured inquiry, or investigations in which the teacher presents the students with a topic, question, and procedure for investigating the question. Guided inquiry, where the teacher provides the topic and question but the students develop a procedure, sits at the middle of this continuum. Finally, open inquiry, in which the teacher provides the topic, but students pick the question and procedure sits at the other end of the inquiry continuum. Though there are varying levels of structure in each of these types of inquiry, each type can provide students with opportunities to engage in creative thinking about scientific questions. Inquiry-based instruction also often offers students opportunities to explore and work collaboratively. Good science instruction incorporates a range of types of scientific inquiries within a classroom. The new framework for science learning suggests scientific inquiry should be coupled with engineering design-type problem solving activities (NRC, 2012). This problem solving design is described as, "problem definition, model development and use, investigation, analysis and interpretation of data, application of mathematics and computational thinking, and determination of solutions." (p. 204). This focus on problem solving forces students to incorporate creative thinking into structured and methodical approaches to understanding phenomena and solving problems.

Collaboration and small group work are critical components of effective reform-based science teaching (NRC, 2012). Working with others in problem solving and inquiry-based settings allows students to consider the perspectives of others, benefit from their knowledge (e.g. Vygotsky's work as cited in Wertsch, 1985), and model authentic scientific practices (NRC, 2012). If the end goal of reforming science instruction is to develop scientists and engineers of the future, then it is essential to use small and large group collaborative settings.

Assessment is a critical component of all educational endeavors; without it, educators would have no way of knowing what their students learned (or did not learn). However, assessment and testing are not synonymous. Science education reform documents such as the NSES and Framework for K-12 Science Education advise that good assessments include multiple data points (rather than a single measure), multiple assessment types, and assessments that are purposefully designed to measure intended learning goals (NRC, 1996, 2012). The NSES encouraged science teachers to use authentic assessments, or "exercises [that] require students to apply scientific knowledge and reasoning to situations similar to those they will encounter in the world outside the classroom, as well as to situations that approximate how scientists do their work" (p. 78). Though assessment is sometimes viewed as the "necessary evil" of education, it can take many forms, and can foster our students to think creatively about science.

In summary, some of the best practices for science instruction remain the same as those recommended many years ago—move away from memorization, facts and formulas, and confirmatory exploration, and replace these experiences with those that allow students to ask questions, solve problems, and work collaboratively to better understand scientific phenomena.

PRACTICES FOR EDUCATING GIFTED STUDENTS WITH ASD

Though many educators agree on best strategies for teaching science, these must be considered within the context of the individual students. Strategies that work with one student may not be successful with another. Since it is a spectrum disorder, autism can manifest itself in a variety of ways and varying levels of intensity. However, there are some characteristics that are shared by the majority of students with autism that can be seen as disabling in the classroom setting. These students generally have difficulties with executive functions such as organization and planning, meaning that they are often very disorganized and have trouble figuring out what they should be doing. This also means that it often takes students with ASD longer to accomplish a task than it would take a typically developing peer. These students tend to be most comfortable when they are following a rigid, predictable schedule, whereas breaks in routine and unfamiliar situations can cause extreme anxiety. Anxiety can also be caused by any extreme stimulus in the environment, such as a loud noise, a bright light, or a potent smell. Many individuals with autism have either hypo- or hypersensitivity, so they can be easily overwhelmed in environments such as these and may even display problem behaviors as a coping mechanism when they are experiencing a sensory overload. Additionally, students with autism have difficulty with communication and social skills, which can pose problems for their interactions with peers (Kluth, 2010).

Yet, students with autism should not be defined merely by what they cannot do. The students we focus on in this chapter – specifically, students on the autism spectrum with above-average IQs – possess many unique skills and abilities that should help them to succeed in school. In addition to their high IQs, many have very strong verbal skills, often accompanied by an advanced vocabulary for their age. They tend to have incredibly strong rote memories and an ability to remember large amounts of factual information, as well as a detailed knowledge in areas of specific interest (Kluth, 2010). They can often be very creative in the sense that they think in ways that are fundamentally different from the way others think (Grandin, 2008). Although there are undoubtedly many challenges that must be faced in educating these students, they are gifted and with the right support, they have the ability to be very successful. The current push in special education is towards inclusion, meaning that students with disabilities should be educated in the general education classroom alongside their typically developing peers (Downing, 2008), and the special education field has a lot of strategies for educating these students with ASD in an inclusive classroom.

One of the most important and helpful things that can be done for a student with ASD is to provide them with a system of organization and structure. This can be done by posting a schedule in the classroom for each school day and following the schedule. If changes are going to be made to the schedule, the student should be prepared in advance about what is going to happen to help reduce anxiety. Classroom rules and routines should also be posted in the room, as students with autism seem to find comfort in being familiar with these types of procedures. Visual aids are especially helpful because the student can refer to them for step-by-step instructions throughout the day. A teacher may want to post instructions for daily routines such as sharpening a pencil, packing up at the end of the day, or turning in homework assignments (Myles, 2006). For in-class tasks and assignments, it can be helpful to provide the student with step-by-step instructions, possibly in the form of a checklist. It can be very overwhelming for a student with autism to receive a large task all at once, but it can be made manageable by breaking it down into smaller pieces. The teacher may even set a timer for each piece to work on time management, but should keep in mind that the student may require additional time or a modified assignment (Kluth, 2010).

Many children with ASD have poor handwriting, so modifications such as allowing them to use a computer to complete an assignment can be very beneficial. They may also get very stressed about test-taking, so for this reason it may be useful to consider alternative forms of assessment. Silverman and Weinfeld (2007) suggest finding other ways for the student to demonstrate his or her understanding, ways that incorporate the student's strengths – such as a project, diagram, or slideshow presentation. Additionally, putting a system of reinforcement in place for the student can help to manage problem behaviors, and finding ways to incorporate the student's personal preferences and special interests into the lesson should encourage the student to be more focused and attentive (Silverman & Weinfeld, 2007). When used effectively, group work can also be a great way to work on social skills. Silverman and Weinfeld (2007) suggest assigning specific roles to each student in the group so that each person has a job to do. Finally, students with ASD tend to interpret everything they hear literally. In order to avoid confusion, instructions should be given concisely and simply, and teachers should say exactly what they mean and what they expect of the student (Myles, 2006).

Students with ASD are often very good at rote memorization and learning facts and formulas – that is to say, they thrive under rules and structure. However, teachers should be encouraging their students to think more creatively. One way to do this is through a focus on problem-solving and real-life applications, which work to develop critical thinking skills.

In conclusion, students with ASD can be very gifted and are capable of achieving incredible success in the science classroom. The teacher needs only to figure out how to best accommodate the student. This might mean providing a system of organization and structure, breaking down tasks to make them less overwhelming, using alternative forms of assessment, incorporating group work into the lesson, utilizing a problem-solving approach, or any other strategy that plays to the student's individual strengths.

COMMON GROUND

Considering the research on best practices in both science education and the education of children with ASD, we can find many areas of common ground, along with areas in which these two fields differ. In Figure 1 below, these areas are depicted in a Venn Diagram.

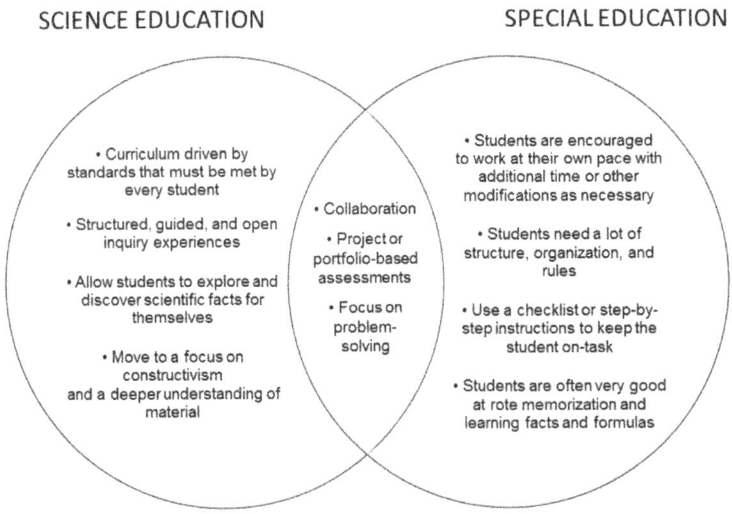

Figure 1. Similarities and differences between practices in Science and Special Education

If our end goal is to develop strategies and suggestions for best practices in teaching science to gifted children with ASD to foster creativity, then we should draw our attention to the areas in which science educators and special educators agree: creating collaborative environments, assessing students authentically, and focusing on problem solving. We can also reflect on the experiences of gifted scientists to help guide our recommendations moving forward.

CREATIVE AND GIFTED AUTISTIC SCIENTISTS

While the scientific and popular literature (e.g. Buchen, 2011; Coghlan, 2011; Tate, 2012) reports on affinity for rules and structure as the main trait shared by scientists with ASD, other similarities in these individuals can also be found. Rawlings and Locarnini (2008) found that scientists with autism scored highly on the Autism Quotient (AQ) subscale associated with both attention to detail (rules) and that of imagination (creativity). Interestingly, this study also found that artists with autism scored higher in other areas, such as schizotopic tendencies. These findings corroborate the speculation by many that some of the most gifted scientists may have had autistic characteristics. A few other examples can be seen in the vignettes below. The first details the experiences of David Finch, and the second that of Dr. Temple Grandin.

David Finch

Engineer-turned-author David Finch, an individual diagnosed with Asperger's Syndrome as an adult, has recently been in the public eye after the publication of his autobiography, *The Journal of Best Practices: A Memoir of Marriage, Asperger Syndrome, and One Man's Quest to Be a Better Dad and Husband,* in 2012. In it, he describes his personal experiences before and after receiving an ASD diagnosis and reveals how this diagnosis helped him to develop coping mechanisms that were useful in everyday life, eventually leading to a better level of self-understanding.

Growing up, Finch's parents helped nurture his interest in science, "My dad regards almost everything through a scientific lens. He and my mom both took time to explain why things happen and how they happen. I would watch my dad analyse a problem from a thousand different angles before approaching the solution. It was cool!" (D. Finch, personal communication, July 30, 2012). Finch also had a life-long love for and fascination with mathematics, and he credits his high school physics teacher, Mr. Anderson, with illuminating the application of mathematics throughout everyday life (D. Finch, personal communication, July 30, 2012). Not surprisingly, he followed his brother's footsteps to pursue a degree and career in music engineering at the University of Miami (Finch, personal communication July 30, 2012). There, under the guidance of Professors Ken Pohlmann and Will Pirkle, he developed an interest in audio and digital signal processing. As he explained, "Besides my music courses, these were the only classes in which my mind didn't wander. I was engaged the entire time and wanted to spend more time learning about these topics." (D. Finch, personal communication, July 30, 2012). After completing his degree, Finch began a career as an audio engineer. He so enjoyed writing software and reports that he often went into a state of flow when doing so. He was able to demonstrate his creativity in designing audio systems. Later, when working in technical marketing on the business side of his profession, he was also able to use his creativity through problem-solving (D. Finch, personal communication, July 30, 2012).

Before his diagnosis, Finch relied on various strategies to "get by," many of which were based on problem solving and rule following, which are typical characteristics of individuals with Asperger's (Finch, 2012). He also relied heavily on mimicking others. After receiving his diagnosis, he was able to better understand how his own mind works and improve his confidence, which then allowed him to perform as an engineer at a higher level (D. Finch, personal communication, July 30, 2012). Despite this newfound understanding, Finch decided to leave the field of engineering to pursue a career as a writer, as it allowed him an opportunity to focus on his creativity. Despite this career change, he maintains a personal interest in the sciences and hopes to pursue scientific hobbies and endeavors with his children as they get older (D. Finch, personal communication, July 30, 2012).

DR. TEMPLE GRANDIN

Temple Grandin, an individual with Asperger's Syndrome, led a childhood marked by frequent temper tantrums, poor grades, and a lack of desire to interact with other people. In fact, she was nonverbal for the first four years of her life. And yet, despite all this, she has gone on to become probably the most well-known person with autism. She holds a Ph.D. in animal science and is now a professor at Colorado State University. She has published many books on autism and frequently lectures on the topic. She is also an incredibly successful engineer; approximately one-half of all the livestock handling facilities in the United States have been designed by her (Grandin, 2008).

Temple Grandin attributes much of her success as an engineer to her ability to think visually. Part of the way her mind works is that she processes information completely in pictures. She is able to design livestock handling systems in her head, in a manner that resembles a 3D design program on a computer. Grandin writes that she is "able to 'see' how all the parts of a project will fit together and also see potential problems" (Grandin, 1986, p. 142). She can visualize designs by taking parts of already existing equipment and piecing them together in her head to create something new. She can "see" this design from many different perspectives and can even rotate images or make them move, much like a computer program would. Grandin can visualize many different test situations, enabling her to "see" how the equipment will work and solve problems and design flaws long before it is ever built.

However, it took a long time to harness and direct these talents of Grandin's. Growing up, she had a very difficult time in school because her teachers did not understand the way her mind worked. She often did poorly on tests and assignments because they were not designed for visual thinkers. The rote memorization tasks that were often required of her were incredibly difficult, and she struggled in working with abstract concepts. Because of this, she was labeled as "brain damaged" for the first few years of her life.

It took the help of a few creative, unorthodox teachers to uncover Grandin's abilities. She could not learn by reading a textbook; instead, Grandin recalls hands-on, real life activities and experiments that encouraged her creativity. She learned about the solar system by drawing it and looking at models, and barometric pressure was something she only understood after her class used milk bottles to make their own barometers. Ever since she was little, Grandin was fascinated with a machine she calls "the squeeze machine," a machine that cattle are placed in before they receive vaccinations. The machine squeezes up against the sides of the cattle and calms them down. Grandin longed for that sort of pressure and tactile stimulation, and she began designing her own squeeze machine that she could get in herself. Most of her teachers and her family discouraged this fixation, but it was all Grandin could think about. It took her high school science teacher, Mr. Carlock, to realize that he could use this fixation to get Grandin interested in schoolwork. He showed her how science could help her to understand how the squeeze machine worked and could give her the ability to build an even better one. This provided the motivation Grandin needed to learn science, and it was at this point that an incredibly successful engineering career was born. If there is one thing to be learned from Grandin's story, it is that the minds of people with autism work differently than the minds of typical students. A good teacher will figure out how to use this to the student's advantage, and with the right support, the student can excel.

EDUCATING FUTURE SCIENTISTS WITH AUTISM

When we consider the literature on best practices for science education as well as those for teaching children with ASD (see Figure 1) alongside the vignettes about David Finch and Temple Grandin, we can conclude that several key strategies can be implemented to help foster scientific creativity in gifted students with ASD.

Collaboration

The image of a scientific genius working alone in a lab is antiquated and inaccurate. The best acts of scientific creativity occur through collaboration, including those that result in acts of genius. In a typical elementary classroom, collaboration tends to manifest itself as group work among students. Group work is certainly a practice which science educators and special educators can agree is beneficial, both to the student with autism and to their typically developing peers. For students with autism, social and communication skills are often a challenge and need to be taught to the student. Group work provides a great opportunity to use language, initiate conversations, respond to the questions and requests of others, and take turns, as students work to meet social skills goals while simultaneously learning the curriculum (Wertsch, 1985). Working collaboratively with peers also teaches gifted students with autism about other people and about how to accommodate differences as the group works together to achieve a common goal. Special educators often look at peers as a very important tool for the inclusion of students with autism, because they tend to be very good at finding ways to involve the student in the lessons and activities. Frequently the creativity and open-mindedness of other students in the classroom means they come up with ideas that even the special educator may have overlooked. Plus, students with autism are generally more engaged and receptive to working and learning when it involves their peers (Downing, 2008).

For science educators, group work encourages the sharing of thoughts and ideas, giving students the opportunity to hear multiple perspectives. This broadens their horizons and enables them to think in new and different ways. It also models science and engineering situations that would be faced in real life, therefore making collaborative work a more authentic way of teaching science, as recommended in the Framework for K-12 Science Education (NRC, 2012). Peer interactions allow students to verbalize their prior conceptions, learn from the experience of others, and see scientific phenomena from novel perspectives, resulting in more creative approaches to scientific understanding.

However, placing gifted students with autism in groups can create very stressful situations if not done correctly. Social interactions often cause anxiety for these students, as do activities that are unstructured or unpredictable. Therefore, group work situations can be incredibly overwhelming and may cause the student to shut down rather than open up. One solution to this problem would be to provide more structure for the group. Open inquiries may not be the best choice for a student with autism, especially not before familiarizing the student with structured and guided inquiries first. Yet, it is not impossible to create successful collaborative scientific activities for classes that contain gifted students with autism—as Temple Grandin explained, teachers can often sense the needs of individual students and in doing so can elect to structure educational experiences based on the needs of these individuals. Silverman and Weinfeld (2007) recommend providing a clear set of goals and expectations for the group, so that the student understands what he or she should be accomplishing. Additionally, they suggest assigning roles for each of the group members that play to their strengths. For example, the gifted student with autism might excel at reading aloud, remembering the steps of the task and making sure they are accomplished, or recalling and recording data. The situation will be much less stressful if the student is familiar with his or her specific responsibilities. The teacher can also help decrease the student's stress by scaffolding social interactions, ensuring that the student has the tools necessary to communicate effectively with the group. As long as teachers are aware of and work to accommodate these challenges, collaborative group work can result in a meaningful and effective learning experience for everyone involved.

Authentic Assessments

Testing can be stressful for all students, especially those with ASD. It can often disrupt the intended schedule in a school day or incite test anxiety in any child. Yet, assessments are not limited to tests alone. Science education reform efforts have pushed to move the focus of assessment away from rote memorization and toward authentic assessments (NRC, 1996, 2012). These authentic assessments can include a range of different formats, and can be tailored based on individual students' needs and teachers' preferences. For example, one teacher might choose to ask students to create models of various phenomena, while another might challenge students to design an instrument or procedure to answer a scientific question. In both of these situations, rather than simply responding to questions or prompts, the students are engaging in scientific practices.

In the NSES, the authors note that assessment and learning are two sides of the same coin (NRC, 1996). Thus, it is critical to engage students in authentic scientific practices throughout instruction, not simply at the end of it. Both David Finch and Temple Grandin reported being most engaged with science instruction that modeled scientific practices—Finch described a high school teacher who pointed out the application of mathematics to physics concepts while Grandin's high school teacher helped her to create a scientific instrument. Providing students with authentic scientific experiences throughout instruction also allows teachers to use an inquiry-based constructivist approach to education (Piaget, 1964). In doing so, students are able to interact with and explore the world around them, and build upon prior knowledge. However, these strategies aren't always the easiest to implement, especially when working with students who fall on the ASD spectrum. Each student comes to their class with a different background and set of experiences and challenges that must be met, thus teachers must approach this type of instruction with flexibility and a variety of teaching strategies.

Project-based instruction and assessment can be a good strategy for educators—both in the science classroom and the special education classroom—to accomplish the goal of creating more authentic instructions and assessments to meet the needs of all students. Project-based instruction and assessment can be done either in collaborative groups or alone, and it often starts with some sort of problem or question that students must work to solve. This type of learning is ideal for gifted students with autism because it allows them to work at their own pace, in a variety of mediums and settings, and on a topic of interest to them. Often, gifted students with ASD have a special area of interest – anything from cars to whales to famous dates in history – on which they are very focused and know a great deal about. Some educators tend to discourage this fixation, but the student's area of interest can actually be a great starting point for project-based learning because of the student's motivation to learn about the topic. For example, a fixation with cars could turn into a physics project in which the student explores concepts such as velocity and acceleration, while an interest in whales could turn into an exploration of marine life, the oceans, or even the organ systems that make up a whale. Reading from a textbook or listening to a teacher lecture can be almost impossible for a student with autism, but being able to do projects that are personally interesting to the students can be very effective. Plus the projects allow the students to explore outside of the classroom, use technology, and possibly interact with others, while also afford teachers the opportunity to modify assignments and assessment techniques, extend deadlines, or otherwise support the student with autism.

Problem Solving

Solving problems means being faced with a novel challenge or obstacle and being able to analyze the situation and develop a solution (Silverman & Weinfeld, 2007). Successful problem solving requires a wide range of analytic and critical thinking skills, as well as creativity. As Temple Grandin (2008) writes, "it involves training the brain to be organized, break down tasks into step-by-step sequences, relate parts to the whole, [and] stay on task" (p. 47). The ability to solve problems is absolutely necessary in order to function in everyday life, and yet all of the aforementioned skills are very difficult for individuals with autism. As such, problem solving is something that people with autism really struggle with, and yet it is a skill that Grandin believes is not incorporated enough into their educations.

The science classroom is an ideal place to teach and practice problem-solving skills. Careers in the STEM fields are based on problem solving; scientists need to be able to do this successfully every single day. Therefore, a good science curriculum should also focus on developing strong problem solving abilities, and it can be done in the context of the lessons.

Grandin says that for her and many others with autism, abstract concepts are very difficult to understand. She learns best from physically doing things. Children with autism (and, in fact, all children) have a natural curiosity about how things work, which can certainly be an advantage in a science classroom. Grandin recalls a windy day when she made a parachute out of a scarf. It took her many, many tries to figure out how to make the parachute fly as far as it could and to keep the strings from tangling, but she continued to try new ideas until the problems were solved. Likewise, Finch found learning by "doing" or solving problems to be the most effective way of learning himself. In his first book, he describes taking apart appliances and searching for order in everyday things. These activities helped him to understand how the world around him worked (Finch, 2012). Good teachers can leverage everyday questions (e.g. how do parachutes work) to structure activities that allow students to utilize problem solving skills. Problem solving skills are something that special educators agree should be taught to students with autism (Kluth, 2010). However, oftentimes teachers do no more than teach the students a list of general problem-solving guidelines (identify the problem, define the problem, organize information, etc.), and this is simply not enough. Gifted students with autism will only become good problem solvers if they are given many real-life opportunities in which to apply these strategies; the more experiences they have, the more they will be able to generalize and apply what they know to novel problems. Science is a perfect application for these skills, and lessons that are taught with a problem-solving focus will not only be good for the gifted student with autism, but will make science memorable, fun, and meaningful for all of the students in the classroom.

CONCLUSIONS

It is not surprising that increased numbers of ASD diagnoses have emerged across areas characterized by high numbers of STEM professionals (Coughlan, 2011). Many of the traits that characterize science—systems, order, organization and classification systems—are also interests of individuals on the ASD spectrum. Yet, these are not the only traits that characterize science or ASD. Science is characterized by exploration, problem solving, creativity, and imagination. Many gifted and creative scientists and other STEM professionals fall on the spectrum of ASD; David Finch and Temple Grandin serve as two examples of successful adults with autism working in STEM careers. Much can be learned from these two examples alongside the literature in both science education and special education about how to best prepare gifted children with ASD for scientific creativity and genius. We believe that modeling authentic scientific practices through collaboration, problem solving, and authentic assessments can be the first steps in developing classroom environments structured to nurture scientific creativity.

NOTE

[1] It should also be noted that the definition of ASD has broadened over the past several decades, and this can also explain some of the increased number of diagnoses.

REFERENCES

Banchi, H., & Bell, R. (2008). The many levels of inquiry. *Science and Children, 46*(2), 26–29.

Buchen, L. (2011). When geeks meet. *Nature, 479*, 25–27.

Coghlan, A. (2011). Childhood autism spikes in geek heartlands. *New Scientist* [Online version]. Retrieved June 22, 2012 from http://www.newscientist.com/article/dn20589-childhood-autism-spikes-in-geek-heartlands.html?full=true&print=true

deBoer, G. E. (1991). *A history of ideas in science education*. New York, NY: Teachers College Press.

Downing, J. E. (2008). *Including students with severe and multiple disabilities in typical classrooms* (3rd ed.). Baltimore, MD: Paul H. Brookes Publishing Co.

Finch, D. (2012). *The journal of best practices: A memoir of marriage, Asperger syndrome, and one man's question to be a better dad and husband*. New York, NY: Scribner.

Fulp, S. (2002). 2000 National survey of science and mathematics education: Status of elementary school science teaching. Chapel Hill, NC: Horizon Research, Inc.

Grandin, T. (1986). *Emergence: Labeled autistic*. New York, NY: Warner Books.

Grandin, T. (1995). *Thinking in pictures: And other reports from my life with Autism*. New York, NY: Doubleday.
Grandin, T. (2008). *The way I see it: A personal look at Autism & Asperger's*. Arlington, TX: Future Horizons.
Kluth, P. (2010). *You're going to love this kid!: Teaching students with autism in the inclusive classroom* (2nd ed.). Baltimore, MD: Paul H. Brookes Publishing Co.
Martin-Hansen, L. (2002). Defining inquiry: Exploring the many types of inquiry in the science classroom. *The Science Teacher*, *2*(69), 34–37.
Myles, B. S., Adreon, D., & Gitlitz, D. (2006). *Simple strategies that work: Helpful hints for all educators of students with Asperger syndrome, high-functioning Autism, and related disabilities*. Shawnee Mission, KA: Autism Asperger Publishing Co.
National Research Council (NRC). (1996). *The national science education standards*. Washington, DC: National Academies Press.
National Research Council (NRC). (2012). *A Framework for K-12 science education: Practices, crosscutting concepts, and core ideas*. Washington, DC: National Academies Press.
Piaget, J. (1964). Development and learning. *Journal of Research in Science Teaching*, *2*, 176–186.
Rawlings, D., & Locarnini, A. (2008). Dimensional schizotypy, autism, and unusual word associations in artists and scientists. *Journal of Research in Personality*, *42*, 465–471.
Silverman, S. M., & Weinfeld, R. (2007). *School success for kids with Asperger's syndrome*. Waco, TX: Prufrock Press.
Tate, R. (2012). *The tech industry's Asperger problem: Affliction or insult?* Retreived June 22, 2012, from http://gawker.com/5885196/the-tech-industrys-asperger-problem-affliction-or-insult
Wertsch, J. (1985). *Vygotsky and the social formation of the mind*. Cambridge, MA: Harvard University Press.

Lauren Madden
Department of Elementary and Early Childhood Education
The College of New Jersey
USA

Kristin Dell'Armo
Ohio State University
USA

SALLY CARSON, STEVE CUTLER AND VICTORIA ROSIN

16. CREATURES, COSTUMES, CRYPTIC CREATIONS

Integrating Creativity in a Secondary Science Gifted Program in Marine Science

INTRODUCTION

In 2001, the New Zealand Government's Ministry of Education formed a Working Party on Gifted Education, which recommended that funding be provided for 3-year programs that could be developed by schools and out-of-school providers. The programs funded by this recommendation, called the Talent Development Initiatives (TDIs), developed new approaches for gifted and talented students aimed at meeting their social and emotional needs as well as providing learning opportunities matched to their learning needs (Riley, Bevan-Brown, Bicknell, Carroll-Lind & Kearney, 2004). The New Zealand Marine Studies Centre was successful in securing a TDI contract in 2006 to develop its Year 10 Gifted and Talented Programs to provide students with an authentic marine science learning experience over multiple days as there was a noticeable lack of gifted programs available to New Zealand schools. The program's over arching theme "making sense of the marine world aids survival and enriches lives" set the stage for integrating creativity in a residential science program.

BACKGROUND

New Zealand Marine Studies Centre

The New Zealand Marine Studies Centre (NZMSC) is a unique educational facility located on the shores of the Otago Harbour, 30 minutes from the city of Dunedin. The Centre is the public outreach arm of the University of Otago's Marine Science Department and operates in association with the Department's Portobello Marine Laboratory. The NZMSC has a long history in delivering science-learning experiences for early childhood, primary and secondary school groups, and the general public. The guiding principles of this TDI program focused on the particular needs of the gifted learner and used creative means for the students to develop and communicate science projects.

M. K. Demetrikopoulos & J. L. Pecore (Eds.), Interplay of Creativity and Giftedness in Science, 281–297.

Program Structure

As the realms of science and technology often hold a negative, uncreative perception to the general public (MoRST, 2001), it is important to change this viewpoint so that secondary students can see the creativity involved in science. The strongest theme that emerged from a study by Lunn and Noble (2008) on how scientists 'do' science, was the creative aspect of science. This finding may conflict with the image of an arts/ science dichotomy that many people hold. The NZMSC understood that elements in the scientific process included creativity which they highlighted through narratives where analogies, and telling the story of science through multiple and varied ways, inspired novel approaches to understanding science. Further descriptions of these various ways are contained below.

The three objectives of the program provided gifted students a continuity of learning to pursue a particular area of interest in marine science, provided better access to expertise in an area of interest, and provided the opportunity to meet and work with like-minded individuals. For this chapter, like-minded individuals refer to other students who are also considered gifted and talented. The objectives listed above were fostered through a variety of methods and guided by the works of Renzulli (1977), Reis and Renzulli (1985), Betts (1985) and the New Zealand Ministry of Education (2000, 2002).

The rationale of the program is based on an integrated curriculum with a general theme that led to more specific small group investigations guided by mentors in a university setting. Using Renzulli's Enrichment Triad (1977), multi-day programs developed; all Year 10 programs eventually contained a residential component that on-going enhanced the creativity of the students in science. Shortly after the program began, it was selected to be part of an active research program, which was also funded by the Ministry of Education TDI, to evaluate and assist in the program's further development. More recently, further research was completed on the program impact on past students as viewed by the parents and teachers. Table 1 illustrates the program's essential elements.

Students and Mentors

Participants in the program are Year 10 secondary school students (14 year olds) and were identified using three forms of evidence including tests, observations, and portfolios. As the characteristics of gifted and talented have broadened and understandings of this type of student have developed within contributing schools, the students who applied to the program did not need to be identified as gifted in science. Since the program's focus was to provide opportunities for all gifted students, students just needed some interest in science. Schools were encouraged to look for: exceptionality, performance and potential, having one or more wide range of abilities, and the recognition of dual exceptionality (may also have special needs/ learning disability). The students could self-select, be teacher or parent nominated,

Table 1. Elements of the Program

Structure	*Tools*	*Process*	*Aim*
Teams and Variable Groupings	* 1-3 students per school. * Team's based on: • Belbin's functional team roles that balance students' strengths and weaknesses. • Mix of males/females and students from different schools. • Residential activity teams formed from a mix of project teams and changed for each residential block.	* Application information from student and teacher. * Belbin's roles disclosed and discussed with students. * Teams review at regular intervals and are asked for details of emerging team members' roles.	* To develop informed awareness of the team's function and student's own role in that team and how team's function may be improved. * To ensure students mix with maximum number of other students on program. * To ensure students share a variety of creative ways of thinking, problem solving and communicating.
Mentors and Science Projects	* Use of laboratory facilities and equipment, boats, seminar room, vans. * Knowledgeable marine science postgraduate students. * Overarching theme that supports and informs projects and creative challenges.	* Mentors' presentations about their research area and their knowledge and experience. * Mentors as scientist role models. * Doing real science in an authentic setting. * Hands-on science with field and laboratory components. * Team problem solving.	* To raise awareness of the possibilities and opportunities in science and demonstrate science as a way of thinking and doing. * To help build engagement and sense of belonging in a knowledge-building community.

(Continued)

Table 1. (Continued)

Structure	*Tools*	*Process*	*Aim*
Residential Component and Linked Recreation Activities	* Island accommodation. * Games and activities. * Teams share the domestic tasks of cleaning and cooking and experience living together.	* A psychological and physical shift with a focus on the social and emotional. * Team challenges that enable marine science ideas and concepts to be used in cross-domain creativity. * Humor and fun emphasised.	* To provide opportunities to develop friendships and for vicarious learning. * To stimulate creative thinking around science ideas. * To provide an environment where it is safe to be smart, unique, odd, inventive and playful with like minds.
Multiple Blocks and Multiple Days	* Scheduling to link with tides, school term, weekends. * 3 blocks, 1 month apart. * Blocks are 3 or 2 days long.	* Time between blocks refreshed the teams and allowed for planning and understanding of projects. * Maximised value for rural students who travel long distance to program. * Multi-block allowed for project development and student's autonomy and mastery of it. * Multi-day allows for in-depth practical work and discovery.	* To allow time for development of academic and interpersonal skills and experiences. * Allows for authentic science experience with reflection, review, critique, and problem solving. * Allows for in-depth development of communication skills. * Helps build engagement and belonging in a knowledge building community.

Table 1. (Continued)

Structure	*Tools*	*Process*	*Aim*
Thinking and Meta-cognition	* Meta-cognitive strategies survey form. * Team reflection and review. * Short keynote talks and discussions around thinking, distributed and augmented cognition and meta-cognition. * Creative challenges linked to marine science. * Science project presentation 'seminars' with peer review and critique. * Use of analogies and modelling.	* Teams use review form with mentor to think about team function, strategies used and progress of their science project. * Analysis of group's meta-cognitive strategies used to frame key questions and to develop meta-cognitive awareness and skills. * Teams challenged creatively and across curriculum. * The fundamental role of analogy in framing, grasping and communicating new ideas and using this to aid creativity and science. * Mentors model peer review and critique. * Participants are encouraged to engage in peer critique and differentiate between science and communication issues.	* To gain a 'world' view and skills to apply to their learning in other situations. * To instil an awareness in students of becoming and being a self-regulated learner. * To develop an awareness in students of the nonlinear path of science thinking and development.

(Continued)

Table 1. (Continued)

Structure	*Tools*	*Process*	*Aim*
Presentations and Communication	* Multiple types of communication challenges – press release, science seminar, poster, novel presentation. * Presentation challenges scaffolded over the multi-block program. * Exemplars, templates and examples are given to teams for press release, PowerPoint presentation and scientific poster. * A final public presentation is given by all teams to communicate their science.	* Teams create a press release to focus on essentials of project and to determine the direction they intend to take the project. * Short PowerPoint presentation by whole team of their project at the end of each block. This allows teams to build their communication skills together and support those less confident in their personal communication skills. * Residential recreational challenges develop skills for teams to develop novel presentations. * Science novel presentations are presented during multiple block days so that ideas can be tried and improved upon. * Peer review and critique supports mastery learning that is meaningful in a science research context. * Poster challenge of research at end of program has team focus on economy of representing the essential aspect of the science.	* To provide students with opportunities for autonomy, mastery and purpose and to share their development with parents and teachers. * To increase students' communication and presentation skills. * To build students' confidence and to have them experience a safe and supportive environment. * To provide a variety of catalysts for team reflection, discussion and knowledge building around research project. * To share the essential role of peer review and critique in science.

or even be nominated by a peer. The students applied to the program as if they were applying for a research position in marine science by using a cover letter and resume. From the early forms of the students' applications appearing in paper, the application process has progressed to the level that some students choose to use technology to create elaborate video presentations with some having a very creative focus.

Programs are divided into 2 different sessions – rural and urban. This is purposefully done to introduce rural students to each other as small rural schools are sometimes in remote locations and do not have access to programs that urban schools do. The desire was to introduce students to like-minded individuals that also shared like experiences in their school environments.

Once selected to attend the program, students are placed into small research teams of 5-6 individuals using a modified version of Belbin's team roles (Belbin, 1981) as described in Table 2. Students engage with this idea during the program and begin to understand how a group functions and how roles change and develop as problems or tasks arise. By using periodic evaluations of how the team is functioning, students discuss how they view each other and who took on which team role for that day.

Table 2. Team member roles

Modified Belbin's (1981) Team Roles	*How Described to Students*
Leader – Makes sure team runs smoothly.	Restaurant Manager – Makes sure team is working together, has what is needed to succeed and deals with any issues that arise.
Innovator – Creative, innovative ideas.	Head Chef – Creates new food dishes, some ideas may be 'out there'.
Practical – Looks at plan and works out what can be done and how.	Assistant Chef – Will make the dishes of Head Chef, but considers if they can be done or should they be modified.
Monitor – Joins pieces together so all parts come together.	Head Waiter – Makes sure restaurant and kitchen staff work together.
Time Manager – Gets project done on time.	Assistant Waiter – Gets food out to table in timely fashion.

A mentor, who is usually a postgraduate student in marine science, leads the team. 'Leading the team' is only in the sense of guiding the students or helping them problem solve rather than directing what is to be done by the group. This is done purposefully and the mentors understand their guiding roles. The students, with their mentor, design a research project by developing questions, hypothesis and methods that then lead to practical hands-on research using the University's facilities and equipment. The mentor proffers the local subject topic to be studied (i.e. camouflage crabs, sea lions, squid) and discusses possible options the team

Table 3. Evening recreational activities

Activity	*Reasoning Behind*
Share Your Sense of Humour – Best team joke, best overall joke, encourage acting out of joke.	Break the ice between students and talk about the foundations of humor being cognitive and its multiple meanings. Also to have fun!
Team Quiz – Questions from what learned at NZMSC and different projects.	Works on distributed cognition idea, as teams have at least 1 member from each research team on them. Fun way to review the generic marine science covered the first day and the base information for each project.
Team Marine Creature Features Challenge – Construct a model of a marine organism that has senses, reproductive system and some internal organ. Its features need to be explained in relation to its ecology, i.e. food, predators, reproductive cycle, etc. The model and narrative need to be cohesive and coherent.	Teams are given limited supplies of kebab sticks, modelling clay, tape, crepe paper, etc. to create model. Imagination and biological principals are applied in a new and novel way, often with humour and use of analogies. Stimulates creative process and students have fun together.
Team Mythical Marine Marvel Challenge – A team member is transformed into a mythical marine god or goddess or superhero. Team explains origins of this myth and how this belief aids in the survival for their 'tribe'.	Teams are given limited supplies of newspaper, crepe paper, tape, and string to use. Imaginative analogizing and another angle of program's overarching theme "making sense of the marine world aids survival and enriches lives". Initiates thinking for novel presentations of science projects by students.
Team Cryptic Clue Orienteering Challenge – Teams use a map of Quarantine Island with directions to find cryptic clues to help the team identify and understand local island features.	Teams leave at timed intervals and have to circumnavigate the island, finding the places or objects hinted at by the clues attached to numbered sites on a map. Teams solve the question posed by the clue with a rhymed response/answer. They are judged on their speed and creative skills with language and thinking.
Team Sea Animal Olympics – A physical competition of 8 activities based on local marine creatures. Each team member must enter at least 1 event and no more than 3. Examples: *Nematode Ninety* – Full body contact with ground, twist and turn in specified manner to finish line; *Stargazer Instant Impact Knack*-sit on chair with hat on head, plate and potato on lap, flip potato in air and catch in hat; *Blue Cod Twister Blister* – Full body contact on floor pad, rotate body around head-feet axis as many times possible in set time.	As these events are new to the teams trying them, usual strategies do not work. Teams have to think about strategy in who enters which event. Lots of laughter and fun and relates to program's theme of "enriching lives".

could investigate. The team then works out the rest of the details together to create their research project.

Mentors volunteer to work with the program and receive training in how to question students, what meta-cognition is and how to develop critical thinking skills in the students. As the mentors are postgraduate students from around the world, many have to learn how to deal with inquisitive students' questions. Mentors learn how to not just give answers but to question students and how to avoid leading questions. They also have to learn that it is ok to admit that they do not know the answer. Mentors learn about collaborative grouping and how to accommodate differences in students' learning needs.

Just as the science extension and enrichment is enabled and supported by capable postgraduate mentors so to the social/emotional recreational activities that feed the creativity needs a creative guiding mentor. This role is filled by the educator who develops and overseas each program.

PROGRAM HIGHLIGHTS

Students attend the program in 3 blocks of 2 or 3 consecutive days. This allows for intense, longer periods of work time that gifted children thrive on when interested in the topic they are studying (Ministry of Education, 2008). They do their research work at the NZMSC and spend the evenings in bunk-style accommodation on Quarantine Island where they continue to engage with each other on creative projects designed to reinforce the science learning.

Once the students' research projects are underway, other communication tasks are assigned. To stimulate and challenge students to look and think in depth about their learning, keynote talks and presentations on aspects of science and meta-cognition are used. Having students 'think about their thinking' is a novel idea to them, as they usually have never stopped to process how they learn best. Students have commented that 'thinking about their thinking' helps in other school areas when they return home to their studies. At the end of each multi-day block, the teams reflect on the progress of their research, communicate their current findings via oral and visual presentations and receive peer reviews on their work just like in the real world of science. Students are shown how to critique work, not just criticize it, and engage in refining their communication skills for a general audience.

The evening residential components consist of a short 2-minute boat ride to Quarantine Island where their accommodation is one of only a few buildings and beyond the caretaker, they are the only inhabitants for the duration of their stay. Placed in work groups that are different from their research teams, they are assigned tasks of cleaning, as well as preparing and serving food for the group of 20–25 students. The evening is filled with informal activities that emphasize creativity by using stories, games, and art to communicate science. The residential component of the program was originally developed for the rural students due to the distance they had to travel to the NZMSC, however the positive feedback from students, parents and teachers led to it becoming part of the urban program as well. Such projects may include designing an imaginary sea creature that has special adaptations to survive and to tell its story or students may dress up a team member as a mythical god or goddess (Fig. 1). As a safe and trusting environment is created during the program, students actively engage in these activities where they may not in other situations. Mentors take their turn helping with the residential component where they interact in a completely different way from the work they are doing in the laboratory. Table 3 describes some of the activities done during the residential part of the program. Some written examples of Year 10 student descriptions of created animals for Team Marine Creature Features Challenge:

Team A – The 'Gillford' fish – Une poisson.

Though a hermaphrodite the full reproductive behaviour and strategy is being openly debated in the ichthyologic community. It is thought to revert to the desperate measure of self-fertilization if in its solitary existence it never comes across another member of its species before a certain age. As a terminal breeder, once the eggs are fertilized and deposited on the sea floor it dies leaving its decomposing body over the eggs both as a protective cover and a source of nutrients to the vulnerable young.

This decomposing body of the fish is thought to be poisonous to any scavenger as well, thus we see few if any disturbances to the nested eggs and early juveniles. With such a strategy leading to a high survival rate in the young it is surprising that more of this species are not seen.

Team E – A SmirkJagger – The female and small globular male.

The male has evolved to an extreme degree. The strong mechanoreceptors have evolved a link to an internal neural and muscular network that is asymmetrical and oblique resulting in a unique means of moving by rapid vibration (not unlike the silent mode of a cell phone) this has recently been termed 'vibroloco'.

The sperm sac is dropped off near a female who, if selecting for the males genetic offering, consumes the sac through the cloacae. Invariably triplets are produced. This somewhat limited reproductive strategy is further restricted by the fact females are terminal breeders, the one set of triplets being their only chance to add to the next generation. One would suspect that some form of parental care would be a necessity for the species to survive but as yet no such behaviour has been observed, or at least not reported in the literature.

The creativeness of the students is encouraged in part by the building of relationships between team members. If the students did not feel safe to reflect and share their creative sides with each other, this portion of the program would not be possible. The freedom to imagine has allowed the students to use their science knowledge to make sense of their creative expression. This provides the students the opportunity to dive into the world of science fiction and use analogies to further explore their creativity. This awareness that the students can tell stories and be creative leads to some exciting finished products to the program's challenges.

Figure 1. Team Mythical Marine Marvel Challenge – Example of students dressed by team members as a marine god/goddess or superhero and explains the origins of this being

In the third and last block, the final component of the groups' research is to prepare a 'novel' presentation that challenges students to think creatively about how to make their research results and projects understandable to a general audience. As students have been presenting their findings to each other throughout their research, this project is seen as an important step of developing their science communication skills. Some of the ways used by teams in the past include puppet shows, dramas, songs, movies, TV parodies or interviews and adaptations of literary works (i.e. Shakespeare). As long as the research and findings are communicated, there are no limitations to how this can be presented other than it is appropriate for the general audience. These presentations, along with the more traditional research presentations, are given to friends and family on the last day of the program. The audience is also invited to ask questions about the teams' projects. The team then have to handle answering the questions with minimal help from the mentors.

Benefits of Program to Students – Views of Students

Students indicated, via team evaluations and individual interviews, that the program opened a window into science they did not know existed. Science was no longer seen as a boring topic where lab notes and step-by-step experiments were the only option. They found themselves in an environment in which they were comfortable to show both their academic ability and their creative talents. Some students commented,

"I'm not a freak here." "It's real." "Less formal (than at school) but we have more responsibility (to define and do the work)." "At school I do all the work whereas here with this team everyone can do the work." Many of the students commented on the enjoyment they received when working on a team with individuals of similar abilities.

Students also encountered, many for the first time, that science is not just facts or correct answers. They were amazed that their mentors were often learning along side of them and did not have all the answers. "Mentors know heaps about the subject yet are learning at the same time as us," a student mused. Another stated, "Before this camp I hated science and saw myself as creative and intelligent at music and stuff, but not science. What changed my mind? Doing it! And also, our mentor - he was so passionate about his work he made us passionate too."

Benefits of Program to Students – Views of Teachers and Parents

When students returned to their own schools and homes, teachers and parents noticed several changes in the individuals who had participated in the program.

The elements of the program that had the most impact on students, as reported by parents and teachers on a survey completed after the gifted and talented program has been running for 6 years, are listed in Table 4 and Table 5.

Table 4. Changes observed in the students by parents and teachers as surveyed one or more years after the program

Changes Observed in Some or All Students After Being Involved in the Program	*% of Time Observed in Some or All Students n = 25–28*	*% of Time Observed in All Students n = 25–28*
Increased communication skills (oral/written/ electronic)	96%	43%
Increased use of specific science language	92%	62%
Increased interest and motivation to engage with science	89%	70%
Increased interest in science careers	96%	39%
Improved self confidence	100%	64%
Improved value of their own identity	96%	57%
Improved teamwork skills	93%	46%
Improved organization and time management skills	93%	37%
Improved ability to carry out science investigations	92%	81%

Table 5. Elements of program that had the most impact on student

Program Elements	*% of Time Medium or High Impact n = 25–28*	*% of Time High Impact n = 25–28*
Setting within an authentic science environment	96%	89%
Development of their own science investigation	100%	68%
Formal presentation challenges	93%	54%
Quarantine Island residential component	88%	54%
Field work (i.e. local shoreline, boats)	100%	82%
Laboratory investigations (i.e. dissections)	100%	89%
Working with live animals	96%	75%
Creative challenges linked to the science (Quarantine Island activities)	88%	60%
Meeting and working with other students	100%	85%
Meeting and working with science mentors	100%	86%
Meeting and working with like minds	96%	89%

Teachers and parents noticed an increase in students' communication skill level including increases in oral, written, and electronic communication capabilities. Teams had challenges of preparing small parts of their presentations throughout each program block and presenting these to the other teams and mentors. Communication skills were developed during the presentation practices and critiquing that happened throughout the program. Electronic skills were enhanced by PowerPoint presentation work and spreadsheet analyses of research data. Postgraduate students, sometimes the mentors but not always, gave talks in a semi-formal or informal fashion about their research to students. These were used as exemplars of how to address an audience.

Other changes indicated by the survey data were improved teamwork skills. Students were placed in teams on day one of the program and this became their research team for the rest of the sessions. Students also were placed in a different team for residential activities that included preparing food, cleaning and the evening game challenges. Mixing of team members introduced students to more individuals in the program and gave them a break from continually interacting with the same 5–6 individuals that they had been placed with for a research team.

In addition to improvement in social skills, the emotional needs of the students were seen as being met in a variety of ways including: improved self-confidence, improved value of their own identity, working and making friends with students who were similar to them or had 'like-minds'. Meeting the emotional needs of the gifted students was an important factor of the gifted program, but the important contribution of the residential portion was not foreseen.

Parents also noted benefits that were apparent at home from their student's attendance at the NZMSC. Students could not stop talking about what they did while attending the program. Developing their own science investigations, working in the field to gather specimens or data and actually using live animals all stimulated the students to see how science is conducted.

Mentors also were noted as important to the program by parents. A parent commented about the mentors, "…the opportunity to interact with a scientist – who turned out to be fun and normal! Exposure to the fact that one can be employed in a field one is passionate about."

The social aspects of spending time with like-minded individuals had a huge impact on the students and, working with like-minded students doing genuine research, were seen as valuable components,, especially for students/families in rural areas. Rural schools can be small, and meeting others who can challenge the gifted student and make them feel 'normal' is not easy. One parent stated,

> My daughter is still friends with other participants on the program. [It is] often difficult to find a range of like-minded motivated "future academic" students in small rural schools. They keep in contact with Facebook and visits. It's great, as they will link up again more than likely next studying first year at Otago.

Benefits of Program to the Teachers and Schools – Teacher Views

Teachers saw the skills the students learned through the program transferring to other subjects and that students now saw science careers as a possibility. "The skills will transfer back into any subject – information skills, information technology. They see possibilities of career opportunities in science. Before they (might have) liked it but now they know what hands on research is all about." As students are with like-minded individuals, teachers noticed, "Their confidence has just gone through the roof. [The experience is] taking away the constraint of being the bright one."

Teachers also benefited as they were invited to spend the day with the students' teams and see them working on their research projects. "For me it has been definitely PD [professional development]: observing how you run things, seeing the kids involved and meeting other teachers of gifted students." "High interest high thinking challenges that could not be accessed back at school to the same level with the same quality environment and same quality of personnel. Best PD for me too – I learnt from observation of the educators [mentors]."

The observation of the student teams in action and observation of how the mentors questioned the students and encouraged them to think out problems was one of the most valuable aspects of the program for some teachers. One mentor explained to his team how he thought out his own Ph.D. marine science research. This out-loud thinking/explaining to the teacher and students helped them to understand a way to process science problems and how some scientists think when doing science. It was an illuminating moment for both teacher and students.

Teachers from smaller rural schools also saw a huge benefit for their gifted and talented students. As a principal stated,

> This is very important in small rural schools as we do not have the critical mass of half a dozen of these students in our individual schools. We do however across our several rural schools. These kids keep the ongoing collaboration going socially through Facebook. We are also challenged to provide the motivational science mentors often available in larger schools. Aspirations come from seeing someone you want to become. Our students are immersed in this passion (of the mentors) and absorb some of it.

SUMMARY

This chapter endeavoured to share the successes noted in a gifted and talented program for Year 10 students at the NZMSC in Dunedin, New Zealand. This program's story highlights the integration of creativeness with the science experience in an authentic environment. It also shows how students use their own creativeness to address the challenges posed.

To meet the learning needs of gifted and talented students, mental challenges are posed during the marine science research projects. In addition to this, the social and emotional needs of these students have been addressed. These needs have been successfully met by the relationships that have developed. Outcomes for students with this program included the fact that building relationships with other students and the team mentors are critical to the success of the program. The relationships were key in creating an atmosphere to facilitate the learning and enjoyment experienced by participants.

Although teachers who observed the program did not always rate the creative challenges and residential components as highly as the science experiences, almost all surveys commented on the friendships that developed and the increase in the gifted students' confidence, self-esteem, and self-belief. The change in participants' confidence, attitudes to science and motivation is strongly connected to the created learning environment. The creative challenges were critical in developing this environment in which the students felt safe, comfortable to be themselves, and that their ideas were valued. The NZMSC believes the creative components of their Gifted and Talented program are instrumental in filling, not only the learning needs for gifted students but also their social and emotional needs. Figure 2 illustrates how the program's elements were triangulated to met students' needs. The program not only helped the students make sense of the marine world but it also helped them make sense of their own world. It aided the students' survival as gifted students and enriched their lives and the lives of others through their own creative science endeavours.

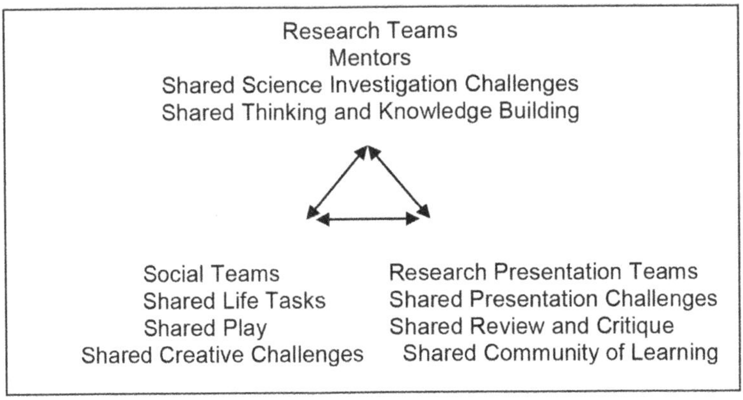

Figure 2. The social/emotional 'Triangulation': Building a safe but dynamic learning community

REFERENCES

Belbin, R. (1981). *Management teams: Why they succeed or fail.* Woburn, MA: Butterworth Heinemann.

Betts, G. (1985). *Autonomous learner model for the gifted and talented.* Greeley, CO: ALPS.

Lunn, M., & Noble, A. (2008). Re-visioning science "Love and passion in the scientific imagination": Art and science. *International Journal of Science Education, 30*(6), 793–805. Retrieved from http://dx.doi.org/10.1080/09500690701264750

Ministry of Education. (2000). *Gifted and talented students: Meeting their needs in New Zealand schools.* Wellington, New Zealand: Learning Media.

Ministry of Education. (2002). *Initiatives in gifted and talented education.* Wellington, New Zealand: Ministry of Education.

Ministry of Education. (2008). *Nurturing gifted and talented children – A parent-teacher partnership.* Wellington, New Zealand: Learning Media.

MoRST (Ministry of Research, Science and Technology). (2001). *Science and society: A policy perspective background.* Wellington, New Zealand: MoRST.

Reis, S., & Renzulli, J. (1985). *The secondary triad model: A practical plan for implementing gifted programs at the junior and senior high school levels.* Mansfield Center, CT: Creative Learning Press.

Renzulli, J. (1977). *The enrichment triad model: A guide for developing defensible programs for the gifted and talented.* Mansfield Center, CT: Creative Learning Press.

Riley, T., Bevan-Brown, J., Bicknell, B., Carroll-Lind, J., & Kearney, A. (2004). *Gifted and talented education in New Zealand schools.* Wellington, New Zealand: Ministry of Education.

Sally Carson
New Zealand Marine Studies Centre
University of Otago
New Zealand

Steve Cutler
New Zealand Marine Studies Centre
University of Otago
New Zealand

Victoria Rosin
New Zealand Marine Studies Centre
University of Otago
New Zealand

SECTION 5

APPROACHES FOR FOSTERING SCIENTIFIC CREATIVITY IN GIFTED LEARNERS

AUDREY C. RULE AND BENJAMIN D. OLSEN

17. USE OF ANALOGY AND COMPARATIVE THINKING IN SCIENTIFIC CREATIVITY AND GIFTED EDUCATION

IMPORTANCE OF COMPARATIVE THINKING

Fundamental Cognitive Processes

Discerning similarities and differences are fundamental cognitive operations for learning. Four important strategies for engaging students in using these foundational operations (Marzano, Pickering, & Pollock, 2001) are (1) comparing similarities and contrasting differences; (2) classifying things into categories based on characteristics; (3) creating analogies that map relationships between pairs of concepts; and (4) creating metaphors that show similar patterns from different domains. A meta-analysis (Apthorp, Dean, & Igel, 2012) of twelve studies from 1998 to 2008 that focused on using similarities and differences, such as analogy, with kindergarten through high school students, indicated that these approaches positively influence student learning. Larger effect sizes were seen when the control group experienced traditional teacher-directed, textbook-based instruction with smaller effects when the control also involved interactive teaching strategies. Student learning improved with the opportunity to reflect and discuss including the systematic guidance of students through analogical reasoning and classification of important concepts along with relationships among and between concepts.

Metaphors, Similes, and Analogies

There are several types of comparisons that people utilize to comprehend new concepts. A *metaphor* is a literary device or figure of speech that is substituted for the concept being examined in order to draw the mind to recognize a resemblance between the two. *Similes* are comparisons that use the words "like" or "as," but generally do not carry as much feeling as metaphors. For instance, saying, "The lion is king" has a stronger emotional impact than "The lion is like a king." Metaphors are often found in poetry and usually have associated value judgments. They are also used in science, sometimes presenting problems of bias; for example, "Cowbirds deposit their eggs in the nests of other species that raise the *freeloaders*," suggests how cowbirds and their young should be viewed (Flannery, 2009). Although the reader may have many associations for cowbirds and the word "freeloader," the

M. K. Demetrikopoulos & J. L. Pecore (Eds.), Interplay of Creativity and Giftedness in Science, 301–320.

mind filters out those that are not similar for the two subjects (Black, 1954). It is the flood of mental associations that helps the learner begin to make sense of the new concept and to connect it to other ideas.

As mentioned previously, metaphors are often used in poetry to convey emotions and values. Poetry has been used as a way to infuse creativity with science learning. For example, Rule, Carnicelli, and Kane (2004) used poetry-writing to motivate high school students in investigating and writing about minerals in an earth science class. An unpublished poem from the study (by the first author of this chapter) that uses similes and metaphor to set the emotional tone and provide positive value judgments about a specific mineral also incorporates accurate scientific information. Students were guided in this process and half of the participants reported improved attitudes toward science after the instructional unit. An illustrative excerpt of the poem follows.

Rose Quartz
Rose quartz is sweet like the nose on a bunny,
It's pretty and nice; sometimes even funny.
It's the colour of roses arching over a gate,
Carnations and clovers painted 'round a cake plate,
A party dress ribbon, pink lemonade punch,
Bubble gum ice cream and jellybeans for lunch,
A fist full of Easter grass, iridescently pink,
Or a powder room rug that matches the sink.
It's smooth and its round – so glass-marble cool,
Carved into statues or cabochon jewels,
Polished as beads or set into rings,
Made into boxes to cherish small things.
It's harder than window glass; difficult to scratch.
Strong Si-O bonding provides a good match.
A substance that's carve-able; yet holds its shine;
The slick, glossy feel of rose quartz is divine!
Its color's mysterious: perhaps colloidal gold,
Or manganese impurities rose quartz may hold.
Titanium could cause its colour within,
Or aligned mineral fibers of unknown origin.
Part of its charm is we don't understand,
The *pink* in that chunk of rose quartz in your hand.
Enjoy its pink radiance, pastel in hue,
Rosy perfection provides a new view.
Calming like rippling pink clouds in the sky,
Rose quartz can ease an over stressed eye.
Rest yours upon it and soon you will know,
Peaceful relaxation of rosy quartz glow.

Analogies, in contrast to metaphors, are used less for drama and more for a direct comparison between concepts to highlight the common relationships of their various features. Inferences can be drawn about the less familiar concept on the basis of what is known about the more familiar concept (Harré, 1972). Deeper understandings of complex concepts can be facilitated by abstracting the most important ideas from the system, delineating its boundaries, and providing appropriate language for presenting a scientific explanation (Arnold & Millar, 1996). Analogies are important to science learning because they often allow the learner to mentally picture a complex concept and may also help in scientific research by alerting the mind to unnoticed, possibly parallel features that may then be explored.

Analogies play several important roles in student learning (Venville & Treagust, 1996): (1) transferring the structure from an unfamiliar domain to a familiar one to aid understanding; (2) motivating students and increasing their self-efficacy in learning science content; (3) facilitating change in mindset of the learner from "matter" to "processes" and (4) supporting memory in recalling features and interactions of a concept. Further evidence of the utility of analogical thinking in memory retrieval comes from several experiments conducted by Gentner, Loewenstein, Thompson, and Forbus (2009). They found that when analogical comparisons were used during learning, later retrieval of information was improved, probably because of the mental representation of the information in abstract comparison categories. They also examined participants' retrieval of a problem-solving strategy when analogy was used at the time of retrieval rather than at the time of learning, showing that this technique enhanced recall and application of the remembered strategy.

Gentner and others (Gentner & Lowenstein, 2002; Gentner & Medina, 1998) have suggested much of children's learning strategies are based on similarity comparisons with analogies being particularly valuable in enabling children to abstract relational knowledge structures. Valle and Callanan (2006) showed that parents effectively used analogies to communicate new science concepts to their four to nine year old children at two science museum displays (topographic maps and a zoetrope exhibit of animation) and through explanation of science in a homework problem about infections. A post-task assessment demonstrated that children whose parents used relational analogies in their explanations performed better.

Assessments Using Analogy

Some classic gifted education creativity tests (e.g., Getzels & Jackson, 1962) asked subjects to generate as many uses for a common object (e.g., pencil, paper clip, brick) as possible, requiring the person to think of properties of the object and how the object might be used differently to exploit one of these properties analogously to another known object (e.g., a heavy brick could be used as a door stop). Similar activities are often used as thinking exercises for gifted students. Intelligence/achievement tests that focus on analogy include the well-known Miller Analogies Test (PsychCorp, 2011) that phrases all items as analogies with a final multiple choice response to complete the analogy and sections of the Graduate Record Exam (Educational Testing Service, 2012). Use of analogy in advanced testing is a testament to the higher levels of thinking addressed by this type of comparative thinking.

USE IN SCIENCE TEACHING

Metaphors in Science Teaching

Metaphors help students make sense of new experiences by connecting them to what they already know. Jakobson and Wickman (2006) examined the spontaneous metaphors and similes of elementary students as they were engaged in science lessons. They noted that children's comparisons of the natural phenomena they were observing to qualities of other things helped them focus on and remember those characteristics. Secondly, they observed that children's spontaneous comparisons were stepping stones to developing the final science concepts rather than endpoints in themselves. However, sometimes, children's metaphors restrict what they observe, resulting in scientific aspects of the phenomenon being ignored. Teachers, therefore, need to interact with students and ask questions that assist them in noticing other important aspects. Children often make comparisons to objects without stating which qualities make these objects similar to the natural phenomena they are exploring. Consequently, it is important for teachers to encourage them to elaborate on how the two are similar. Jakobson and Wickman also noted that because understandings of metaphors and similes rely on prior experiences, all comparisons will not be equally effective for all students. Additionally, some metaphors contain negative aesthetic or value judgments that hinder students from exploring the phenomena further. The researchers of this study suggested that teachers might make a game of students trying to think of positive metaphors when a conversation turns negative. Additionally, the teacher can mediate metaphors that appeal mostly to one gender or culture by suggesting more universal ones.

Teaching with Analogies

Analogies serve as early mental models that connect prior knowledge to developing understandings, but they may be used ineffectively when a learner interprets unshared attributes as valid or when learners are not familiar with the analogy (Harrison & Treagust, 1993). Therefore, teachers need to guide students in mapping the relevant features of the analogy and in identifying its limits (Adúriz-Bravo, Bonan, Galli, Chion, & Meinardi, 2005).

The Teaching with Analogies Model helps students avoid some of the problems associated with using analogies to explain complex concepts (Glynn, 2007, 2004; Glynn, Duit, & Thiele, 1995). This model has six steps: (1) introduce the new, unfamiliar concept called the *target* concept; (2) remind students to think about what they know about the *analogue* concept, a familiar concept to which the target will be compared; (3) identify the most important features of both the target and analogue concepts; (4) connect the ideas from the two concepts that have the same types of relationships through mapping – drawing a diagram or making a chart; (5) identify areas in which the comparison breaks down; and (6) draw conclusions about the target concept – what do students now understand about this new idea since comparing it to a familiar idea? For teaching through analogies to work well, both target and analogue need to have a number of similar features; the more features shared, the better the analogy.

The pairing of components from the target and analogue that have similar roles in each system is called *structural alignment* and is accomplished through mapping. Mapping can be made visible through a chart that connects the two features (one from each system or domain). Gentner and Markman (1997) identified three psychological constraints on the alignment of an analogy: (1) structural consistency, (2) relational focus, and (3) systematicity. One-to-one correspondence of features of the target and analogue with matching relationships within their respective systems constitutes structural consistency. Relational focus means that the paired elements do not have to have similar visual or surface appearances; they just need to have similar relationships in their systems. Systematicity refers to the fact that analogies are comparisons between systems of related elements.

Bridging analogies (Brown & Clement, 1989) can also be used to assist students in understanding difficult concepts. There are four steps to this process, illustrated here with a physical science case: (1) Student ideas that are inconsistent with scientific knowledge are made clear by using a target question. For instance, the teacher may ask if a table exerts a force on the book resting on its surface. A student responding, "No, because the table is not moving," is exhibiting an idea that does not match scientific understandings. (2) The teacher suggests an analogous case that students find intuitively acceptable, such as considering a hand pressing down on a spring. This is called the *anchoring analogy*, because it forms the strongly accepted end of a chain of ideas that will connect to the disputed idea. The teacher asks, "Does the spring exert a force on the hand? (3) The teacher asks students to compare the two analogies: the book on the table and the hand on the spring. (4) The teacher supplies another analogy that is closer to the book on the table, because it is easier to understand a close analogy than a distant one. This is the *bridging analogy*. In fact, several close analogies many be provided to bridge the gap between the target case and the anchoring analogy. In this situation, the teacher may ask students to consider a book resting on a foam cushion or suspended by a flexible strip. A study of Turkish high school students using bridging analogies to study physics concepts demonstrated that both male and female students learned more compared to a control group (Yilmaz, Eryilmaz, & Geban, 2006).

Active student engagement in acting out an analogy can assist students in better understanding the science. For example, upper elementary students learned how an electric circuit works by forming a loop with a table (representing the battery) as part of the loop (Ashmann, 2009). At the end of the table designated as negative, a large pile of pennies was placed. A few pennies were also placed at the opposite, "positive" end of the table to represent the electrons present there. Each student, standing shoulder to shoulder, had one hand behind the back and one in front, grasping a single penny. The circuit operated as the student next to the positive end of the battery felt an imagined attraction of that end of the battery for her negatively-charged electron-penny and placed it on the table. This allowed her to take the penny from the person next to her and to continue the chain reaction of penny movement around the circuit. She continued to place pennies at the positive end of the battery, taking pennies one at a time from the person next to her. The person at the negative end of the battery took single pennies from that end of the table to feed the current until the pile of pennies at the negative end of the battery was exhausted and the battery was "dead." Students were also able to act out how an insulator stops current, how a light bulb operates, and the differences between series circuit and a parallel circuit.

Generative analogies are so-called because these analogies are created and modified by students as they explore a concept (Wong, 1993). Students who generate their own analogies think deeply about concepts and tend to ask important questions. Generative analogies are effective because they originate from a students' base of understanding, thereby avoiding analogies not understood by students. Students activate and connect to previous knowledge as they attempt to devise the analogy. The process of devising an analogy pushes the student to probe and question their current understandings of the topic.

Many successful science analogy lessons have been documented in the professional literature. For example, Orgill and Bodners (2007) used a two by two pane of postage stamps as an analogy for the binding of oxygen to hemoglobin. To remove the first stamp, two perforated sides must be torn, just as the binding of an initial oxygen molecule to hemoglobin requires quite a bit of energy. The second and third stamps removed requires less tearing and less energy just as binding of successive oxygen molecules requires less energy. In their study, students reported better understanding of course information, increased ability to visualize biochemical concepts, and improved recall of content. Additionally, the results of the investigation indicated instructional analogies increased student motivation and enhanced communication of science ideas.

Using analogies can help students conceptualize relationships that exist between structure and function within a complex system rather than merely memorizing information. Student-created analogies facilitate students' higher levels of thinking and actively involve them in the process (Marzano et al., 2001). Middle school learners who created models of cells as cities, restaurants, baseball games, or homes (Grady & Jeanpierre, 2011) showed increased test scores compared to previous groups who did not engage in such work, indicating their improved understanding of cell parts and functions. However, it is important to note that use of analogies in teaching a science concept may not be enough; additional opportunities to understand, discuss, and apply the new ideas are important (Guerra-Ramos, 2011).

Models

Mental representations of physical phenomena or systems that have analogous structures (similar spatial arrangement and relationships between components) are mental models (Gilbert, Boulter, & Elmer, 2000). If the relationships between the parts are causal in nature, then one can mentally "run" or conduct a simulation of the model to make predictions and explanations (Nersessian, 2008). A person's mental models of phenomena continuously evolve throughout the lifetime, changing as the individual encounters new situations. However, although a mental model may evolve to be a scientifically accepted one, individuals may fall back on past ideas or experiences to generate predictions and explanations, rather than actively engage to mentally manipulate the model they espouse as correct. Because of this potential fallback to earlier naive conceptions, it is important to provide in-depth explanations of the underlying mechanisms of how the physical processes work (Chiou & Anderson, 2009).

Physical models are another type of analogy. Beads woven together with nylon thread can be used to model many chemical structures, such as fullerenes, in which each spherical bead represents the electron density of a carbon-carbon bond (Chuang, Jin, Tsoo, Tang, Cheung, & Cuccia, 2012). The construction of models helps students notice the symmetry of the molecule in the three-dimensional representation of its configuration. Another activity described by Nassiff and Czerwinski (2012) showed how students in a high school chemistry class used large and small paperclips to model the Law of Conservation of Mass. Students took different quantities of large, then small paperclips, adding them together to be weighed. Then they linked one small to each large paperclip to form the compound LgSm (Large-Small). They weighed the product LgSm and leftover small and large paperclips, comparing it to the initial weight of the original quantities to verify the law.

PROBLEM-SOLVING AND CREATIVITY

Combination of Elements

Many eminent scientists have also been artists such as Leonardo da Vinci and Richard Feynman. Root-Bernstein (2003, p. 267) states that ".many scientists and engineers employ the arts as scientific tools and that various artistic insights have actually preceded and made possible subsequent scientific discoveries and their practical applications." He outlined four ways that the arts assist scientists: (1) new phenomena are often invented or discovered by the arts before being investigated by science; (2) the arts supply non-traditional physical and mental tools including models and analogies for problem-solving; (3) words, images, and models used to communicate scientific ideas and results often come from the arts; and (4) fantasy and the generation of possible worlds for exploration and testing according to real-world constraints contribute to scientific discovery and invention. Perrine and Brodersen (2005) found that both artists and scientists share the personality trait of openness to experience, but in artists this is best seen as openness to aesthetics, while in scientists it is openness to ideas.

Michalko (1998) postulated that successful ideas of creative geniuses come from having a rich pool of alternatives and conjectures from which to choose the best, likening it to the blind mutation pool in biological evolution from which only the best adapted changes survive. He studied eminent scientists, inventors, and artists, compiling their creative thinking strategies for generating ideas, summarizing them into two categories: (1) ways to see what no one else is seeing, and (2) ways to think what no one else is thinking. The first category addresses perception, including defining the problem at different levels of scope or perspective, attending to different aspects of the issue, and making thoughts visible through a large variety of diagrams. The second category focuses on ways to generate "blind" ideas that are shaped by chance or random factors. A strategy that supports this category is determining major parameters of a challenge and listing possibilities for each. Then the possible solutions are created by randomly selecting one possibility for each parameter to create a whole that combines different elements. According to Michalko, Leonardo da Vinci used this technique in drawing a large variety of grotesque heads or caricatures by making a chart of major features of the head (overall shape, eyes, nose, mouth, chin) and listing possibilities for each (e.g., for the chin: double-chinned, slack-jawed, sagging, angular, receding, projecting). After providing several techniques for combining and connecting ideas, Michalko suggested, among other strategies, that one can use similarities, differences, and analogy between domains to produce new ideas.

Thagard (2010) evaluated the *combinatorial conjecture* that all creativity results from combinations of mental representations. He studied two existing lists to avoid any personal bias, not arguing whether the selected one hundred were actually the ultimate "best", but assuming that they all were, indeed, important: (1) one hundred important scientific discoveries (Haven, 2007) and (2) one hundred great technological inventions (Philbin, 2003). He concluded from his analysis that all of the hundred scientific discoveries show evidence of combination of concepts leading to the discovery. New concepts (evidenced by newly–coined words) occurred in only 60 of the 100 cases, while analogy was used in 14 discoveries, with all but two of these involving comparisons across different domains. Forty-one of these involved visual representations while 87 of the 100 technological inventions used visual representation. Twelve of the inventions used analogy with seven of these being across domains.

Analogies in Problem-Solving and Innovation

"[A] problem occurs when there is an obstacle between a present state and a goal and it is not obvious how to get around the obstacle" (Goldstein 2005 p. 388). Problems can be well-defined or ill-defined (Kahney, 1994). A well-defined problem provides the solver with all the information necessary to solve the problem. This information falls into these four categories: the initial state, the goal state, legal operations, and operator restrictions. An ill-defined problem is missing one or more of these types of information. Analogies can be useful in solving problems if the solver recognizes the similarities between two analogous problems and can also recall the solution to the problem (Condell, Wade, Galway, McBride, Gormley, Brennan, & Somasundram, 2010). Comparing two similar problems helps people develop a general schema that operates across domains, making them more able to think of the problem in broad terms and use analogous thinking to solve it. However, functional fixedness, the inability to perceive new relationships or uses for objects, and mechanization of thought (using the same problem-solving steps for all problems) inhibits the problem-solving process (Anderson, 2005); therefore, it is important for problem-solvers to recall possible applicable methods while remaining open to new approaches.

Visual analogies (analogical reasoning with visual knowledge) are important in architecture (Casakin, 2004) and other types of design (Ferguson, 1992). Davies, Goel, and Nersessian, (2009) analyzed the sequence of drawings made by undergraduate college students who were presented with the problem of designing a weed trimmer that extends on a pole from a truck to trim the roadside but needs to be able to "pass through" traffic signposts. A diagram and a description of an airlock vestibule separating a clean room from the rest of the building were provided for students to use as an analogy in solving the problem. Davies and colleagues determined how the new designs for the weed trimmer may have been produced by incremental transfer from the provided airlock example and developed a computer program that simulated the visual input and output of several of the participants. They concluded that designers can create new solutions by transferring ideas from prior, analogous models by using visuospatial representations of the stages of the design organized in chronological order. Their computer modeling work indicates that analogical transfer can occur using only visuospatial knowledge. This reinforces the importance of using visuals such as diagrams in creative thinking, as previously discussed when mentioning Michalko's creative thinking strategies.

The use of analogy assists scientists in making structured connections between different domains to better understand how they work and to exploit well-known relationships in one domain for innovations in another. Many scientists and inventors have used analogy to assist them in making conceptual breakthroughs. For example, James Dyson, while looking for ways to make vacuum cleaners more effective, observed the whirling action of a sawmill cyclone sucking sawdust without becoming clogged. His first vacuum cleaner prototype was based on this analogy (Foreman & Drummond, 2008). Similarly, Hans Krebs defined the citric acid cycle, later named the Krebs Cycle, by recognizing the similarities of parts of the chain to components in other cyclic processes (Lightman, 2005). Likewise, Charles Darwin compared evolution to a tree, connecting budding twigs to existing species and older growth as the long succession of extinct organisms. He noticed that new growth overtops older branches, blocking the light from them in the same way that new species may outcompete others in the struggle for resources. This analogy helped Darwin notice other aspects of evolution to investigate (Darwin, 1859; Marcelos & Nagem, 2012).

Analogies have been used by scientists and engineers to develop new theories and experimental approaches. For example, in the area of artificial intelligence, Brooks (1999) used an analogy to convince many members of his field that concentrating on general aspects of intelligence such as vision and movement were more important to the development of artificial intelligence than the then-current approach of focusing on specialized intelligence like solving difficult mathematics or playing chess. The analogy he employed in this argument was that nature required over three billion years for life to evolve from single cells to insects, but only 450 million years to evolve from insects to humans. The conclusion to be drawn is that basic properties of life are more difficult engineering problems and should form the foundation for artificial intelligence in a bottom-up manner, rather than trying to reproduce more specialized aspects first (a top-down approach) (Gibson, 2008).

Experts Compared to Novices

It has been posited that it takes about ten thousand hours of concentrated effort to reach expert level in most academic fields, sports, and games (Ericsson, 1996). This information is important to educators of gifted and talented students, as it indicates that preparation in a student's area of interest should begin early.

"An expert is a person with special knowledge or ability to perform an allocated task skilfully," whereas a "novice is someone who is new to the field or activity" (Condell et al., 2010, p. 232). Several studies have delineated the differences between the ways experts and novices solve problems. Experts work backward from the unknown to the given information in a "means-end" approach, while novices write down the given information and try to make connections to the unknown (Larkin, McDermott, Simon, & Simon, 1980). Experts search a greater breadth of possibilities than novices and consider the consequences of each step (deGroot, 1965). They are able to recognize crucial configurations of information and their implications. Experts in physics arrange their knowledge in a hierarchical fashion, producing specific solutions to the problem; whereas novices are less organized and more general in their solutions (Larkin et al., 1980). Physics experts also used underlying science concepts rather than surface features and use multiple representations of the problem.

Schenk, Vitalari, and Davis (1998) identified five differences between experts and novices: (1) novices have less knowledge about the domain, which limits their ability to ask important questions to gather information and therefore generate good solutions; (2) Experts tend to involve users in the system development stage; (3) Experts react to specific information rather than general triggers; (4) Experts generate more hypotheses and goals for the problem; (5) Experts notice features and patterns differently than novices.

Easton and Ormerod (2001) found that both experts and novices spend the same amount of time working on a problem, but experts provide more alternative recommendations, critical issues, evaluation criteria, and more quantitative rather than qualitative solutions. Experts are able to activate and retrieve previous knowledge related to the problem but may take more time in solving the problem because they spend more time understanding it (Hung, 2003). Experts, however, only outperform novices when solving problems in their area of expertise. Otherwise, they perform similar to novices because their advantages are based on their store of previous experience and knowledge in the field (Goldstein, 2005).

Experts are more likely than novices to recognize analogical relationships between different situations and to encode these into memory, allowing later retrieval (Blanchette & Dunbar, 2001). Memory retrieval of this type of relational information is very important to effective problem-solving and functioning in many educational and workplace situations (Pfeffer & Sutton, 2000). The most effective means of relational transfer is for the person to compare analogous examples during learning (Gentner, Loewenstein, & Thompson, 2003).

Form and Function Analogies

Form and function is a unifying concept of science noted in the National Science Education Standards (National Committee on Science Education Standards and Assessment and National Research Council, 1996) that can be applied to both the natural and designed world, therefore allowing analogies between these domains. *Forms*, physical properties that include shape, colour, pattern, texture, motion, and configuration, support the *functions* of manufactured objects or natural organisms such as animal body parts and plant parts. Research studies have shown the efficacy of high school students using form and function analogies to learn human body systems (Rule & Furletti, 2004), and of second graders learning animal adaptations (Rule, Baldwin, & Schell, 2008).

The two studies just mentioned utilized a unique instructional material called an “object box,” which was a set of small manufactured items (the “objects”), each representing an analogue, and a set of corresponding two-sided cards housed in a plastic shoebox (the “box”). The front of each card described the form and function of an animal body part (second grade study on animal adaptations) or a component of a human body system (high school study). The student’s first task was to take a card, read about the form and function, and then search through the objects to locate one that had a similar form and function. This activity had the advantage of being hands-on and of having concrete examples of the analogues used in the analogies for students to examine. The reverse side of each card presented the name of the correct analogous object and an explanation of how its form and function matched that of the target. Figure 1 shows example card fronts and backs from a form and function analogy object box about the saguaro cactus. Although not used in either of the mentioned studies, this example shows the form and function relationships well and was similar to the sets used in the study. The fronts of the cards are shown on the left and the backs of the cards (with the corresponding answers) are shown on the right.

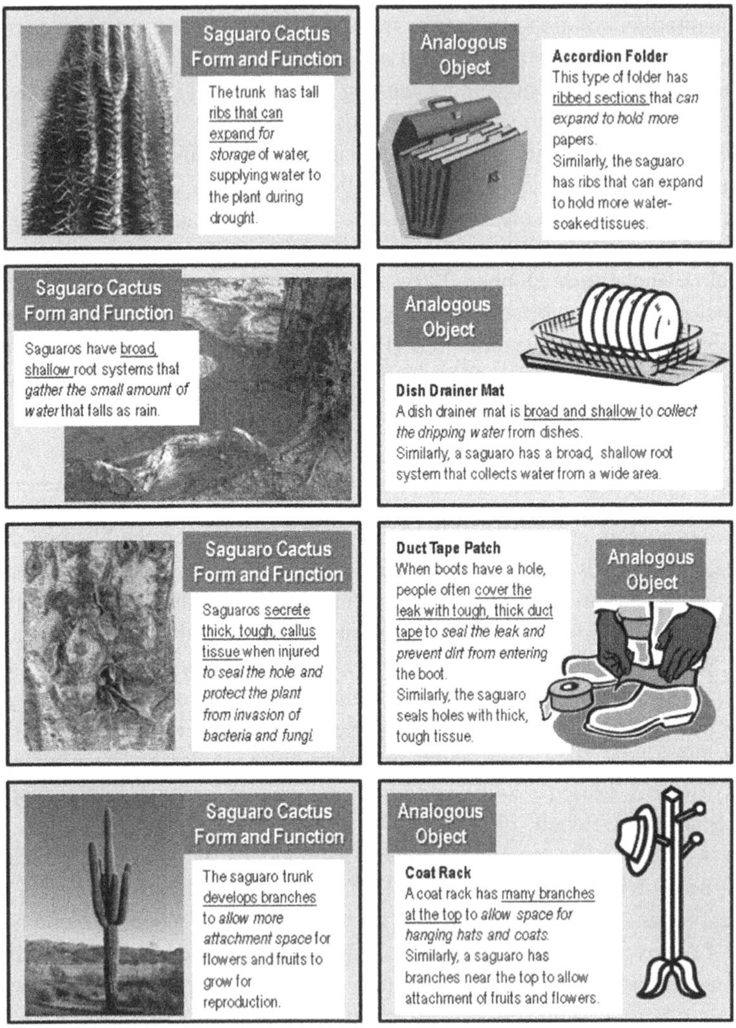

Figure 1. Example form and function analogy cards for the Saguaro cactus

A sequence of activities to enhance learning through deeper analysis of the analogies was used in both studies. First, students matched the cards to objects as just described. Next, students chose one of the objects and mapped the relationships between the target and the analogous object, using a chart like the example in Table 1 that maps the analogies between a saguaro cactus and an accordion folder. The limits of the analogy were noted on the bottom of the chart.

Table 1. Mapping of the analogy of a saguaro cactus and an accordion folder

Saguaro Cactus	*Similarities*	*Accordion Folder*
Saguaro ribs	Both are expandable	Accordion-like sections
Widens to store water	Widens when fuller	Widens to store paper
Stores water	Storage mechanism	Stores papers
Contracts during drought	Can contract	Contracts when few papers
Waxy skin to seal out organisms	Protective skin	Tough paper and clasp to protect documents
Saguaro Cactus	Limits	Accordion Folder
Living plant	Different materials	Thick paper
Green	Different colouration	Brown or variety of colours
Naturally growing	Different origin	Manufactured item
Stores water	Stores different items	Stores paper
In desert areas	Different location	Found in offices

Third, students considered other objects that might be used as alternative analogues for the target animal body part or human body system component. These objects needed to have the same form and function relationships as the target concept. For example, other items that could be used as analogies to the expanding nature of the cactus are: an elastic waistband, a knitted hat, blacksmith bellows, a pleated skirt, certain pleated vacuum cleaner bags, and some suitcases that can expand by unzipping a section. In the final activity, students were given a new animal body part or human body system component written on an index card. Students generated their own form and function analogy, drawing a sketch of the analogue object function on another index card. These were then mixed and students worked to match these new analogies devised by classmates, discussing issues and strengths of that work.

Form and function analogies have been combined successfully with the SCAMPER method to create new inventions or innovations of manufactured items (Rule, Baldwin, & Schell, 2009). This creative thinking technique's name, SCAMPER (Eberle, 1972), is an acronym for various operations that can produce changes for innovations: Substitute, Combine, Adapt, Modify-Minify-Maximize, Put-to-another-use, Eliminate, and Rearrange. These ideas were developed from

Osborne's checklist (1963) of tactics for producing creative transformations. First an item is identified to which innovation or invention will be applied. In work with second graders, Rule et al. (2009) used simple items such as an envelope, plastic spoon or paper cup. A chart is used to implement this technique, as shown in Table 2. The first column has the creative SCAMPER operations that will be applied to ideas; the second column is used to note a form and function relationship present in one or more organisms that will be applied to the item in conjunction with the SCAMPER operation to generate ideas for innovation. The combination of disparate ideas in this manner is called forced relationships, an effective strategy for producing novel ideas (Guilford, 1986). The last column shows ideas for innovation of the product, in this case, a canvas tennis shoe.

Table 2. Applying the SCAMPER technique in conjunction with form and function to generate ideas for improving canvas tennis shoes

SCAMPER Operation	*Saguaro Form and Function Idea*	*Idea for Improving Tennis Shoe*
Substitute	Saguaros have broad shallow root systems	Substitute broad woven mats for soles to walk easily on sand.
Combine	Saguaros have branches to support colourful flowers and fruits	Attach flowers and baubles to the shoestrings to make them more attractive.
Adapt	Saguaros have a waxy skin to prevent water loss and keep out pathogens.	Sell a waterproofing gel that can be applied for walking in wet grass or puddles.
Modify, Minimize, Maximize	Saguaros have ribs so they can expand.	Have pleated fabric along the side so that a swollen or growing foot is easily accommodated.
Put to Another Use	Saguaros have spines to protect them from browsing animals.	Fill old shoes with cement and use as a self defence weapon.
Eliminate	Saguaros develop tough tissue to seal wounds.	Have a tube of gel that can be applied to worn spots in the canvas to seal and eliminate them.
Rearrange	Saguaros have wooden rings inside to support the trunk and branches.	Take the insole out and use it as padding for the ankle with a higher, more supportive shoe top.

USING ANALOGY IN PREPARING GIFTED INDIVIDUALS

Use of Analogies in Gifted Education Programs

Many problem-solving activities for gifted students require students to reuse or re-purpose common items such as plastic lids, cardboard trays, plastic bottles, and Popsicle sticks to make a project. For example, the Future City problem-based program with computer simulation (Gardiner, 2007; National Engineers Week Future City® Competition, 2011) requires students to design a model of their planned city with recycled materials after using simulation software to design their city. Another problem-solving exercise using analogical thinking for gifted students is one described by Rule et al. (2011; 2012). In this activity, participants are each given an identical set of recycled and craft items, given a theme, and asked to make a scene or object related to the theme in a limited amount of time. Participants must envision the various recycled items as new analogous parts of the construction.

Synectics, a Greek word meaning binding together of seemingly unrelated elements, was the term used by Gordon (1961) for his program of strategies aimed at uncovering the psychological mechanisms for creative activity. He first developed these skills for use in industry while he was a member of a consulting group that helped businesses develop new product ideas, but later they were applied to developing workbooks for gifted education (Gordon, 1974). The synectics program utilized analogical thinking strategies (among other strategies) including personal analogy and direct analogy. In *personal analogy,* the participant puts himself or herself in the place of one of the objects involved in the problem to be solved – *becoming* the object. How the object is feeling, moving, wishing, and interacting with other objects is verbalized using imagination, emotions, and the senses. *Direct analogy* uses animals, appliances, everyday items, or systems to make analogies to the problem or parts of it to enhance idea production for a solution.

Synectics also offered an additional strategy for problem-solving. First, everyone is reminded to postpone judgment of ideas. A facilitator asks the participant to state the problem. Group members translate the problem into wish form, "I wish that." and the participant chooses a few that seem to represent the problem best. The participant explains what words or phrases made the chosen wishes most appealing. The facilitator leads the class in imagining an excursion to a distant place, describing the scenes and events encountered there for a few minutes and making connections to the favoured words and phrases. Then, class members use that excursion to form connections and analogies to the problem at hand, often generating unusual perspectives that assist in finding a solution. For example, a second grade student may choose this statement: "I wish I could make a book of shadows that presents mysteries." The facilitator may take the class on an imagined cave tour with many formations that are likened to common objects when viewed from different perspectives and which reveals mysterious cave inhabitants such as blind fish and crickets. The class may connect mysteries in the cave darkness to guessing the object that made a shadow and viewing objects from different perspectives to matching different shadows of the same object created by moving the light source.

Another successful approach for school-wide enrichment and gifted programming is the Talents Unlimited Thinking Skills Program (Schlichter & Palmer, 1993), which provides a set of thinking skills to be used in kindergarten through high school education. This system consists of the following talents (to be combined with specific academic talent such as science): productive thinking, planning, decision-making, forecasting causes and effects, and six additional communication talents. One of the communication talents is generating similes using the words "like" or "as" and adding details to portray the situation in which the similarity is at its most magnified point. For example instead of saying, "The butterfly was as colourful as a painting," the simile would be taken to the limit by saying, "The butterfly was as colourful as a bold Mondrian painting of black, white, and primary colours."

CHAPTER SUMMARY

Comparative thinking is a fundamental cognitive process that positively affects student learning. Metaphors, similes, and analogies use mental associations that assist the learner in making sense of new concepts and in connecting them to existing knowledge. Poetry containing metaphors mixed with science information can infuse creativity with science while conveying emotions and values that motivate students. Analogies allow learners to draw inferences about a less familiar concept through what is known about the more familiar one, thereby aiding understanding, self-efficacy, and memory along with moving the learner to focus on process. Gifted education creativity tests often assess quantity and quality of idea generation through analogy tasks while college entrance or achievement tests use analogy to assess vocabulary and fine divisions of concept understanding.

Analogies are useful in science teaching and learning. Children often make spontaneous analogies when discussing observations of natural phenomena; discussions with adults can facilitate, enrich, and guide their use of analogies. Effective analogies make reference to familiar analogues and clearly define the limits of the comparison. Mapping the paired components from the new target idea to the familiar analogy, a process called structural alignment, helps in this process. A series of analogies can assist students in understanding difficult concepts by starting with one easily understood and then moving to others that are closer to the unfamiliar concept, bridging analogies. Kinesthetically dramatizing an abstract science process through analogy with concrete objects or actions facilitates student understanding. Asking students to generate and demonstrate analogies of science processes is effective because learners must think deeply about the concepts, often asking important questions, while operating from the learner's base of understanding. Many teachers successfully use analogy or physical models made with common items such as beads and paper clips to present difficult science concepts to students.

Creative scientists have used arts ideas such as exploration of imagined worlds and use of words, images and models derived from the arts to fuel their ingenuity. Eminent scientists, inventors, and artists examined problems from different perspectives and tried many combinations of ideas. Analogies can be helpful in problem-solving by allowing the individual to apply a previous solution to a problem in another domain to the present problem. Drawing sketches of problem steps can assist in translating and applying this knowledge. Many scientists have had breakthrough ideas that resulted from analogies to other domains. Experts differ from novices regarding solving problems in the domain of their expertise. They search a greater breadth of possibilities, generating a hierarchical set of problem solutions with critiques. They are more likely to recognize and use analogical relationships. Form and function analogies are useful for comparing natural to manufactured objects and systems and have resulted in many innovations. Hands-on sets of materials that include objects and analogy explanations on cards, along with analogy mapping and generation activities can assist students in deeper understanding of concepts such as human body systems or animal adaptations. Combining these ideas through creative operations such as substitute, combine, or adapt and applying them to a given objects can produce innovative ideas.

Analogy has been used in gifted education programs in many ways. Re-purposing of materials to design models, scenes, or objects provides hands-on, engrossing activities. The Synectics program utilized personal analogies of becoming the object, direct analogies, and fantasy analogical excursions to develop solutions to problems. Another popular thinking skill program encouraged students to develop similes that used descriptive phrases that magnified the comparison being made, thereby communicating the idea well. The wealth of applications of analogy to idea generation and science understanding make this approach an essential component of science and gifted education.

REFERENCES

Adúriz-Bravo, A., Bonan, L., Galli, L. G., Chion, A. R., & Meinardi, E. (2005). Scientific argumentation in pre-service biology teacher education. *Eurasia Journal of Mathematics, Science and Technology Education, 1*(1), 76–83.

Anderson, J. R. (2005). *Cognitive psychology and its implications* (6th ed.). New York, NY: Worth Publishers.

Apthorp, H. S., Dean, C., & Igel, C. (2012). Using similarities and differences: A meta-analysis of its effect and emergent patterns. *School Science and Mathematics, 112*(4), 204–216.

Arnold, M., & Millar, R. (1996). Exploring the use of analogy in the teaching of heat, temperature and thermal equilibrium. In G. Welford, J. Osborne, & P. Scott (Eds.), *Research in science education in Europe: Current issues and themes*. London, England: Farmer Press.

Ashmann, S. (2009). The pennies-as-electrons analogy: An engaging model helps upper elementary students understand the flow of electricity. *Science and Children, 47*(4), 24–27.

Black, M. (1954). *Metaphor: Proceedings of the Aristotelian Society, 55*, 273–294.

Blanchette, I., & Dunbar, K. (2001). Analogy use in naturalistic settings: The influence of audience, emotion, and goals. *Memory & Cognition, 29*, 730–735.

Brooks, R. A. (1999). *Intelligence without representation. Cambrian intelligence: The early story of the new AI.* Cambridge, MA: MIT Press.

Brown, D. E., & Clement, J. (1989). Overcoming misconceptions via analogical reasoning: Abstract transfer versus explanatory model construction. *Instructional Science*, *18*(4), 237–261.
Casakin, H. (2004). Visual analogy as a cognitive strategy in the design process: Expert versus novice performance. *Journal of Design Research*, *4*(2).
Chiou, G.-L., & Anderson, O. R. (2009). A study of undergraduate physics students' understanding of heat conduction based on mental model theory and an ontology-process analysis. *Science Education*, *94*(5), 825–854.
Chuang, C., Jin, B.-Y., Tsoo, C.-C., Tang, N. Y.-W., Cheung, P. S. M., & Cuccia, L. A. (2012). Molecular modeling of fullerenes with beads. *Journal of Chemical Education*, *89*, 414–416.
Condell, J., Wade, J., Galway, L., McBride, M., Gormley, P., Brennan, J., … Somasundram, T. (2010). Problem solving techniques in cognitive science. *Artificial Intelligence Review*, *34*, 221–234.
Darwin, C. R. (1859). *On the origin of species by means of natural selection, or the preservation of favoured races in the struggle for life* (1st ed.). London, England: John Murray.
Davies, J., Goel, A. K., & Nersessian, N. J. (2009). A computational model of visual analogies in design. *Cognitive Systems Research*, *10*, 204–215.
deGroot, A. D. (1965). *Thought and choice in chess*. The Hague, The Netherlands: Mouton.
Easton, G., & Ormerod, T. C. (2001). *Expert/Novice difference in case analysis: Final report to the European case clearing house*. Retrieved from http://www.ecch.com/files/downloads/research/RP0303M.pdf
Eberle, R. F. (1972). Developing imagination through Scamper. *Journal of Creative Behavior*, *6*, 199–203.
Educational Testing Service. (2012). *G. R. E.* Retrieved from http://www.ets.org/gre
Ericsson, K. A. (1996). The acquisition of expert performance: An introduction to some of the issues. In K. A. Ericsson (Ed.), *The road to excellence: The acquisition of expert performance in the arts, science and games* (pp. 1–50). Mahwah, NJ: Lawrence Erlbaum.
Flannery, M. C. (2009). Mirrors and maps: Two sides of metaphor. *The American Biology Teacher*, *71*(6), 371–374.
Ferguson, E. S. (1992). *Engineering and the mind's eye*. Cambridge, MA: MIT Press.
Foreman, L., & Drummond, M. (2008). Rebel, rebel: James Dyson bucked the odds – Now his vacuum business is cleaning up. *Inventor's Digest*, *24*(6), 18–24.
Gardiner, K. M. (2007).*The future city competition – A successful stem experience for middle school students. Proceedings*, ASEE New England Section, Spring Meeting, University of Rhode Island, Kingston, RI.
Gentner, D., & Lowenstein, J. (2002). Relational language and relational thought. In E. Amsel & J. P Byrnes (Eds.), *Language, literacy, and cognitive development: The development and consequences of symbolic communication* (pp. 87–120). Mahwah, NJ: Erlbaum.
Gentner, D., Loewenstein, J., & Thompson, L. (2003). Learning and transfer: A general role for analogical encoding. *Journal of Educational Psychology*, *95*, 393–408.
Gentner, D., Loewenstein, J., Thompson, L., & Forbus, K. D. (2009). Reviving inert knowledge: Analogical abstraction supports relational retrieval of past events. *Cognitive Science*, *33*, 1343–1382.
Gentner, D., & Markman, A. B. (1997). Structure mapping in analogy and similarity. *American Psychologist*, *52*, 45–56.
Gentner, D., & Medina, J. (1998). Similarity and the development of rules. *Cognition*, *65*, 263–297.
Getzels, J. W., & Jackson, P. W. (1962). *Creativity and intelligence: Explorations with gifted students*. London, England: John Wiley & Sons, Inc.
Gibson, K. (2008). Analogy in scientific argumentation. *Technical Communication Quarterly*, *17*(2), 202–219.
Gilbert, J. K., Boulter, C. J., & Elmer, R. (2000). Positioning models in science education and in design and technology education. In J. K. Gilbert & C. J. Boulter (Eds.), *Developing models in science education* (pp. 3 – 17). Dordrecht, The Netherlands: Kluwer.
Glynn, S. M. (2004). Connect concepts with questions and analogies. In T. R. Koballa & D. J. Tippins (Eds.), *Cases in middles ad secondary science education* (pp. 136–142). Upper Saddle River, NJ: Pearson Education.

Glynn, S. (2007). The teaching with analogies model: Build conceptual bridges with mental models. *Science and Children, 44*(8), 52–55.
Glynn, S. M., Duit, R., & Thiele, R. B. (1995). Teaching science with analogies: A strategy for constructing knowledge. In S. M. Glynn & R. Duit (Eds.), *Learning science in the schools: Research reforming practice* (pp. 247–273). Mahwah, NJ: Erlbaum.
Goldstein, E. B. (2005) *Cognitive psychology: Connecting mind, research, and everyday experience*. Belmont, CA: Thomson Wadsworth.
Gordon, W. J. (1961). *Synectics: The development of creative capacity*. New York, NY: Harper and Row.
Gordon, W. J. J. (1974). *Making it strange* (Books 1–4). New York, NY: Harper and Row.
Grady, K., & Jeanpierre, B. (2011). Population 75 trillion: Cells, organelles, and their functions. *Science Scope, 34*(5), 64–69.
Guerre-Ramos, M. T. (2011). Analogies as tools for meaning making in elementary science education: How do they work in classroom settings? *Eurasian Journal of Mathematics, Science and Technology Education,7*(1), 29–39.
Guilford, J. P. (1986). *Creative talents: Their nature, uses, and development*. Buffalo, NY: Bearly Limited.
Harrison, A., & Treagust, D. (1993). Teaching with analogies: A case study in grade-10 optics. *Journal of Research in Science Teaching, 30*(10), 1291–1307.
Harré, R (1972). *The philosophies of science: An introductory survey*. London, England: Oxford University Press.
Haven, K. (2007). *100 greatest science discoveries of all time*. Westport, CT: Libraries Unlimited.
Hung, S. Y. (2003). Expert versus novice use of executive support systems: An empirical study. *Information and Management, 40*(3), 177–189.
Jakobson, B., & Wickman, P.-O. (2007). Transformation through language use: Children's spontaneous metaphors in elementary school science. *Science and Education, 16*, 267–289.
Kahney, H. (1994). *Problem solving current issues* (2nd ed.). Buckingham, England: Open University Press.
Larkin, J. H., McDermott, J., Simon, D. P., & Simon, H. A. (1980). Models of competence in solving physics problems. *Cognitive Science, 4*, 317–345.
Lightman, A. (2005). Moments of truth. *New Scientist, 188*(2526), 36–41.
Marcelos, M. F., & Nagem, R. L. (2012). Use of the "tree" analogy in evolution teaching by biology teachers. *Science and Education, 21*, 507–541.
Marzano, R., Pickering, D., & Pollock, J. (2001). *Classroom instruction that works: Research-based strategies for increasing student achievement*. Alexandria, VA: Association for Supervision and Curriculum Development.
Michalko, M. (1998). *Cracking creativity: The secrets of creative genius*. Berkeley, CA: Ten Speed Press.
Nassiff, P., & Czerwinski, W. (2012). Modeling atoms using paperclips. *Journal of Chemical Education, 89*, 370–372.
National Committee on Science Education Standards and Assessment and National Research Council. (1996). *National science education standards: Observe, interact, change, learn*. Washington, DC: National Academy Press.
National Engineers Week Future City® Competition. (2011). *Future city competition*. Retrieved from http://futurecity.org/
Nersessian, N. J. (2008). Mental modeling in conceptual change. In S. Vosniadou (Ed.), *International handbook of research in conceptual change* (pp. 391–416). New York, NY: Routledge.
Orgill, M., & Bodners, G. (2007). Locks and keys: An analysis of biochemistry students' use of analogies. *Biochemistry and Molecular Biology Education, 35*(4), 244–254.
Osborn, A. F. (1963). *Applied imagination* (3rd ed.). New York, NY: Scribner.
Perrine, N. E., & Broderden, R. M. (2005). Artistic and scientific creative behavior: Openness and the mediating role of interests. *Journal of Creative Behavior, 39*(4), 217–236.
Pfeffer, J., & Sutton, R. I. (2000). *The knowing–doing gap*. Boston, MA: Harvard Business School Press.
Philbin, T. (2003). *The 100 greatest inventions of all time*. New York, NY: Citadel Press.
PsychCorp. (2011). *Pearson clinical assessment: Miller analogies test*. Retrieved from http://psychcorp.pearsonassessments.com/haiweb/Cultures/en-US/site/Community/PostSecondary/Products/MAT/mathome.htm

Root-Bernstein, R. (2003). The art of innovation: Polymaths and the universality of the creative process. In L. V. Shavinina (Ed.), *The international handbook on innovation* (pp. 267–278). Oxford, England: Elsevier Science, Ltd.
Rule, A. C., & Furletti, C. (2004). Using form and function analogy object boxes to teach human body systems. *School Science and Mathematics, 104*(4), 155–169.
Rule, A. C., Carnicelli, L., & Kane, S. S. (2004). Using poetry to teach about minerals in earth science class. *Journal of Geoscience Education, 52*(1), 10–14.
Rule, A. C., Baldwin, S., & Schell, R. (2008). Second graders learn animal adaptations through form and function analogy object boxes. *International Journal of Science Education, 30* (9), 1159–1182.
Rule, A. C., Baldwin, S., & Schell, R. (2009). Trick-or-treat candy-getters and hornet scare devices: Second graders make creative inventions related to animal adaptations. *Journal of Creative Behavior, 43*(3), 149–168.
Rule, A. C., Zhbanova, K., Hileman Webb, A., Evans, J., Schneider, J. S., Parpucu, H., … Ruan B. (2011). Creative product problem-solving game: Exploring Torrance's creative strengths by making an object from a set of given materials. *ERIC Document Reproduction Service No. ED527045.*
Rule, A. C., Alkouri, Z. A., Criswell, S. J., Evans, J. L., Hileman Webb, A. N., Parpucu, H., … Zhbanova, K. S. (2012). Practicing creative thinking skills by making creative products related to economic issues. *Social Studies Research and Practice*, *7*(3), 47–67.
Schenk, K. D., Vitalari, N. P., & Davis, S. (1998). Differences between novice and expert system analysis: What do we know and what do we do? *Journal of Management of Information Systems,15*(1), 9–50.
Schlichter, C. L., & Palmer, W. R. (1993). *Thinking smart: A primer of the talents unlimited model.* Mansfield Center, CT: Creative Learning Press.
Thagard, P. (2010). Creative combination of representations: Scientific discovery and technological invention. In R. Proctor & E. J. Capaldi (Eds.), *Psychology of science*. Oxford, England: Oxford University Press.
Valle, A., & Callanan, M. A. (2006). Similarity comparisons and relational analogies in parent-child conversations about science topics. *Merrill-Palmer Quarterly, 52*(1), 96–124.
Venville, G., & Treagust, D. (1996). The role of analogies in promoting conceptual change in biology. *Instructional Science, 24*, 295–320.
Wong, E. D. (1993). Understanding the generative capacity of analogies as a tool for explanation. *Journal of Research in Science Teaching, 30*, 1259–1272.
Yilmaz, S., Eryilmas, A., & Geban, O. (2006). Assessing the impact of bridging analogies in mechanics. *School Science and Mathematics, 106*(6), 220–230.

Audrey C. Rule
Department of Curriculum and Instruction
University of Northern Iowa
USA

Benjamin D. Olsen
Center for Educational Transformation
University of Northern Iowa
USA

KEITH S. TABER

18. 'CHEMICAL REACTIONS ARE LIKE HELL BECAUSE…'

Asking Gifted Science Learners to be Creative in a Curriculum Context that Encourages Convergent Thinking

INTRODUCTION

The focus of this chapter is an activity, a science analogy game, introduced as part of a science enrichment programme for 14–15 year old gifted students attending English state schools. The 'game' was designed to be fun, but had a serious rationale. The activity was intended to encourage students to think divergently around school science concepts, and thus to be creative in a science learning context. Creativity is an essential part of the scientific process, and is an important area for development for all learners; but is arguably of particular relevance in identifying and developing those learners who may be labeled as gifted (Kim, 2008; Sternberg, 2010). Yet, arguably, since a mandatory National Curriculum was introduced in England over twenty years ago, English school science provides very limited opportunities for students to demonstrate their creativity, or indeed to even appreciate that science is a creative endeavour (Osborne & Collins, 2001).

The chapter will offer the reader some background regarding the English curriculum context, and the national 'gifted and talented' policy issued to guide schools in working with their most able learners. This provided the context in which the ASCEND ('Able Scientists Collectively Exploring New Demands') project was conceived as an after-school enrichment programme intended to challenge secondary age students. ASCEND was designed around a number of principles relating to a focus on the nature of science, learning through collaborative group work, and encouraging a metacognitive approach to science learning. The rationale and structure of the ASCEND programme will be outlined, before the analogy game, and student responses to the activity, are considered in more detail.

Terms such as 'gifted', 'high achiever', 'high ability' do not have an agreed meaning internationally, and are not always used consistently even within the English context that is the setting for this chapter (Taber, 2007c). The present chapter tends to refer to gifted science learners, meaning those perceived by their schools as being of notable (but not necessarily exceptional) high ability in science, as this is how the term is generally used in the English system.

M. K. Demetrikopoulos & J. L. Pecore (Eds.), Interplay of Creativity and Giftedness in Science, 321–349.

The intention behind the analogy game will be explored, and its basic structure explained. The use of the analogy game within the ASCEND programme will be illustrated through samples of student dialogue. Publication of the ASCEND teaching materials (Taber, 2007b) means that the analogy game is available for adoption, or adaption, by teachers. However, even more importantly, it offers an example of how creativity can be encouraged in the teaching of science at secondary (middle/high) school level.

CREATIVITY IN SCIENCE AND SCIENCE LEARNING

Creating New Scientific Knowledge

Science is a creative process. Scientific discovery relies upon the construction of new ideas, new mental models, new hypotheses, new explanations, new techniques, new instrumentation, new analytical tools, new theories, and so forth. This is widely recognised in the cases of major scientific breakthroughs, and great scientists are seen as creative geniuses alongside composers, novelists, creative artists and so forth (Csikszentmihalyi, 1997; Simonton, 2004). However, it is not just the Newtons, Darwins and Einsteins that are creative. The basic process of science relies upon the power of the human imagination to consider new possibilities that can be conceived, and – if judged promising – tested. The scientific research paper is often a rational reconstruction of the discovery process that focuses on the context of justification – the argument and evidence that supports a new knowledge claim (Medawar, 1963/1990). This might easily lead to the impression that it is the demonstration of a 'proof' of discovery that matters, not how the discovery came about. It is the logical rigour of the argument from observations and measurements to a theoretical claim that is privileged in the traditional scientific report.

This is fine as far as it goes if we recognise that the given purpose of research reports is to make new knowledge claims and that they necessarily focus on the justification of those claims. So although a seed for Einstein's theory of special relativity may well have germinated from the paradox encountered when imagining how light should be understood by a traveler who moved at the same velocity as light (Gutting, 1972), it is the calculations and their eventual relevance to real measurements that persuaded others that Einstien's revolutionary ideas should be taken seriously. It is irrelevant whether August Kekulé, as he claimed (Rothenberg, 1995), really dreamt up the ring structure of benzene, as long as the evidence supported his proposed structure. It matters not if Barbara McClintock could only explain how she came up with the idea of jumping genes in the vague terms of how her brain was able to 'integrate' her observations (Keller, 1983): what matters is whether she offered persuasive evidence that genes do sometimes self-transpose within a genome. Whilst the scientific community was (for some time) unconvinced by her arguments, she could be considered something of an eccentric crank; but once her arguments from research evidence were considered convincing, she was promoted to the status of visionary and Nobel laureate.

Perhaps part of this attitude links to the notion of science as an objective activity: where in principle scientific observers are interchangeable without changing the outcomes of scientific investigations. What one scientist claims is worth little until other researchers can reproduce the original results. The claimed discovery of cold fusion, with its immense potential to overcome a world 'energy crisis' generally ceased to be considered a major scientific breakthrough as it become clear that it could not be replicated reliably in other laboratories.

Unfortunately, just as history is said to be written by the winners, science texts books generally tend to report only what is now generally accepted. Not only do school and college texts usually focus on the 'winners', but they often offer reconstructions of scientists' work that are tided up on the basis of hindsight (Niaz & Rodriguez, 2000). So what is presented is often what we now consider to be the case, based on the decades of careful work that followed (what in retrospect is considered to be) a major breakthrough. So all the uncertainty, controversy and the flaws of early results are ignored in the process of offering students a concise and simple account of what we *now* think, and why we *now* think it. This may be a sensible approach to pedagogy if what we wish to teach is the products of science (the laws, the theories, the models of consensus science): but it leaves a lot to be desired if we hope to teach students something of scientific processes (Lawson, 2010) – let alone suggest something of the thrill and drama of research.

'Admitting' the Subjective into Science

A problem with the creative step is its lack of objectivity. We cannot directly share our imaginings (which draw upon pre-conscious thinking that we cannot even access ourselves by introspection), and we cannot readily replicate the creative process. Major breakthroughs have relied on creative insight that derives from a nexus of the problem-context, the institutional and professional environment, and the cognitive resources of a unique individual thinker (Gardner, 1998; Sternberg, 1993). As Pasteur noted, chance favours the prepared mind, and the experience and learning of each scientist is unique, preparing them to notice specific things and understand them in particular ways. If Benjamin Franklin and Rosalind Franklin had been swapped at birth (using a suitable time machine of course), then it seems very likely that discoveries about both the nature of lightning and the structure of DNA would have been delayed. Interestingly, where most people find special relativity counter-intuitive when they first meet it, Einstein's own path to the discovery was to follow-through on the consequences of his own intuition of the invariance of the speed of light. It has been argued that the personal, implicit knowledge of the individual scientists plays a key role in scientific work as the source of such intuitions, and that this necessarily amounts to an unaccountable subjective element in scientific discovery (Polanyi, 1962/1969).

Yet, even though we do not fully understand the creative process, and even though it is subjective and unpredictable, every scientific discovery ever made (whether of the revolutionary or more typical routine variety) has involved someone imagining a new possibility. The creative process is essential to science, and indeed creativity is the necessary complement to logic, for science to make progress. Kuhn referred to the way progress in science depends upon both tradition and innovation existing in an 'essential tension' (Kuhn, 1959/1977). It may be logical and rational thinking which is so often recognised as essential to science, but without creativity and imagination, logic would have no scientific work to do (Taber, 2011b).

It has long been suggested that school science attracts convergent, rather than divergent, thinkers (Hudson, 1967), but that does not reflect the nature of science itself. We do not need to admit creativity, with its subjective nature, into science, in the sense of letting it in; as it is already 'in' science: being an inherent and essential foundation for any kind of productive science to exist. However, when we teach science, we do need to admit creativity in the sense of acknowledging to our students that creativity plays a major role. We need *to confess* that science has a strong subjective component at the individual level, which provides the raw material that objectivity and logical argument need to work on. If nothing else, this should help our students realise that science is an ongoing and unfinished human adventure with plenty of scope for them to make unique individual contributions.

Creating New Scientific Learning

Now in a sense, what is true of science is true of science learning. The science student, at least at school level, is not expected to help develop new scientific knowledge claims, and rather is generally charged with recapitulating the scientific discoveries that others have already made: developing new personal understanding. However, learning is still about developing new knowledge (albeit personal knowledge), and this requires the ability to imagine possibilities not considered before. The constructivist perspective on learning suggests that each individual learner is set the task of reconstructing, from their own existing mental resources, the ideas valued in the wider community (Taber, 2011a). Yet, of course, if each learner has unique cognitive resources, and constructs their personal knowledge within a unique conceptual ecology, each reconstruction is actually not a replica of the science represented in the curriculum, but more a pastiche or homage.

Indeed, one of the criticisms addressed to learning theorists is to explain the so-called learning paradox (Scardamalia & Bereiter, 2006), that is, how any individual can construct new learning, given that they would not have had the understanding in place to know what they were constructing until they actually formed the new learning. I have never been convinced there is a problem here as the natural world is full of systems that build up iteratively, without following any 'deliberate' (i.e., consciously developed) plan: but even if learning something new is not a paradox, it is certainly a personal achievement.

Learners' potential for creativity in science is hardly in doubt, given the immense challenge faced by science teachers of children who commonly come to class already holding intuitive ideas at odds with science, and then proceed to interpret teaching in all sorts of imaginative, if non-canonical, ways (Duit, 2009). Unfortunately this creativity, often recognised in terms of misconceptions, alternative conceptions, learning difficulties etc, is commonly only seen as a barrier to learning the 'right' ideas (Larkin, 2012). Sometimes this may be a reasonable stance to take, but it has long been argued that learners' non-canonical thinking can often be more helpfully seen as the available resources for new learning (Smith, diSessa, & Roschelle, 1993). If the vast research effort exploring students' ideas in science has shown anything, it is that it is a mistake to assume that learners' alternative ideas about science topics all have the same nature, and play the same potential role in learning (Taber, 2009, 2014). Rather, students' ideas in science vary along a wide range of dimensions, including their openness to modification, and their potential as starting point for learning the scientifically accepted ideas.

Any one who spends time talking to children and young people about their scientific understanding will come across all sorts of creative suggestions. So, Bert, a Y10 (14–15 year old student) suggested to me that bonding was something that arose during the evolution of atoms. Amy, at the same age, suggested that differences in electrical conductivity depended on the density of particles in a material because the densest materials did not leave space for electron flow. A younger student, Jim in Y7, hypothesised that the air hole in the base of a Bunsen burner was there to let dirt out. (These and many other examples of student thinking about science topics can be found at http://www.educ.cam.ac.uk/research/projects/eclipse/index.html).

In each case, it would be easy to simply dismiss the idea as 'wrong' – but such ideas potentially offer the raw material for scientific testing (although Bert's idea might be rather difficult to test in practice). The philosopher of science, Karl Popper, suggested science proceeded through bold conjectures (Popper, 1989): 'bold' because there is little risk in hypothesising something that already seems very likely in terms of existing scientific theories; and so there is likely to be little learned by testing timid hypotheses. When *our students* make bold conjectures, it is too easy to simply see this as getting the science wrong, but we should perhaps do more to acknowledge and encourage their thinking when they actively engage with possible reasons and mechanisms to explain their observations (Taber, 2007a). Creative scientists often have lots of bold ideas, only a few of which bear fruit. We therefore need to encourage science learners to see having ideas as intrinsically a good thing in science, even when those ideas are later found not to be right. Indeed, à la Popper, science is meant to proceed through bold conjectures, *and their refutations* - and the scientist is supposed to celebrate the process that being proved wrong represents. Whilst that prescription seems a little idealistic, it is certainly the case that successful scientists have to produce creative suggestions, and then accept that most will be pruned away in identifying the few that actually take forward our understanding of the world.

If we teach science in a way which implies students' creative ideas are unwelcome distractions or, worse, simply wrong, and so suggest to students that the job of learning science is to just receive the already demonstrated correct ideas, then we do little to either encourage the creative thinkers to become scientists, or to encourage future scientists to think creatively. The present chapter reports on a study from England, where the curriculum has been considered as too crowded with prescribed topics that teachers are required to 'cover' to offer scope for creative activities. Moreover, gifted education in English schools is generally conceptualised in simplistic terms (Taber, 2007c), and, for historical reasons, expertise in gifted science education is rather thinly spread in the state schools.

REPRODUCTION RATHER THAN CREATION: LEARNING SCIENCE IN ENGLISH SECONDARY SCHOOLS

England has a national Education system, with a good deal of local variation. The wider UK context is more complex, with Northern Ireland and especially Wales having much in common with England, but Scotland having quite a different school system. The project discussed here, ASCEND, took place in England, and discussion here is limited to that country. The vast majority of school-age students in England attend state maintained schools, that is schools funded from the public purse. All children of school-age are entitled to a place at a state school. Approximately 7% of children are educated outside of the state system in independent, 'fee-paying', schools that are subject to some of the same regulations as state schools, but are exempt from others. So, for example, whilst anyone working as a teacher in any school in England has to be vetted to check whether they have a criminal record involving an offence which would deem them unsuitable to work with young people, independent schools are otherwise free to employ who they see fit as teachers, whereas state schools are usually expected to employ qualified teachers who have completed a nationally recognised course of training. At the time of the ASCEND project, state schools were required to follow a national curriculum established by legal statute, whereas independent schools were largely free to set their own curriculum (Since the project was carried out, the requirement to follow the National Curriculum has been relaxed for some, but not all, state schools and increasingly some state schools are being allowed to put aside the requirement to only).

Whilst the independent sector educates a minority of students nationally, it is worth reflecting that traditionally a disproportionate number of those who achieve high office in the UK are educated in the independent sector. For example, a very high proportion of cabinet ministers and senior civil servants, and high achievers in other public spheres, are privately educated. A simple interpretation might be that independent schools offer a 'better' education, but that of course is over-simplistic. Certainly some well-established private schools have excellent records of getting students into Oxbridge (i.e. Oxford and Cambridge) and other prestigious Universities, which is often a major step to later career success. But, of course, these top independent schools are highly selective. Students generally obtain places at independent schools because either their parents are paying a hefty fee, or because they have won competitive scholarships demonstrating they are very high achievers who will likely attain at high levels in public examinations. Arguably, disposable income to send your children to private school, and aspiring to have your children educated away from the proletariat, are both factors linked to the remnants of a class system that dominated social structures in Britain for many centuries.

Gifted Education Moving Out of, and Back into, Educational Fashion

This 'political' aside is necessary to understand the nature of the state school system in England, for it has two consequences. One of these related to 'creaming': that in many state schools a disproportionate number of the most able students who might have attended that school have moved out of the state system, leaving a somewhat skewed distribution of ability among the school's student population. In most cases this is not extreme, yet if the most able students (who by definition only make up a small proportion of the cohort) benefit from time working with their similarly able peers, then even the loss of small numbers of such youngsters can significantly reduce the cadre of 'gifted' students in any year group in a state school.

The second issue concerns the acceptability of focusing attention and other resource on 'gifted' learners in state schools. In the 1960s, gifted education was a focus of some research attention (Fisher, 1969), and in the last decade or so, gifted education has become a policy issue that all state schools are required to respond to (DfES, 2002; The National Strategies, 2008). However, between these periods there was tendency for this issue to become unfashionable, and even to be seen as 'politically incorrect'.

The secondary school system established in England at the end of the 1939-1945 'World War' had set up discrete types of schools for students with different aptitudes, determined by an academic test taken by all students at eleven years of age. Passing that test meant admission to what was termed a grammar school, preparation for public examinations, and a good chance of progression to university education - and so to the professions and other well-paid 'white-collar' positions. For many years, failing a test taken on one day at primary school meant attending a secondary school where public examinations were not taken and there was no provision for study to university entrance level. Over time it became very clear not only that failing the 'eleven-plus' made it extremely unlikely a student could ever progress to higher education or the professions unless the family could afford private education; but also that testing pupils at age eleven meant sorting them by family background and associated social capital as much as sorting them in terms of true academic potential.

This led to a major shift, largely in the 1970s, in most parts of England, to 'comprehensive' secondary schools that accepted all local students regardless of primary school achievement (Crook, 2002). Moreover, given the dangers of early (or perhaps any) labeling of students by ability, there was a strong movement toward teaching largely in 'mixed-ability' classes, rather than streaming students within schools according to perceived ability. Awareness of the potential implications of premature labeling of students by ability, and the dangers of Hawthorne-type (Rosenthal & Jacobson, 1970) effects (with ability labels becoming self-fulfilling prophecies), made many working in the comprehensive school system wary of any kind of sorting of students by ability (White, 1987).

More recently the comprehensive system has suffered a great deal of tinkering by successive governments attempting to offer parental choice between schools, and significantly, setting students into ability-based classes within specific subjects has now become the norm in most school subjects, at least in the senior secondary years. Indeed the party in government has suggested that they wish to offer 'guidance' to the school inspectorate "to ensure that schools – particularly those not performing at high levels – set all academic subjects by ability" (Conservatives, not dated, p. 33). Just as any kind of ability labelling became suspect for some decades, the once ideologically correct notion of mixed-ability teaching has now come to be seen as something radical and almost subversive: too liberal; too progressive; too egalitarian; too relaxed. Such are the whims within an education system directed centrally by whichever ideologically motivated political party is in government.

There are clearly some important and complex issues here which cannot be treated in depth in the present chapter, but it was against the background of 'comprehensivisation' and the widespread adoption of (what was then considered) ideologically sound mixed-ability teaching that an interest in the learning of gifted students in state schools seems to have been widely seen as unsound (Boaler, Wiliam, & Brown, 2000). After all, it was sometimes suggested, all students have their gifts. And surely the most able learners already have advantages, so limited resources should be focused on those who will not succeed without extra support. No doubt, for some, there was also a sense that the 'gifts' of gifted students were often largely down to luck in their upbringing, and therefore there was a social imperative to attempt to make up for such inequity by offering some form of compensation for the less lucky.

This is not the place to consider the merits of these arguments in any depth, but there was clearly a rather major flaw in the social equity stance when the most able students in state schools were not getting special attention, but would have to later compete for university places and employment with students attending prestigious fee paying schools (with their smaller classes and myriad extra curricular opportunities), which were largely able to ensure their student body consisted only of those youngsters judged to be academically strong. An alternative egalitarian argument might be that education should offer all learners the best possibility of meeting their true potential, and gifted learners are unlikely to be supported by being assumed to have the same needs as all their classmates (Reis & Renzulli, 2010; Taber, 2015). There is of course also an economic argument for public investment in meeting the needs of gifted learners, if it considered that those gifted learners will potentially offer the greatest return on public spending on their education, because they may pay back to society in disproportionate ways through wealth creation and other significant contributions.

A National Policy on 'Gifted and Talented'

By the time of the ASCEND project, there had been a significant shift in attitudes to the gifted. The government had introduced a 'gifted and talented' (G&T) policy, first in particular geographical areas (generally cities with relatively high levels of social deprivation), and then nationally. In the language of G&T, students could be gifted in one or more academic subjects, such as science, and/or talented in subjects such as the creative arts or sport. The G&T policy was both simplistic and simple. It was simple in that all state schools would be expected to show that they had a register of G&T students and could account for how they were providing suitable provision for their G&T cohort. The policy was simplistic in that it defined students as gifted if they were in the top 5-10% of students in that school, according to whatever measures might be seen as appropriate. Given the near-dearth of decent research into gifted provision in England for many years, and the lack of capacity in gifted education within the state sector, a good deal of guidance was provided on how to identify gifted learners, but much of this was vague, general, aspirational, and untested (Taber, 2007c).

Moreover, having identified the gifted students, there was limited expertise among science teachers about what to do with them, apart from expect them to be performing especially well when studying the National Curriculum. Yet that National Curriculum was based on a 'one-size-fits-most' approach that assumed the same basic curriculum was appropriate for the majority of students, with just a small proportion of secondary students adding a bit more of the same (even more topics, rather than seeking more advanced understanding) and another small proportion taking a reduced spread of topics to either address limited attainment in the subject, or to make room for extra study elsewhere (DfEE/QCA, 1999).

The curriculum was a fairly comprehensive tour of most important areas of biology, chemistry, physics and some earth/space science. It offered a broad view of the scope of the sciences, but limited opportunities for studying any particular topics in depth - and virtually no opportunities for extended enquiry work within normal curriculum time, as commonly happens in some parts of the United States for example (Eilam, 2008). The National Curriculum was linked to an increasingly specified, high stakes, public examination system that encouraged teaching for the test, but had little scope for questions that might expect creativity from students – and so require markers to think beyond a highly structured mark scheme (Collins, Reiss, & Stobart, 2010). This pattern has also been noted in other National contexts, such as the US (Longo, 2010).

Whilst schemes had long existed to support teachers in helping students to undertake creative project work in science (Taber & Cole, 2010; West, 2007), the perceptions of the requirements of the curriculum, and its associated assessment regime, meant that these were often only adopted for occasional use, or as an extra-curricular option where teachers were especially keen to offer additional opportunities.

It was also increasingly recognised that although the national curriculum for science had been intended to help learners appreciate the nature of science, as well as learn about specific science topics, in practice students generally learnt very little about the processes of science beyond a formulaic and simplistic approach to fair testing (Taber, 2008). In a 'investigation' for their school science examination course work, a student learning science under the English National Curriculum might typically have decided to test whether the concentration or temperature of an acid influences the rate of its reaction with some magnesium ribbon – knowing full well that both factors should make a difference. Already knowing the answer hardly provides an authentic experience of enquiry, even if it helps students get good marks for their practical work. Attempts to modify curriculum and assessment to address this issue were recognised as making limited progress – partly because teachers generally felt they lacked subject knowledge in this area themselves, and because they felt unsupported by available school text books or suitable teaching resources. Inadvertently, a system seems to have developed which provides a science education where the conscientious and studious can do well, but where the creative gifted learner with a tendency to think divergently is likely to both struggle to perform to their potential, and indeed struggle to find the kind of challenge likely to engage their interest.

Meeting the Needs of the Most Able in Science

This set of circumstances left many science teachers in state schools concerned about what they should be doing to support their most able learners, beyond helping them obtain good grades in the formal examinations. This set of circumstances motivated a project (organised with Prof. John Gilbert, then at Reading University, and Prof. Mike Watts, then at Roehampton University, now at Brunel University) labelled as APECS (Able Pupils Experiencing Challenging Science), to explore these issues. With some modest funding from the University of Cambridge Faculty of Education's Research Development Fund, a seminar series was established on *Meeting the Needs of the Most Able in Science.*

The seminars involved teachers as well as academics and students, and explored various aspects of teaching science to the most able pupils, with a particular focus on how such learners could be challenged in the context of state schools charged with 'covering' the national curriculum requirements. This led to the publication of an edited volume, with contributions primarily based upon the ideas explored in the seminar series (Taber, 2007d). An opportunity to secure a small award to fund a project on 'teaching ideas and evidence in science' for the 11–14 year age group supported an initiative to help trainee teachers develop teaching about an aspect of the nature of science whilst on placements in schools, with a particular (but not exclusive) focus on the most able learners. This project was partially funded by a government agency, and partly by an educational charity, the Gatsby Science Enhancement Programme (SEP). SEP were open to considering other similar projects, and this coincided with the possibility of working with the state secondary schools in Cambridge (England), who had developed a confederation to share ideas and experience. This was the background for ASCEND.

ASCEND: A PARTNERSHIP PROJECT TO SUPPORT SCHOOLS IN PROVISION FOR THEIR GIFTED SCIENCE STUDENTS

ASCEND was conceived as a way to support the local schools in working together, by offering a programme of after-school extra-curricular science sessions in a University context for 14–15 year old students. The invitation to schools to join the project clearly indicated that this was seen as enrichment for gifted science learners, but it was left to each school to nominate the students considered most likely to benefit from attending. This was in keeping with national policy that required the schools to identify their own 'gifted' cohorts. The funding from SEP supported operation of the programme (employing graduate students to support the sessions), and developing materials which could then be made available to the partner schools, as well as to any other schools and teachers who might wish to use them (Taber, 2007b). By working in partnership with four local schools it was possible to allow the students to work in the context of a relatively large group of like-minded individuals. The ASCEND cohort was about the size of a typical secondary school science class, whereas within each school the number of students considered 'gifted' in science in any year group was necessarily limited. The programme was designed to incorporate a number of features:

- Group work
- Metacognition
- The nature of science
- Conference format

Group Work

Nearly all of the activities were designed to be carried-out in groups of about 4 students. We also encouraged the students to work in groups composed of students from more than one school, so they would meet and work with new potential friends. This approach gave students opportunities both to explain their own ideas to an audience likely to be thinking at a similar level, and to have their ideas questioned and challenged in ways that were less likely to occur when students were normally working with less able peers.

Metacognition

Gifted students should be expected to demonstrate high levels of metacognitive ability, although in this, as in other aspects of cognitive development, learners are only likely to reach their potential when given suitable opportunities to practice their skills. Given the nature of school science in England (see above), and the normal pattern of lessons being broken into short, structured activities (something that tended to be seen as required by an inspection regime much concerned with lesson 'pace') it was anticipated that students would not be regularly asked to be independent (or even collaborative) learners in their normal science classes. This was confirmed in the feedback from the students attending the sessions (Taber & Riga, 2006).

Most ASCEND activities were designed to require groups to plan their actions, and to then regulate their work over a period of about an hour. Plenty of support was available if needed, but the 'staff' (graduate students who were either preparing for teaching or undertaking research degrees) were briefed to act primarily as observers, and to only intervene when asked, or if it was clear a group was making no progress. Again, the feedback from the students made it clear they were not used to being given this level of responsibility for learning in their science lessons and nor were they used to being allowed to develop and explore ideas for any extended period without regular scrutiny and guidance from teachers (Taber & Riga, 2006).

Most of the ASCEND activities were designed to have no single right answer, so students were expected to be able to evaluate their own progress and achievements. One of the activities asked the groups to work with a large amount of learning resource material to develop their own model/representation related to studying and learning. The time available did not allow a close study of all the source material, especially if all the students wished to read everything provided (rather than adopt a strategy to divide up the material).

Another activity asked students to determine which activities and occupations counted as scientific: the focus was on the use of criteria and argumentation, not on the actual decisions reached. Another activity offered three simplified models of the processes by which science progresses, and asked students to identify aspects of these 'philosophies' of science in brief vignettes of the work of well known scientists. Most of the examples did not clearly fit just one of the models. Another activity asked groups to act as an interdisciplinary team of scientists building up a synthetic model of plant nutrition drawing upon information for biology, chemistry and physics, and allowed the groups freedom in how they chose to represent their model. The idea that gifted learners appreciated having some level of choice in learning activities had been something highlighted in the earlier APECS project (Taber, 2007a). These, and the other activities, are detailed in the book and teaching materials published by SEP (Taber, 2007b). It was hoped that this type of approach would offer students the opportunity to demonstrate their creativity within the context of science learning.

The Nature of Science

The theme of the nature of science (NOS) was selected as a suitable theme for most of the activities (Taber & Riga, 2006). This was selected partly because it was known that this was generally a weak area in student learning in England (as suggested above), but also because it was felt to offer contexts for challenging tasks, suitable for gifted learners. Contexts included: the demarcation of science, the nature of scientific method, the nature of scientific laws, the criteria for a good scientific explanation, the nature of scientific models and so forth, all linked with the school science curriculum, whilst offering the opportunity to tackle tasks that would be considered too complex, too advanced, too abstract for use in classrooms with many students of 14–15.

It also allowed the selection of material that would not be met within the normal curriculum, whilst still being relevant to the objectives of secondary school science such as introducing historical examples that would not normally be met, or taking challenging topics that were featured in the curriculum (chemical bonding, natural selection, photosynthesis), but asking students to discuss in depth abstract or complex features that would be only touched upon in school. That is, the sessions offered enrichment not in terms of these topics, but in terms of the abstract, theoretical treatment of the topics, and so the level of intellectual demand.

Conference Format

A final feature of the programme was having the students be treated as adults. These 14–15 year olds were accustomed to being considered as minors in school, but they were referred to as 'delegates' at ASCEND. Anyone who has taught students of this age knows most have considerable potential to behave as children or as adults depending upon the circumstances they find themselves in. The delegates arrived (often by bus, foot or on their bikes) to be met by a conference style registration, and to be given their delegate badge and conference pack (with materials they would need for that day's activities). They would then be able to claim refreshments from the Faculty cafeteria on showing their delegate badge. As most students were coming straight from school (and would arrive at different times depending on the school) this was a pragmatic approach, as well as a deliberate attempt to get the delegates to feel they were in an adult study environment.

THE ANALOGY GAME

The analogy game (Taber, 2007b) was designed to be played during the registration/refreshments period at the last session in the programme, before moving to the Science Education Centre for the main activity. It was meant to link to the NOS theme by indicating, in a fun way, that creativity was something to be valued in science, and to give delegates the opportunity to offer lateral, including humorous and perhaps even irreverent, thoughts relating to science concepts, in a context where this was seen as something positive. It also built upon the idea of analogy being important in science, that had been incorporated in an earlier ASCEND session.

Analogy, Discovery and Learning

Analogy is at the heart of major processes by which scientists develop new ideas, and by which learners acquire new understandings. In science, the importance of analogy in making discoveries has been recognised (Nersessian, 2008). If we notice that a system in which we are interested is somewhat like another system we already know about, then we can produce a mental mapping relating the two: drawing upon the known structure of the familiar to conjecture possible relationships about the system we are enquiring into.

In effect there is a multi-stage process here, the first part of which is to form a simile (to notice something is in a sense like something else). So, for example, Lise Meitner and her nephew Otto Robert Frisch, 're-cognised' the fission of a heavy nucleus as being like a water drop breaking into two smaller drops (Frisch, 1979). The nucleus was not a water drop, but that simile had potential for developing constructive hypotheses about nuclear fission. The simile of a nucleus being like a liquid particle in the way they could both divide into two smaller parts was a productive starting point for developing new ideas about the nucleus that could then be tested against empirical data (Meitner & Frisch, 1939).

In a similar way, teachers use such comparisons to help learners become familiar with the unfamiliar. A classic, if flawed, example is the teaching analogy of the atom being like a tiny solar system (Taber, 2013). The logic is that the student may not know what an atom is like, but may be familiar with the general structure of the solar system. The analogy is in terms of structural similarities between the two systems: e.g. smaller bodies considered to orbit a large central mass (Nakiboglu & Taber, 2013). The analogy goes beyond a simple similarity – but allows a mapping of structural features of the two systems (i.e. the relationships between parts of the systems). For example, if we know that a force is needed to keep the planets in orbit around a star, we might conjecture that similarly a force might be needed to keep electrons in orbit about a nucleus, in the planetary model of the atom. Of course, we might then conjecture – as many students seem to – that gravity provides centripetal force in the atom, as it does in the solar system (Taber, 2013). Analogy offers possibilities to consider: it does not assure us of drawing correct deductions. However, it can be fruitful in science: if we know that oscillations in a liquid drop can lead to the drop dividing, we may ask the question 'is it possible that some kind of similar oscillation process is going on in the nucleus that leads to it dividing?' (see Figure 1).

In an earlier ASCEND session students had undertaken some simple laboratory work to explore the patterns of discharge of a capacitor, water level evening out between two connected burettes, and cooling of hot water. They were also given

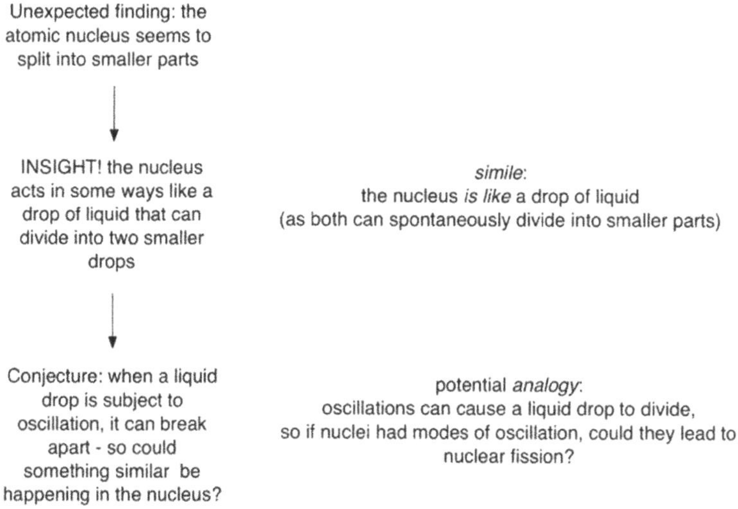

Figure 1. An example of the role of analogy as a creative process in science

information about the pattern of radioactive decay, and about negative feedback in a system where the driver for change is itself reduced by the change it drives. (For example, the rate of cooling of hot water depends upon the temperature difference between the water and the surroundings. The greater the temperature difference, the faster the water cools – but as the water cools the temperature difference is reduced, which reduces the rate at which the water cools, etc.) That is, the ASCEND delegates were in effect introduced to a number of systems that exhibited exponential decay characteristics, and where the pattern of changes observed could be explained in analogous terms (Taber, 2011c). Having experienced how analogies between physical systems can be productive ways of thinking in science, the analogy game offered the students a chance to generate their own analogies.

Generating Similes for Scientific Concepts

The analogy game is a card game in which each player is dealt cards, and seeks to win the game by laying down their cards first. There are two types of cards: science concept cards (with the name of a science concept that should be familiar from school science), and analog cards (which had the name of everyday objects, abstract ideas, or then current celebrities – as well as some 'wild' cards where students could offer their own analog term). Players took turns in playing, and on their turn were challenged to make an analogy between one of their concept cards and one of the analog cards. Players who could not make an analogy on their turn were able to swap some cards with the pack, but had to wait for the others to play before it was their turn again.

The materials used in the ASCEND session are included in the SEP publication (Taber, 2007b), but the activity could readily be modified (e.g. numbers of card originally dealt; precise terms used on the two decks of cards etc) to make it suitable for students following a different curriculum or of different ages. What was essential to the spirit of the game was that in laying down an analogy, the players had to have a reason as the other players could challenge a proposed analogy. Students were asked to explain "the [concept] is like the [analog] because…". So a student could not pair, say electricity with love without explaining their analogy: e.g. electricity is like love because both can give you a tingly feeling.

So the game incorporated a degree of peer review (something that might be considered an implicit NOS feature perhaps) in that it was for the player wishing to lay down cards to persuade the other players in their game to accept the analogy. Now, of course, this opens up possibilities for the game to be undermined in several ways. Friendship ties and perceived social status could influence how such justifications are achieved, and there is also scope for game strategies à la prisoner's dilemma. As the game was intended as a fun activity, with little at stake, these 'threats to validity' were not considered to be serious concerns. The two key features were that students were being asked to explicitly think analogically in a science education context (providing a context for highlighting the importance of thinking creatively in the processes of science itself), and that they were being asked to justify their suggestions (potentially linking to the issue of what makes a good explanation, something that had been the focus of another earlier session in the ASCEND programme). In this simple fun activity, there was potential to reflect on both the context of discovery (imagining a novel similarity) and the context of justification (explaining why a scientific idea is like an everyday idea or entity): in the same way that the creative and the logical are both essential in science itself (Taber, 2011b).

Playing the Analogy Game

Digital voice recorders were used to record two of the groups playing the game (with the knowledge and permission of the players – recorders were not set up where delegates had reservations). Modern digital voice recorders are very effective at recording sound, but when collecting group talk, especially in a public place like a cafeteria where there are various conversations underway, it is often not possible to obtain complete transcripts from recordings. That was the case here. However, the partial transcripts produced from the two groups certainly gave a strong flavour of the activity, and are drawn upon in the account below.

IMAGINING AND JUSTIFYING SIMILARITY

Observations of the ASCEND delegates playing the analogy game, supported by transcriptions from the two recorded groups, suggested that most of these 'gifted' science students appreciated the rationale of the game quite quickly, and they were generally very willing to enter into the spirit of finding points of similarity between science concepts and everyday entities. Whether or not these students would have questioned the value or point of the activity had it been presented in the context of a formal science class, they seemed to take to the challenge in the context of the ASCEND enrichment programme. The similarities that were suggested, at least those identified from the clearly audible parts of the recordings, were quite diverse in nature – and often not well thought through. However, students were being asked to produce similarities spontaneously without preparation, and without guidance on what kind of similarity they might look for – in others words, in common with most of the ASCEND activities, there was considerably less imposed structure than they would expect in a science learning activity they were asked to undertake in their school science. The analogy game had been set up with necessary structure in terms of procedure (as there was not time for the students to devise and negotiate their own rules), but without the use of examples or indeed any formal teacher introduction.

To some extent this fitted the general approach of ASCEND to provide a learning experience that contrasted with the high level of scaffolding of tasks typical of school science, and rather to ask students to take more initiative. However, there was also the pragmatic consideration, that delegates were arriving unevenly, and the game was being played in the public context of a Faculty cafeteria that prevented any formal or extended introduction to the activity. This setting for the activity was therefore not ideal, and likely impacted the quality of the productions the delegates were able to offer.

Sharing the Creative Act

The first comparison offered in one of the recorded groups was that *cola is like a chemical reaction.* After this suggestion was offered by one of the girls, another player questioned whether saying 'cola is like a chemical reaction' was the same as 'a chemical reaction is like cola' (i.e. the task was to find analogies for the scientific concepts, not scientific analogies for the everyday items), but another player adjudicated that 'it's the same thing'. The justification for this first stab at a creative analogy was:

> cola is like a chemical reaction because it's fizzy and it bubbles...and some people can drink absolutely loads of it and some people can have absolutely loads of it and other people can't and also ...um people...and people can never...people can never tell whether they've got diet coke or real coke and (unclear) in chemical reactions you've got...and in chemical reactions you can never quite tell what elements are there unless you're really, really good at it.

There seem to be three phases to this proposed analogy. The first comparison (fizzing and bubbling) was a similarity, but was not developed. The delegate quickly moved on to a new suggestion related to how much cola a person can drink, but this did not seem to have been thought through (we might say the suggestion fizzled out), and she shifted to a third aspect of the justification. So a chemical reaction was like cola because people could not tell what type of cola they were drinking, and in a similar way one had to be very good at chemistry to know which elements were present during a chemical reaction.

The rest of her group seemed happy enough with this suggestion, and the group moved on. The next suggestion was that "the light is like money because you never know how much you've got till you're sitting...till you're sitting in the dark without it". The next suggestion, prefixed by the qualification "okay, this is slightly rubbish", was that "a molecule is like Africa because...Africa is one little thing in a mass of other countries and big things that makes up the world".

After a short pause one of the girls suggested that "DNA is like high-heels... um...every woman's got a pair and...okay...an ionic bond is like love because it's all about two things being joined". This was greeted with an exclamation of "that's good!" from one of the boys in the group, leading to the modest self-evaluation that "I got the cushy one, that was easy". The other players accepted the bond-love analogy for the bond, perhaps filling in the mapping (the elements as the pair of lovers) for themselves (see Figure 2), whereas the DNA-shoes suggestion did not seem to offer the player potential for moving beyond finding a similarity:

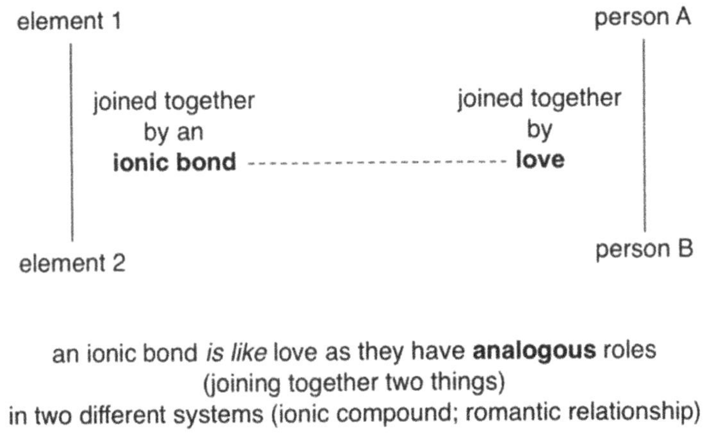

Figure 2. An analogy offered by ASCEND participant

The very next suggestion offered a more explicit mapping: "the nucleus is...like the brain because the nucleus controls what the cell (does) and the brain controls what we (do)". This is a familiar teaching analogy, but was received as though novel to these students ("yes...sure!"). Later in the game, another analogy familiar to teachers was offered ("a cell is like a brick...they're used to build up the body"), but even if these examples might have been ones the delegates had met in their science lessons (perhaps recalled without being consciously aware of this), most of their suggestions seemed original.

The next comparison was that "reproduction is like Versace because...I presume Versace started as something very small...and has grown and multiplied, reproduced, big". One of the girls in the group then suggested that bicarbonate and alkali were analogous, based on bicarbonate being used in the same way as an alkali. At this point, the author intervened

Girl taking her turn:	Bicarbonate...I think you put bicarbonate of soda on a wasp's sting so the wasp's sting must be acid.
Keith:	But isn't that saying something *is* an alkali, not *like* an alkali?
Players:	Mm.
Keith:	What do you know about alkalis?
Girl taking her turn:	They neutralise...
Keith:	Okay, can you think of something else that neutralises something?
	(No response.)
Keith:	...doesn't have to be a chemical substance...what might neutralise something?
Other girl:	Aah...a peacemaker.
	[Some chatter and laughter...]
Other girl:	The United Nations are like an alkali...they neutralise...disagreement in the world...not always successfully.
Next girl taking her turn:	Okay...an electric current is like a smile because from the right charge and through the right battery it ends up brightening up your day. [Then after group laughter]. Sorry that was very cheesy, I know!

This new suggestion was not spelt out in detail, but seemed to be appreciated by the others in the group, and seemed to play upon the literal and metaphorical meanings of 'brighten'. The group continued to play the game in this spirit. Some of their suggestions' continued to be underdeveloped: "velocity is like a courier...velocity is like speed with direction so it's like couriering the...whatever...in the direction of the speed"; an unclear suggestion along the lines that carbohydrates are like sandals because carbohydrates provide energy, and energy is needed to walk around in sandals; an apparent attempt to suggest waste is like bananas as many bananas are wasted in their countries of origin. However, other offerings were both structured as analogies, and creative:

- "condensing is like death...it's where you change from one state to another... from living to dead, from gas to liquid...I win...ha ha!"
- "acceleration is like...holidays...because acceleration is a change of speed... [and] holidays is a change of...place"

Prioritising Justification

If anything, the group's progress was perhaps limited by their apparent reluctance to challenge each other to explain their proposed analogies. The situation was very different in a second group, where the players were much more argumentative, and seemed more competitive in playing the game. This is clear from the very beginning of the transcript where one student is arguing that he can justify his mooted analogy "because they're both written on little rectangles of paper". There then followed debate about whether this was allowed within the rules of the game. The author was asked to arbitrate and limited input to pointing out that it could be a very boring game if it proceeded on the basis of accepting any two terms as analogous simply because they were both written on rectangular cards. The next suggestion was more in the spirit of the game, if not presented as a comparison: "I've got one – you need Energy to run a Control Center". This suggestion was rightly challenged on the grounds that "you've got to say why [it's] like the Control Center". An initial attempt at a response to this challenge was interrupted before it could be developed. After some squabbling over whether they were now experiencing some tit-for-tat challenges, the group went back to the instruction card and re-read (or perhaps, read) the rules. This led to a reformulation of the analogy as "Control Centers make things work and Energy makes things work". This was accepted by the other players.

The next suggestion was that 'electric current is a flow of electrons' The group considered whether to accept this suggestion, and one of the other players agreed, before another then challenged: "well, no, what's the analogy?" This led to a disagreement over whether *being the same* counted as *being alike*. It was eventually decided that identity did not count as similarity in the context of the game. The next mooted analogy was that 'Smith is like an atom because both are extremely common'. After this straightforward comparison, the next suggestion was somewhat more involved: that 'string is like acceleration [because] when things accelerate and you look at them through a camera they go like blurred and long and string is blurred and long'. Whilst somewhat convoluted (as with many of the suggestions this had the flavour of ideas being developed as spoken), this example revealed a creative link between two quite different ideas (acceleration and string).

The next comparison suggested is of particular interest given the competitive way the group had entered into the activity, a one player offered a suggestion that another member of the group developed.

Player 1: Here we go here we go…a bible…is like a molecule…both contain (lots of) information
Player 2: How does a molecule contain information?
Player 1: Molecules contain loads of atoms…proven!
Player 2: I've got a good one I've got a good one…for that.
Player 1: Yeah, go on then.

Player 2: A molecule is a complex arrangement of atoms and a bible is a complex arrangement of stories...and books and things.
Player 1: That's quite a good analogy.

Here there seemed to be a genuine co-construction of an analogy, coinciding with a portion of the activity where the students worked together rather than argued. Unfortunately, the next mooted idea that "rules are like light because...rules make up light basically." became bogged down in a dispute over whether rules were the same as laws, before ending in a claimed similarity that "you can break light and you can break rules". Further similarities suggested were that "combustion is like a fire because fire is a type of combustion"; that "evaporation is like fuming [as] they both make steam"; "adrenalin is like a brick because when they get up too high they start falling"; and "string is like carbohydrate [as they were both] long...strands". The latter suggestion was challenged, and in response was further specified as "starch is a carbohydrate and starch is a long thin strand and string is a long thin strand".

There was then an extended argument about whether falling could be compared to acceleration. Interestingly, the point of dispute seemed to relate to whether acceleration was inherently limited. One of the teaching assistants intervened in the argument to ask the groups about their understanding of analogy, and was told they were looking for "a similarity" where "something's like something". Spurred on by this the boy proposing the acceleration-falling analogy explained: "I said they're similar because in both of them you're accelerating to a point...I'm sure when you're accelerating you'll stop eventually...and when you're falling". At one level acceleration and falling may seem too similar for this to be considered as a creative link, but it seemed the intended comparison was between terminal motion when falling, and the limits on the velocity of a massive body due to relativistic effects, which potentially at least made an interesting comparison: "if you accelerate fast enough, don't you go as fast as the speed of light and then you go backwards which means that you must've stopped accelerating at some point and reversed direction", and that at this point "some scientists think that everything goes black". Whilst the latter part of this suggestion appeared to reflect an alternative conception, the initial basic comparison seemed sound.

After this rather extended discussion, one of the delegates made a new suggestion that "a cell is like an ant...small things but when they work together they can make up organs...the ants when they work together...". The suggestion was interrupted, but accepted, and it was observed that "the [thing] about this game is that you can make a link with basically anything". The potentially strong analogy here (cells are like ants as cells work together in an organ, and ants work together - in a colony presumably), was not explored further, as one of the delegates had moved on to suggestion that testosterone was like an agent because they both made things happen. The group had time for one more suggestion before the activity was wound up to allow delegates to move from the cafeteria to the Science Education Centre for the main activity for the evening. It was suggested that "chemical reactions are like hell because the common theory of hell is lots of fire…and burning", as "many chemical reactions were caused by a burning heat". In response to a query from another player, the proposing student agreed that he was indeed "saying that hell is exothermic".

DISCUSSION

Generally the students responded to the activity with enthusiasm, and most groups became engaged in the game. As well as several groups of students, a group of accompanying teachers played the game with one of the graduate assistants at one table in the café, experiencing first hand the challenge involved. He later reported:

> The students enjoyed it a lot and it made them question their understanding of a topic. It's very easy to think analogies are singular and absolute. Electricity is like water, chemical reactions like competing for friends etc. It was good to hear some really creative new analogies. Even at the [teacher's] table, we were really stretched and forced to examine some of the concepts more closely than we had before.

The dialogues recorded as the two groups of 14–15 years played the analogy game are certainly intriguing. An issue raised earlier in the Chapter was the nature of 'giftedness' in the English curriculum context. English schools are required to identify a certain proportion of their students as gifted, but largely left to devise or adopt whatever local criteria they feel fit. The students who attended ASCEND were certainly above-average attainers in science, but probably reflected a much broader range of ability or attainment than would be signified by the 'gifted' label in many other educational contexts. However, there are hints here of the kinds of exchanges expected of gifted learners. In both groups there were attempts to clarify the scope of what was allowed within the game: was the analogy commutable (if cola is like a chemical reaction, can we assume that a chemical reaction is like cola)? Could identity ('electric current is a flow of electrons') or class membership ('fire is a type of combustion') be counted as a suitable kind of similarity? The suggestion that *being represented by being written on cards* might be a sufficient ground for analogy itself displayed an ability to 'think out of the box'. Some of the suggestions certainly demonstrated a degree of ingenuity: such as considering how things appeared when filmed whilst moving fast as the basis for a comparison.

Evaluating and Developing the Activity

Some of the suggestions offered by the students seem far from ideal, and this raises an interesting point about the activity. The students were provided with both the science concepts for which analogies were to be found, and a pool of potential analogues. Although some 'wildcards' were included for students to make their own suggestions, most of the time the delegates were trying to force some kind of comparison from their hand of five science concepts, and seven potential analogies they had been dealt. This explains the odd nature of some of the choices, and raises the issue of what students might have thought of if they were required simply to offer their own everyday comparisons. That would perhaps encourage more originality and so creativity (although also admit disputes about what counted as an everyday idea or entity), and so might be a more productive learning activity. In the context of meeting the game for the first time, the inclusion of a pool of potential analogues was probably sensible, but a development suitable for those who have played the game this way would be to require players to offer their own analogues.

A clear limitation of the game as presented was that it had not been sufficiently set up to ensure that only clear analogies, rather than similarities, would be seen as acceptable moves. Whilst spotting a previously unnoticed similarity between a scientific concept and an everyday idea certainly called upon students' creativity, it could be a more valuable activity to ask players to then develop the similarities by analysing the structure of the target and analogue to make parallels explicit. This would be valuable in terms of teaching about the nature of science, as just as logic has nothing to work on without creativity; creativity is only of value when creations are subject to analysis and critique (Taber, 2011b). This further stage certainly happened in some of the examples presented above, but not all. Some mooted similarities were not developed ("cola is like a chemical reaction because...some people can drink absolutely loads of it and some people can have absolutely loads of it and other people can't"), and some were left only partially explained ("an ionic bond is like love because it's all about two things being joined"; "a molecule is like Africa because...Africa is one little thing in a mass of other countries and big things that makes up the world"). In some cases, this may have been because the ASCEND delegates were not able to take the ideas further: but perhaps they saw no need to if they understood the task as simply look for, and justify, a "a similarity" where "something's like something".

This could have been addressed by more detailed instructions making it necessary to show a structural similarity between target and analogue, and having a formal teacher-led introduction to the game (not viable in the particular context where we used the game in ASCEND, but certainly possible when the game is used in a more formal classroom context). Again, there is the issue of how much to challenge those new to the activity: transcripts may suggest that such a specification could have made the task too challenging for some of this group of students. Whilst the games generally moved along with similes being accepted, it seems likely that a formal requirement to map out an explicit analogy might have made the activity a little too difficult for some players on first meeting the game, and so perhaps not the fun activity it was meant to be. My recommendation to other teachers who wish to adopt or adapt the game with high ability students of this age (14–15 year olds) would be that unless the students were already familiar with analysing analogies it would make sense to initially allow any similarities justified to, and accepted by, the group of players, as occurred in ASCEND. It would then be possible to revisit the game later after considering some examples of analogies used by scientists, but now with the more demanding expectation that only formally explained analogies would be allowed:

- an ionic bond is like love because two different elements are joined together
- a molecule is like a bible because a molecule is a complex arrangement of atoms and a bible is a complex arrangement of stories

Despite this reservation, the analogy game did prove successful at engaging students' interest, and encouraging them to apply their imaginations in a way that (though no fault of their teachers) they seldom experienced in their science lessons. Although some of the imaginative suggestions offered by the ASCEND delegates left something to be desired - many were not well thought through, and a number were discarded even as they were being suggested - this was not necessarily a bad thing. Indeed, the context of the game format seemed to elicit thinking in progress: the 'mental cogs' moving though cola as fizzy, to cola as differentially tolerated, to colas as being hard to distinguish. Having creative ideas in science is hard work. Most such ideas transpire to be of limited potential, making it even more important we encourage creative youngsters capable of divergent thinking into science to make sure there is plenty of raw material for the convergent and logical thinking to work on. First there must be ideas. Then the ideas that are worth working with are selected. Then these ideas are developed into something suitably operationalised for testing. Then the most promising ideas are tested to see if they prove to support a better understanding of nature. Only a small proportion of creative ideas will transpire to be genuinely productive in moving science forward – but that is why it is so important to help students to see that it is important to think divergently in science; and to be prepared to share 'brave' ideas that may seem a little bizarre; and not to worry about generating ideas that may prove to be 'dead-ends'.

Building Upon Learners' Ideas

In this context, it is worth noting that the creative science teacher could look to adopt some of the ideas produced by learners in this activity, as raw material for further development, thus working with students' ideas in a dialogic way (Mortimer & Scott, 2003) and showing learners that imaginative ideas can have real value in science learning. Of course, this would require some careful eavesdropping during the activity, or asking students to keep records of their accepted analogies (especially if this could be done in a simple format which did not interrupt the spontaneity of the activity).

The example from our transcripts of the comparison between falling and acceleration will be used to illustrate this process. As suggested above, it would be easy to dismiss this suggestion as not very imaginative: when we fall, we accelerate, so that does not seem a very creative link. However, the crux of this delegate's suggestion was that falling (by implication, in an everyday context, through a resistive fluid such as air) is limited, and that acceleration is also inherently limited in principle according to special relativity. There is certainly a structural analogy that could be drawn upon here (see Figure 3).

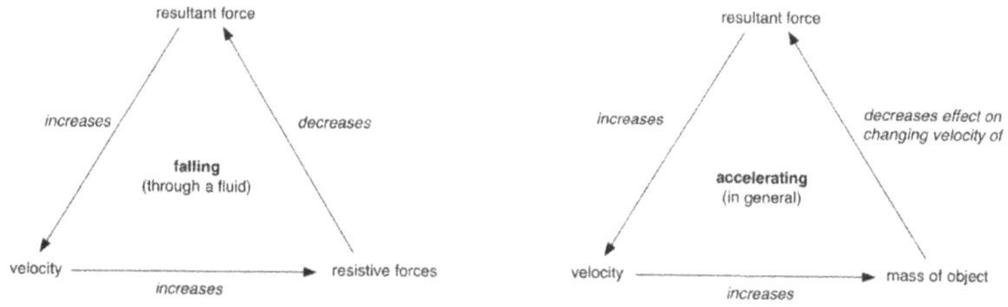

Figure 3. Relativistic increases in mass puts a limit on acceleration analogous to how resistive forces limit terminal velocity of a falling object

Although learning about terminal velocity is part of the secondary school curriculum, it is unlikely that special relativity had been met in formal science lessons (it was certainly not part of the curriculum at this age), so this suggestion seems to draw upon learning outside school. Again, this is a feature that might be considered typical of (though by no means restricted to) gifted learners. It is possible to represent the target and analogue in structurally very similar ways (as in Figure 3), which not only highlights the analogy itself, but makes it clear that this is an analogy between relationships rather than a superficial similarity. Being able to draw analogies at such a highly abstract level would again seem to be something we might associate with gifted learners. It was also very intriguing that the pattern of relationships here took the form of a negative feedback cycle (where the driver for a change brings about a change that diminishes its own effect), which had been a focus of an earlier ASCEND session (Taber, 2011c). That may be no more than coincidence, but perhaps at some level learning about the analogy between other examples of negative feedback in physical systems may have cued this particular insight in one ASCEND delegate.

IN CONCLUSION

The analogy game was one of a programme of science learning activities developed in ASCEND to counter a perceived imbalance in aspects of school science in England, especially when considering the needs of the most capable learners. In effect, science under the English National Curriculum was providing a very limited, and uninspiring, image of science, and one that lacked the opportunities to engage in depth with the complexity of important scientific concepts that were most likely to motivate and challenge gifted learners. Table 1 summarises the types of changes needed to provide suitable science teaching for the most able learners, and which various ASCEND activities attempted to address, at least to provide some enrichment

Table 1. A basis for enriching, or if possible modifying, school science for gifted learners

Shift	*Notes*
From product to process	School science education needs to balance learning about the outputs of science (laws, theories etc) with learning about the processes of science
From justification to discovery	School science education needs to emphasise the creative process of imagining new ideas as well as the logical process of testing them
From breadth to depth	School science education needs to allow students opportunities to engage with ideas in depth in an exploratory mode of study, as well as opportunities to learn key established ideas from a range of important topics across the sciences
From cognition to metacognition	School science should offer opportunities for learners to develop their self-knowledge of the strengths and limits of existing knowledge, and how it can be used as a starting point for development
From analysis to synthesis	School science education should provide opportunities for learners to demonstrate divergent thinking, and find new perspectives and linkages, as well as opportunities to deconstruct, analyse and critique existing arguments and thinking
From facts to possibilities	School science should reflect post-positivist views of science as offering robust but provisional knowledge that is always open to being revisited in the light of new evidence
From prescription to responsibility	Schools science should provide learners with opportunities to take responsibility for planning, monitoring and evaluating their learning in science, as well as will opportunities to learn well-established procedures from structured teaching.

for one group of students, and to show what might be possible in a more flexible curriculum context.

School science in England has put a great emphasis on teaching children to understand fair testing, but has all but ignored the creative process that produces the ideas that might be worth testing. Gifted science students (indeed, all students) are given an impression of science as a discipline that makes heavy demands on memory, but has little use for imagination. Despite this, the analogy game, and other ASCEND activities, demonstrated that students can be creative in science when the opportunity presents itself. Activities like the analogy game also provide an opportunity for teachers to legitimise and value students' creative ideas in science. That, of course, is not just something that is important in the English context, but should be an aim for all those working to support the development of gifted young scientists in all educational contexts.

ACKNOWLEDGEMENTS

Thanks are due to Gatsby SEP for funding the ASCEND project, and publishing the teaching materials. Thanks are also due to the staff in the partner schools that joined the project (Chesterton Community College, Netherhall School and Sixth Form College, Parkside Community College, St. Bede's Inter-church School), and the team of graduate assistants who supported the sessions. Fran Riga was the project research assistant, and produced the transcripts that are drawn upon in this Chapter.

REFERENCES

Boaler, J., Wiliam, D., & Brown, M. (2000). Students' Experiences of Ability Grouping – disaffection, polarisation and the construction of failure. *British Educational Research Journal, 26*(5), 631–648. doi: 10.1080/713651583

Collins, S., Reiss, M., & Stobart, G. (2010). What happens when high-stakes testing stops? Teachers' perceptions of the impact of compulsory national testing in science of 11-year-olds in England and its abolition in Wales. *Assessment in Education: Principles, Policy & Practice, 17*(3), 273–286. doi: 10.1080/0969594x.2010.496205

Conservatives. (n.d.). *Raising the bar, closing the gap; An action plan for schools to raise standards, create more good school places and make opportunity more equal.* No place of publication given: Alan Mabbutt on behalf of the Conservative Party.

Crook, D. (2002). Local authorities and comprehensivisation in England and Wales, 1944–1974. *Oxford Review of Education, 28*(2–3), 247–260. doi: 10.1080/03054980220143405

Csikszentmihalyi, M. (1997). *Creativity: Flow and the psychology of discovery and invention.* New York, NY: HarperPerennial.

DfEE/QCA. (1999). *Science: The national curriculum for England, key stages* (pp. 1–4). London, England: Department for Education and Employment/Qualifications and Curriculum Authority.

DfES. (2002). *Teaching able, gifted and talented pupils: Overview, 2002.* Retrieved January 14, 2004, from http://www.standards.dfes.gov.uk/midbins/keystage3/

Duit, R. (2009). *Bibliography – students' and teachers' conceptions and science education.* Retrieved from Kiel, http://www.ipn.uni-kiel.de/aktuell/stcse/stcse.html

Eilam, B. (2008). Long-term laboratory enquiry: Promoting understanding of ecology. In D. W. Sunal, E. L. Wright, & C. Sundberg (Eds.), *The impact of the laboratory and technology on learning and teaching science K-16* (pp. 77–109). Charlotte, NC: Information Age Publishing.

Fisher, S. G. (1969). Working with gifted children in science. In S. A. Bridges (Ed.), *Gifted children and the brentwood experiment* (pp. 128–135). Bath, England: The Pitman Press.

Frisch, O. R. (1979). *What little I remember.* Cambridge, England: Cambridge University Press.

Gardner, H. (1998). *Extraordinary minds.* London, England: Phoenix.

Gutting, G. (1972). Einstein's discovery of special relativity. *Philosophy of science, 39*(1), 51–68.

Hudson, L. (1967). *Contrary imaginations: Psychological study of the english schoolboy.* Harmondsworth, England: Penguin.

Keller, E. F. (1983). *A feeling for the organism: The life and work of Barbara McClintock.* New York, NY: W H Freeman and Company.

Kim, K. H. (2008). Underachievement and creativity: Are gifted underachievers highly creative? *Creativity Research Journal, 20*(2), 234–242. doi: 10.1080/10400410802060232

Kuhn, T. S. (1959/1977). The essential tension: Tradition and innovation in scientific research. In T. S. Kuhn (Ed.), *The essential tension: Selected studies in scientific tradition and change* (pp. 225–239). Chicago, IL: University of Chicago Press.

Larkin, D. (2012). Misconceptions about "misconceptions": Preservice secondary science teachers' views on the value and role of student ideas. *Science Education, 96*(5), 927–959. doi: 10.1002/sce.21022

Lawson, A. E. (2010). *Teaching inquiry science in middle and secondary schools.* Thousand Oaks, CA: Sage Publications.

Longo, C. (2010). Fostering creativity or teaching to the test? Implications of state testing on the delivery of science instruction. *The Clearing House: A Journal of Educational Strategies, Issues and Ideas, 83*(2), 54–57. doi: 10.1080/00098650903505399

Medawar, P. B. (1963/1990). Is the scientific paper a fraud? In P. B. Medawar (Ed.), *The threat and the glory* (pp. 228–233). New York, NY: Harper Collins. (Reprinted from: The Listener, Volume 70: 12th September, 1963.)

Meitner, L., & Frisch, O. R. (1939). Disintegration of uranium by neutrons: A new type of nuclear reaction. *Nature, 143,* 239–240.

Mortimer, E. F., & Scott, P. H. (2003). *Meaning making in secondary science classrooms.* Maidenhead, England: Open University Press.

Nakiboglu, C., & Taber, K. S. (2013). The atom as a tiny solar system: Turkish high school students' understanding of the atom in relation to a common teaching analogy In G. Tsaparlis & H. Sevian (Eds.), *Concepts of matter in science education* (pp. 169–198). Dordecht, The Netherlands: Springer.

Nersessian, N. J. (2008). *Creating scientific concepts.* Cambridge, MA: The MIT Press.

Niaz, M., & Rodriguez, M. A. (2000). Teaching chemistry as a rhetoric of conclusions or heuristic principles – a history and philosophy of science perspective. *Chemistry Education: Research and Practice in Europe, 1*(3), 315–322.

Osborne, J., & Collins, S. (2001). Pupils' views of the role and value of the science curriculum: A focus-group study. *International Journal of Science Education, 23*(5), 441–467. doi: 10.1080/09500690010006518

Polanyi, M. (1962/1969). The unaccountable element in science. In M. Greene (Ed.), *Knowing and being: Essays by Michael Polanyi* (pp. 105–120). Chicago, IL: University of Chicago.

Popper, K. R. (1989). *Conjectures and refutations: The growth of scientific knowledge* (5th ed.). London, England: Routledge.

Reis, S. M., & Renzulli, J. S. (2010). Is there still a need for gifted education? An examination of current research. *Learning and Individual Differences, 20*(4), 308–317. doi: 10.1016/j.lindif.2009.10.012

Rosenthal, R., & Jacobson, L. (1970). Teacher's expectations. In L. Hudson (Ed.), *The ecology of human intelligence* (pp. 177–181). Harmondsworth, England: Penguin.

Rothenberg, A. (1995). Creative cognitive processes in Kekulé's discovery of the structure of the benzene molecule. *The American Journal of Psychology, 108*(3), 419–438.

Scardamalia, M., & Bereiter, C. (2006). Knowledge building: Theory, pedagogy, and technology. In R. K. Sawyer (Ed.), *The Cambridge handbook of the learning sciences* (pp. 97–115). Cambridge, England: Cambridge University Press.

Simonton, D. K. (2004). *Creativity in Science: Chance, genius, logic and zeitgeist.* Cambridge, England: Cambridge University Press.

Smith, J. P., diSessa, A. A., & Roschelle, J. (1993). Misconceptions reconceived: A constructivist analysis of knowledge in transition. *The Journal of the Learning Sciences, 3*(2), 115–163.

Sternberg, R. J. (1993). The concept of 'giftedness': A pentagonal implicit theory. *The origins and development of high ability* (pp. 5–21). Chichester, England: John Wiley & Sons.

Sternberg, R. J. (2010). Assessment of gifted students for identification purposes: New techniques for a new millennium. *Learning and Individual Differences, 20*(4), 327–336. doi: 10.1016/j.lindif.2009.08.003

Taber, K. S. (2001). When the analogy breaks down: Modelling the atom on the solar system. *Physics Education, 36*(3), 222–226.

Taber, K. S. (2007a). Choice for the gifted: Lessons from teaching about scientific explanations. In K. S. Taber (Ed.), *Science education for gifted learners* (pp. 158–171). London, England: Routledge.

Taber, K. S. (2007b). *Enriching school science for the gifted learner*. London, England: Gatsby Science Enhancement Programme.

Taber, K. S. (2007c). Science education for gifted learners? In K. S. Taber (Ed.), *Science education for gifted learners* (pp. 1–14). London, England: Routledge.

Taber, K. S. (Ed.). (2007d). *Science education for gifted learners*. London, England: Routledge.

Taber, K. S. (2008). Towards a curricular model of the nature of science. *Science & Education, 17*(2–3), 179–218. doi: 10.1007/s11191-006-9056-4

Taber, K. S. (2009). *Progressing science education: Constructing the scientific research programme into the contingent nature of learning science*. Dordrecht, The Netherlands: Springer.

Taber, K. S. (2011a). Constructivism as educational theory: Contingency in learning, and optimally guided instruction. In J. Hassaskhah (Ed.), *Educational theory* (pp. 39–61). New York, NY: Nova. Retrieved from https://camtools.cam.ac.uk/wiki/eclipse/Constructivism.html

Taber, K. S. (2011b). The natures of scientific thinking: Creativity as the handmaiden to logic in the development of public and personal knowledge. In M. S. Khine (Ed.), *Advances in the nature of science research – Concepts and methodologies* (pp. 51–74). Dordrecht, The Netherlands: Springer.

Taber, K. S. (2011c). Patterns in nature: Challenging secondary students to learn about physical laws. *Physics Education, 46*(1), 80–89.

Taber, K. S. (2013). Upper secondary students' understanding of the basic physical interactions in analogous atomic and solar systems. *Research in Science Education, 43*(4), 1107–1136.

Taber, K. S. (2014). *Student thinking and learning in science: Perspectives on the nature and development of learners' ideas*. New York, NY: Routledge.

Taber, K. S. (2015). Developing a research programme in science education for gifted learners. In N. L. Yates (Ed.), *New developments in science education research* (pp. 1–29). New York, NY: Nova Science Publishers.

Taber, K. S., & Cole, J. (2010). The CREST awards scheme: Challenging gifted and talented students through creative STEM project work. *School Science Review, 92*(339), 117–126.

Taber, K. S., & Riga, F. (2006). Lessons form the ASCEND project: Able pupils' responses to an enrichment programme exploring the nature of science. *School Science Review, 87*(321), 97–106.

The National Strategies. (2008). Gifted and talented education. *Guidance on preventing underachievement: A focus on excpetionally able pupils* [00066-2008KT-EN]. Nottingham, England: Department for Children, Schools and Families.

West, A. (2007). Practical work for the gifted in science. In K. S. Taber (Ed.), *Science education for gifted learners* (pp. 172–181). London, England: Routledge.

White, J. (1987). The comprehensive ideal and the rejection of theory. *British Journal of Educational Studies, 35*(3), 196–210.

Keith S. Taber
Faculty of Education
University of Cambridge
United Kingdom

KYUNG HEE KIM AND STEVE V. COXON

19. FOSTERING CREATIVITY USING ROBOTICS AMONG STUDENTS IN STEM FIELDS TO REVERSE THE CREATIVITY CRISIS

INTRODUCTION

Creativity is required for innovation in STEM fields. The success of South Korea on the Programme for International Student Assessment (PISA) (Center on International Educational Benchmarking, 2012), contrasted with that nation's lack of success in measures of innovation including few patents per capita and no Nobel Prizes in the sciences, suggests that exceptional test scores are insufficient to produce exceptional innovating. Ironically, as South Korea has recognized the need for creativity in science education, the U.S. has become hyper-focused on reading and math standardized test scores to the extent that science education and creative opportunities have been greatly reduced in schools. This trend must be reversed with the inclusion of science-focused programs in schools that are likely to foster creativity in children. In particular, STEM programs that include demands on children's creativity are warranted. Robotics programs likely meet this need and are discussed later in this chapter after an overview of creativity, measuring creativity, and a synthesis of the research demonstrating the Creativity Crisis.

DIVERGENT, CONVERGENT, AND EMERGENT THINKING

Creativity is making or doing something useful and new or better in the arts, science, business, or other worthwhile endeavor. The creative thinking process requires three types of thinking: divergent, convergent, and emergent thinking, as Table 1 shows. Based on years of analyses of divergent thinking tests and creativity test scores and investigation into creativity, three types of thinking have emerged: divergent, convergent, and emergent. For the creative process to be successful, the three types of creative thinking must work well together.

Divergent thinkers are quick or loose thinkers. They generate original ideas. Divergent thinking processes include expanding and spreading like having a wide-range lens on a camera. Divergent thinking involves fluency, flexibility, and originality, whereas convergent thinking involves analysis and logic (Guilford, 1956, 1959, 1986).

Convergent thinkers are tight or narrow thinkers and can decide whether the original ideas are valuable and worth pursuing. To continue the metaphor, convergent

M. K. Demetrikopoulos & J. L. Pecore (Eds.), Interplay of Creativity and Giftedness in Science, 351–365.

thinking processes include focusing and narrowing down like having a narrow-angle lens on the camera. Analyzing, evaluating, and narrowing down divergent ideas with convergent thinking process is necessary for the creative process to be successful.

Table 1. Descriptions of the TTCT measures in relation to creative thinking and attitude

	Creative Thinking & Creative Attitude	*Name of the TTCT measure/ Description*
Divergent Thinking	Fluency (Generating many ideas)	*Fluency*—The number of ideas generated
	Originality (Generating unusual ideas)	*Originality*—The number of unique ideas generated
	Flexibility (Having another perspective or using another sense)	*Unusual Visualization*—Looking with another angle *Internal Visualization*—Seeing through the hidden *Colorfulness of Imagery*—Using the five senses *Movement or Action*—Using body movement
Emergent Thinking	Abstract mindset (Enjoying the complex & ambiguous)	*Abstractness of Titles*—Thinking beyond what is seen
	Persistence & elaboration (Working on details or describing with imagination)	*Elaboration*—The degree of detail & persistence *Storytelling Articulateness*—The skill to tell a story *Expressiveness of Titles*—The skill to be expressive *Richness of Imagery*—The skill to visualize
	Integration (Unconventional & connecting between the seemingly irrelevant)	*Extending or Breaking Boundaries*—Nonconforming *Synthesis of Lines or Circles*—Reorganizing *Synthesis of Incomplete Figures*—Connecting the different
Creative Attitude	Open-mindedness	*Resistance to Premature Closure*—Deferring judgment
	Emotional sensitivity	*Emotional Expressiveness*—Emotional & sensitive
	Humor	*Humor*—Playful, childlike, & humorous
	Fantasy	*Fantasy*—Future-oriented and enjoying fantasy, daydreaming, and the unknown
✓Convergent Thinking	Logical	✓*Analytical/evaluative/logical thinking*—Part of intelligence

Emergent thinkers are deep or stretched thinkers. They develop ideas to final, useful products. Emergent thinking involves integration and imaginative elaboration. To extend the metaphor, emergent thinking is developing the pictures taken by these camera lenses using artistic technique, completing the picture, and presenting with a frame that best enhances the beauty of the image. Emergent thinking to refine and implement the idea into a final product is critical for the creative process to be successful.

Measuring Creativity with the Torrance Tests of Creative Thinking (TTCT)

Creativity has been measured in hundreds of thousands of children and adults in the U.S. over nearly five decades. As indicated by Torrance's 40-year longitudinal study (Torrance, 2002), scores on the TTCT are good predictors of adult creative performance. The Torrance Tests of Creative Thinking (TTCT) was developed in the 1950s and has been regularly re-normed. There are verbal and figural versions of the TTCT; the Figural was used in the Creativity Crisis study and is therefore the focus of this chapter. The TTCT has been translated into over 35 languages and is the most widely used test of creativity. Research shows that among all of the creativity tests, the TTCT predicts creative achievement the best (Kim, 2008). Unlike other creativity tests, the TTCT-Figural measures creative thinking and not merely divergent-thinking. Torrance designed the test to score responses for Guilford's four divergent thinking factors of Fluency, Flexibility, Originality, and Elaboration (Torrance, 1966). The TTCT can be administered to kindergarteners through adults in 30 minutes as an individual or group test. While test takers are required to draw, artistic quality is not required to receive credit.

The TTCT–Figural has two parallel forms, A and B, and consists of three activities: picture construction, picture completion, and repeated figures of lines or circles. The TTCT-Figural is comprised of five norm-referenced measures so that the numbers of points earned are relative to the norm group: fluency, originality, elaboration, abstractness of titles, and resistance to premature closure. In addition, there are 13 criterion-referenced measures of Creative Strengths so that credit is given depending on whether the criterion appears in the responses (see Kim, 2006 for details).

The TTCT measures divergent thinking, emergent thinking, and creative attitudes, as Table 1 shows. Divergent thinking skills are assessed in the categories of fluency, originality, and flexibility. The *Fluency* (Generating many ideas) and *Originality* subscales (Generating unusual ideas) are each assessed. Flexibility (Having another perspective or using another sense) is assessed by four of the 13 checklists of *Creative Strengths* subscales: *Unusual Visualization* (Looking with another angle), *Internal Visualization* (Seeing through the hidden), *Colorfulness of Imagery* (Using the five senses), and *Movement or Action* (Using body movement).

Emergent thinking skills are assessed through abstract mindset, persistence and elaboration, and integration. Abstract mindset (Enjoying the complex and ambiguous) is assessed with the *Abstractness of Titles* subscale (Thinking beyond what is seen). Persistence and elaboration (Working on details or describing with imagination) are assessed by the *Elaboration* subscale (The degree of detail and persistence) and by three of the 13 checklists of *Creative Strengths* subscales: the *Storytelling Articulateness* subscale (The skill to tell a story), the *Expressiveness of Titles* subscale (The skill to be expressive), and the *Richness of Imagery* subscale (The skill to visualize). Integration (Unconventional and connecting between the seemingly irrelevant) is also assessed by three of the 13 checklists of Creative Strengths subscales: the *Extending or Breaking Boundaries* subscale (Nonconforming), the *Synthesis of Lines or Circles* subscale (Reorganizing), and the *Synthesis of Incomplete Figures* subscale (Connecting the different).

Creative attitude is assessed through open-mindedness, emotional sensitivity, humor, and fantasy. Open-mindedness is assessed by the *Resistance to Premature Closure* subscale (Deferring judgment). Emotional sensitivity is assessed by the *Emotional Expressiveness* subscale (Emotional and sensitive), one of the 13 *Creative Strengths*. Humor is assessed by the *Humor* subscale (Playful, childlike, and humorous), another of the 13 *Creative Strengths*. Fantasy is assessed by the *Fantasy* subscale (Future-oriented and enjoying fantasy, daydreaming, and the unknown), also one of the 13 *Creative Strengths*.

RESULTS OF THE CREATIVITY CRISIS STUDY

Emergent Thinking

Elaboration scores decreased by 19% from 1984 to 1990, by 25% from 1984 to 1998, and by 37% from 1984 to 2008. Elaboration scores decreased the earliest, starting in 1984, which indicates that individuals are less able to elaborate ideas with imagination and think reflectively and that they are less persistent. Creativity is more than just coming up with an idea. Hard work, persistence, and endurance are required to produce a final product. In our view, creativity does not exist without a final, useful product. Imagining a new invention is different than developing it and getting it to market. IQ continues to increase steadily (Flynn, 1984, 2007). Elaboration is usually correlated to IQ (Torrance, 2000). Therefore, as IQ continues to increase, elaboration should have increased. So, to decrease while IQ has increased suggests that the divergent component of Elaboration actually decreased even more than the gross scores indicate.

Abstractness of Titles scores, another element of emergent thinking, decreased since 1998. Abstractness of Titles refers to thinking beyond the obvious and what is seen. This is what allows some people to recognize and describe patterns and the essence of problems without distorting the information. Abstractness of Titles scores decreased by 7% from 1998 to 2008, a little later than the decreases of other TTCT subscales, which started in 1984 (Elaboration) or in 1990 (Fluency, Originality, & Creative Strengths). Abstractness of Titles scores are also inflated because they relate to IQ (Torrance, 2000).

Divergent Thinking

Fluency. The more original ideas one generates, the better the chance of having an idea that can be developed into a useful product becomes. Divergent thinking skills had the next greatest decrease. *Fluency* scores have decreased since 1990 by nearly 5% from 1990 to 1998 and by 7% from 1990 to 2008. The biggest decrease in Fluency scores was for kindergartners through third graders, and the second biggest decrease was for fourth through sixth graders, which indicate that younger children's ability to produce many ideas decreased the most gravely since 1990.

Originality, another divergent thinking skill, is one of the most critical elements of creative thinking. *Originality* scores also decreased since 1990, indicating that we are less able to generate unusual ideas. Originality scores actually continued to increase until 1990, but decreased by nearly 4% from 1990 to 1998, and remained static from 1998 to 2008. Originality is the only TTCT subscale that is reflective of different cultures and time. The originality lists are periodically updated due to changes over time. For example, drawing a cell phone in the repeated lines task was once original, but is now quite common. Thus, Kim (2006) questioned the credibility of Originality scores of the TTCT based on the Originality Lists that Torrance developed in 1984. The continued use of 1984 Originality Lists leads to an expectation that the Originality scores should go up artificially the longer the Originality Lists are not updated. However, the results indicated that the Originality scores also decreased from 1990 to 1998 and remained static from 1998 to 2008. Therefore, Originality scores may have actually decreased even more than these results demonstrate.

Examining each age group separately revealed that the biggest decrease in Originality scores from 1990 to 2008 was for kindergartners through third graders. It can be concluded that younger children's ability to produce statistically infrequent, unique, and unusual ideas has significantly decreased since 1990.

The decrease in Originality scores indicate that the decrease resulted from a climate that facilitates creativity less well than in the past and continues to grow less tolerant of creative expression. Almost everyone expresses positive views of creativity, but very few understand the challenges that come with creative work. Most people are uncomfortable with the change, uncertainty, new ideas, challenges, and risk that accompany creativity and creative behavior. In order for thinkers to present original ideas, the climate needs to be receptive, or at least not hostile, to expression and consideration of unusual ideas. The proponent of an original idea starts out as a *minority of one*. In an intolerant climate, many ideas are dismissed immediately. A healthy creative climate resists closure and respects new ideas for the possibilities they offer. A healthy climate considers how new ideas may work, instead of dismissing ideas because of reasons they may not work. The decrease in Originality scores is an indirect measure of growing social pressures toward conformity and increasing intolerance for new ideas.

Creative Attitude

Resistance to premature closure. While decreasing less than divergent and emergent thinking, creative attitudes also decreased. *Resistance to Premature Closure* scores decreased since 1998. Resistance to Premature Closure scores decreased by about 2% from 1998 to 2008. Again, these scores are likely inflated because they correlate to IQ (Torrance, 2000). This indicates that individuals are less able to defer judgment. An open mind is needed to understand the problem and to consider many possible solutions. Considering various viewpoints as well as accepting and celebrating diversity are means of fostering creative thinking.

Creative Strengths

Emergent thinking, divergent thinking, and creative attitudes have decreased continuously since 1990. The 13 Checklists of Creative Strengths scores decreased by 3% from 1990 to 1998 and by nearly 6% from 1990 to 2008. Unfortunately the 13 scores for this subscale are combined by the test scores, so they could not be analyzed further. The decrease of Strengths scores after 1990 likely indicate that over the last 20 years, that children display less emergent thinking, divergent thinking, and creative attitudes.

Emergent thinking. Based on the decrease of the combined scores, it appears that individuals have lower emergent thinking skills. Less elaboration, including skills to tell a story, to be expressive, and to visualize, is evident among Americans. Less integration, such as skills to think unconventionally, to reorganize and synthesize, and to connect between the seemingly irrelevant things or concepts, are also evident.

Divergent thinking. Individuals display lower divergent thinking skills. Less flexibility, including such skills as looking at things through another angle, seeing through the hidden aspects of a problem or a situation, and using the five senses including expressing by body movement sense, are evident.

Creative Attitudes. Creative attitudes have decreased in the population. Fewer emotions and emotional sensitivity are displayed. Americans are less playful, childlike, and humorous. We are less future-oriented and less likely to enjoy fantasy, daydreaming, and the unknown.

Results summary

The Creativity Crisis results (Kim, 2011) demonstrate that creativity scores decreased significantly, starting before 1990.

- Elaboration decreased by 17%
- Abstractness of Titles decreased by 7%

- Fluency decreased by 7%
- Creative Strengths decreased by almost 6%
- Originality decreased by almost 4%
- Resistance to Premature Closure decreased by almost 2%

Americans are losing their ability to elaborate upon ideas, are less capable of detailed and reflective thinking, and are less motivated to be creative. Why this has occurred remains unknown, but perhaps home, school, and society facilitate and encourage creativity less. Children in kindergarten through third grade suffered the greatest decline, followed by those in fourth through sixth grades. The findings may also mean that younger children are becoming less capable of synthesis and organization, important critical thinking processes. They may be less able to determine what is important in a complex problem.

We do not yet know what the consequences of this broad decrease of creativity among U.S. students will be. Given America's dependence on innovations in STEM fields for economic and quality of life improvements, the risks seem very high especially if other nations make efforts in their own school systems. At the same time, creativity is improvable through educational experiences. Among those possibilities, robotics competitions appear to be one means by which schools can increase creativity in STEM education. However, an overall paucity of research in these areas leaves many questions to be answered.

CREATVIITY RESEARCH IN STEM EDUCATION

Even without regard to creativity, little high quality experimental or quasi-experimental research has focused on elementary or middle school STEM education. Science education research makes a particularly apropos example of the problem. A recent meta-analysis of effective science teaching strategies failed to locate many studies concerning elementary science instruction where control groups were used (Schroeder, Scott, Tolson, Huang, & Lee, 2007). Only 61 of nearly 400 science education studies identified were deemed appropriate for meta-analysis, and only 16 of those 61 studies were conducted with K-8 participants. However, while individual studies were not reported on separately by grade level and science intervention, the mean effect size of the K-8 studies of various science education strategies was .68 (Schroeder et al., 2007). That is, science education at the elementary level is a meaningful endeavor in terms of student achievement and failing to provide it represents a significant failure to develop science talents in young students.

In fact, the longitudinal research of Novak (2005) suggests that students are harmed when science education does not begin early in their schooling. He found that children taught science concepts beginning in second grade continued to perform better throughout high school compared to students who did not receive science instruction until sixth grade. Not only did the students with earlier instruction learn more, but they were wrong about less. Novak also found that students receiving earlier instruction had fewer misconceptions about science in the twelfth grade than those whose instruction was delayed.

While a paucity of research exists in science education generally, research conducted on elementary science education specifically in terms of creativity gains is almost absent. Much of what has been done has been conducted outside of the U.S. in contexts that may not relate to U.S. schools. Perhaps this lack of interest in the U.S. stems from a widespread myth that creativity is solely useful in the arts (Newton, 2010) while, in reality, creativity is necessary for success in all fields including the sciences. We live longer and higher quality lives because of successful creativity in scientific fields such as medicine and technology (National Science Board, 2010). More research is needed to improve science education and creativity, which we argue can and should go hand-in-hand.

In some of the few such studies that have been conducted, significant creativity gains have been made by treatment groups, suggesting that creativity can be improved through at least some STEM education treatments. Technology, in particular, takes a predominate role in this small pool of research. For example, Eow, Ali, Mahmud, and Baki (2010) reported on an experimental study of 69 Malaysian 13 and 14-year-old students developing computer games with differ strategies and found that students in the experimental group made significantly greater gains than the control group. Younger children also appear to benefit. Shawareb (2011) reports on a quasi-experimental study of 76 Jordanian kindergarten children where the treatment group used educational software thought to facilitate creative thinking and found that they demonstrated significantly greater gains on the TTCT. Much more research is needed to illuminate STEM education programs that facilitate growth in creativity, especially in the context of U.S. schools.

Along with fostering creativity, the National Science Board (2010) offered several recommendations for future STEM programming in their report, *Preparing the Next Generation of STEM Innovators*, including early intervention and a focus on the most able students. The next section of this chapter discusses promising research on robotics programs that respond to these three recommendations.

ROBOTICS PROGRAMS

Robotics programs are a potential means of improving creativity through STEM education. In particular, the non-profit group, For Inspiration and Recognition in Science and Technology (FIRST) programs such as the FIRST LEGO League (FLL) and Junior FIRST LEGO League (Jr.FLL) appear to increase STEM interest among children (Melchior, Cutter, & Cohen, 2004), to be challenging enough to meet the needs of the gifted (Coxon, 2010; 2012a), and to offer opportunities to be creative within STEM fields (Geeter, Golder, & Nordin, 2002). Robotics competitions contain a wealth of open-ended problem solving opportunities with unlimited possible solutions. Performing successfully requires participants to generate many ideas through divergent thinking, to move toward a solution through convergent thinking, and to actually engineer and program a robot to carry out the chosen solution through emergent thinking. Robotics competitions involve each aspect of the STEM acronym. Both the FLL and Jr.FLL competitions have an annual, relevant science theme. Technology is included as students program the robots with a computer. Engineering is required as students create working robots from LEGO elements including gears and motors. Math is required in the logic of programming, in manipulating the robot to go specific distances and directions, and in many other aspects.

The Jr.FLL is an academic program centered on a real-world science theme, such as food safety, that changes annually. The program engages teams of up to six children ages 6-9 in developing a research poster and building a LEGO model, that may use the LEGO WeDo robotics kit, to help solve a problem associated with the science theme. The LEGO WeDo robotics kit allows children to build a working robot with LEGO bricks including a special motor and two sensors. The robot is programmable with a drag-and-drop programming language. Both the model and poster are shared at events with other teams and the participating children receive non-competitive awards from experts such as engineers and university faculty volunteering as judges.

The FLL also has an annual, real-world science theme. Recent themes include medical technology, the Mars rovers, and improving the lives of seniors. Children and adolescents ages 9-14 compete on teams of up to ten. Teams build robots with the LEGO NXT or EV3 robotics kits that include multiple sensors, a rechargeable battery, and a computerized brick into which programs written on a computer can be downloaded. The robots complete pre-determined tasks on a field based on the year's theme such as placing a LEGO solar panel on a LEGO house or repairing a LEGO leg bone with a LEGO pin. Teams also do a research project on an aspect of the year's theme. The program is highly competitive, with qualifying and state-level events leading to an annual World Championship. Each level becomes more competitive, making the competition challenging even for the most advanced and gifted students (Coxon, 2010, 2012a). In 2012, more than 200,000 children and adolescents from more than 60 countries participated in the FLL (US FIRST, 2012).

We are not aware of any completed research on robotics' effect on creativity using pre- and post-assessments such as the TTCT other than the study reported here. However, some experimental research has been conducted on the effects of robotics use on children's spatial ability, often also called visualization or visual-spatial thinking. Spatial ability is "the ability to generate, retain, retrieve, and transform well-structured visual images" (Lohman, 1993, p. 1). Spatial ability is highly predictive of success in STEM fields (Wai, Lubinski, & Benbow, 2009). It is also likely important for creativity in fields where it is necessary to form and manipulate a mental picture of something that is novel, such as a new technology (Coxon, 2012b). Spatial ability is likely related to creativity (Coxon, 2012b; Liben, 2009). Although research demonstrating this is currently lacking, the biographies of eminent creators, particularly Michael Faraday, Francis Galton, Benjamin Franklin, and Albert Einstein, provide evidence of this relationship. Each of these creators reported on the need for visual-spatial thinking to facilitate their creativity with Einstein noting that he did not use verbal processes in his creative thinking at all, preferring to rely on visualization (Lohman, 1993). Research into a possible link between spatial ability and creativity is currently underway by the second author.

In an experimental interventional study of 75 public school gifted children ages 9-14, the treatment group completed a 20-hour simulation of the FLL competition and evidenced significant gains on the Project Talent Spatial Battery (Coxon, 2012a). The Spatial Battery is a set of four assessments each focused on a different aspect of spatial ability (Flanagan, 1979). In particular, males made very large gains with an effect size of .87. While there were few participants from groups traditionally underrepresented in gifted programs represented in the study, those present also made large gains. As the program only involved children and adolescents scoring within the top 10% on an ability test, it appears that the FLL program is challenging enough to meet the needs of the gifted participants. However, more research is needed to affirm the gains made by participants from groups traditionally underrepresented in gifted programs and to find means to raise spatial ability in females.

Creativity Research in Robotics Education

Both the Jr.FLL and the FLL offer an early intervention in STEM education and are likely challenging enough for the most able learners. It seems likely that building original robots to create novel solutions to real-world science problems also facilitates creative growth. The research on spatial gains is also suggestive of this possibility. However, while some researchers have suggested this likelihood (Coxon, 2012b; Geeter et al., 2002), we were unable to locate any other studies using robotics with a pre- and post-measure of creativity at the time of this writing other than the one we report on here along with two promising studies nearing completion.

The completed study involved three Jr.FLL teams at high poverty public elementary schools in a Midwestern metropolitan area. Each team consisted of six children ages 6–9 years for a total of 18 children. Unfortunately, six participants did not return their parental informed consent forms, a common challenge in working in high poverty settings. The remaining participants were largely in areas commonly deficient in STEM education programs: 83% African American, 67% on free or reduced lunch, and 67% female.

All participants with parental informed consent took the TTCT-Figural at the onset of the program, then participated in the 2012 Jr.FLL challenge, Snack Attack. Two participants did not complete the post-assessment. Teams researched food safety issues and selected one on which to focus. Over the two months prior to their culminating event, teams built models using a LEGO kit specially designed for the Jr.FLL competition and the LEGO WeDo robotics kit. The models were entirely built by children and focused on demonstrating a solution that the food safety issue each team selected. As the model was both a novel creation made without the use of instructions and built for a useful purpose, it was hypothesized that participants' creativity would improve and that this would be reflected when they are post-assessed with the TTCT-Figural.

All 12 participants with parental informed consent and their parents completed a select portion of the survey created by Melchior and his collaborators at Brandeis to evaluate FIRST programs focused on interest in STEM and school engagement (Melchior, Cohen, Cutter, & Leavitt, 2005; Melchior, Cutter, & Cohen, 2004). As the participants in this study were younger than in the previous research, it was determined that the original survey was too long and it was honed to 15 questions about interest in STEM fields, motivation for school, and interest in attending college. As FIRST programs appear to be highly motivating for children, it was hypothesized that participants' interest in STEM fields, in school, and in attending college would improve according to the survey results.

Results

As Table 2 indicates, participants saw improvements in all three types of creative thinking: divergent increased by about 19 points; creative attitude increased more than 11 points; and emergent thinking increased about 5 points. The divergent thinking score was statistically significantly increased. Creative attitude and emergent thinking scores also increased, but did not reach to a statistically significant level.

Table 2. The torrance tests of creative thinking-figural pre- and post-assessment results of Jr.FLL

TTCT	*Pretest Mean (SD)*	*Posttest Mean (SD)*	*t*	*p*
Divergent thinking	197.2 (39.0)	216.0 (44.7)	2.32*	.046
Creative attitude	89.1 (26.7)	100.5 (11.3)	1.65	.133
Emergent thinking	210.9 (46.6)	215.4 (55.26)	0.39	.704

Participants (N=10)
Note: * $p < 0.5$

Of the 12 surveys returned from Jr.FLL participants in Spring 2012 (67% response rate), 100% indicated an increased interest in going to college and doing well in school. Additionally, 83% agreed that they wanted to become a scientist or engineer because of the program. For parents, 92% believed the program increased their children's interest in going to college and doing well in school. The same number reported that their children were more interested in math and science and how those subjects can be used to solve problems in the real world.

Discussion

Although only divergent thinking increased significantly from the pre- to post-assessment, it is likely that a study with a larger sample size would demonstrate significance in all three areas. Thus, although the increased scores for emergent thinking and creative attitude of this study might not be statistically important, considering the fact that 1) divergent thinking, emergent thinking, and creative attitude are increased after the use of LEGO robotics; 2) this study is the first study that examined the effects of the use of LEGO robotics on creativity test scores because no similar studies have been conducted before; and 3) the participants represent minority groups usually underrepresented in STEM fields, the results of the present study are educationally important. A larger study has just been completed and results will be available soon.

Even if Jr.FLL participation only increases divergent thinking, generating many ideas is foundational to later convergent and emergent thinking. Thus, the increase in divergent thinking is an important finding. School policy makers should be encouraged to incorporate Jr.FLL into regular programming to help combat the Creativity Crisis. It is likely that the use of LEGO robotics generally in academic work will increase divergent thinking and possibly other aspects as well. Future research that may demonstrate this is discussed below.

The survey results were also positive. They indicated improvement in interest in pursuing STEM fields, motivation for school, and interest in attending college among participants as reported both by participants' parents as well as the participants in regards to themselves. This is concomitant with previous findings about FIRST programs in research using the same surveys (Melchior, Cohen, Cutter, & Leavitt, 2005; Melchior, Cutter, & Cohen, 2004). It is important to note that this finding is among a group that is largely African American, from low socio-economic backgrounds, and female: all groups traditionally underrepresented in post-secondary STEM programs. This suggest that incorporating Jr.FLL in school programming for these groups will lead to an increase of these populations in post-secondary STEM programs.

Future Research

Important questions remain regarding robotics in education. In a larger study, will convergent and/or emergent thinking be significantly improved under treatment with LEGO WeDo robotics (the kit used in the Jr.FLL)? Is creativity improved by the use of LEGO NXT or EV3 robotics (the kits used in the FLL)? Are these improvements limited to participants in the competition or can creativity be similarly improved in academic classes involving robotics? Two studies are currently underway that will help answer these questions. In a nearly completed study, more than 100 high ability children participating in a summer robotics program on a university campus have taken the TTCT-figural and the Project Talent Spatial Battery as pre- and post-assessments to both explore possible gains on both measures and to look for a possible correlation between both the measures as well as any possible gains on both measures. About half of participants took classes using the LEGO WeDo robotics kit and approximately half used the LEGO NXT robotics kit. If a significant, positive correlation exists between the rise of spatial ability and creativity, this will open up more activities, projects, and programs for use by schools to combat declining creativity scores in the U.S. Most importantly, with the larger number of participants, it is likely that significant gains will be found for the expected improvements on convergent and emergent thinking.

Another study now underway involves three FLL teams at high poverty public elementary and middle schools in a Midwestern metropolitan area with a similarly diverse group of 8-10 participants per team ages 9-14. Teams participated in the 2012 FLL challenge, Senior Solutions, which focuses on solving problems faced by seniors as they age to help them live more independent and higher quality lives. All participants took both the TTCT-figural and the Project Talent Spatial Battery as pre- and post-assessments. This study has the potential to add further evidence to the earlier finding that groups traditionally underrepresented in STEM programs can make large spatial gains and will help to determine if females make gains when the treatment is longer (Coxon, 2012a). Furthermore, it will determine if participation in the FLL competition improves creativity scores, a finding that would provide further support to encourage schools to become involved in the program.

However, classroom teachers, school leaders, and education policy makers should not wait until research catches up to the national need: The U.S. cannot afford to wait complacently. The Business Roundtable (2005) warns of a "slow withering, a gradual decline, a widening gap between a complacent America and countries with the drive, commitment and vision to take our place" (p. 5). In general, engaging students in creative activities, projects, and programs works to raise their creativity (Coxon, 2012b; Davis & Rimm, 1998; Erez, 2004; Sternberg, 1984; Treffinger, Isaksen, & Dorval, 2006). Participating in the FLL improves spatial ability. Participating in either competition improves interest in STEM fields, motivation for school, and interest in attending college. Most importantly in light of the Creativity Crisis, participating in the Jr.FLL improves children's divergent thinking. Schools should move to include more creative programs such as the Jr.FLL as a means to help combat the Creativity Crisis.

REFERENCES

Business Roundtable. (2008). *Tapping America's potential: The education for innovation initiative* (p. 5). Washington, DC: Author. Retrieved from: http://www.eric.ed.gov/ERICDocs/data/ericdocs2sql/content_storage_01/0000019b/80/1b/ad/71.pdf

Center on International Educational Benchmarking. (2012). *South Korea.* Retrieved from http://www.ncee.org/programs-affiliates/center-on-international-education-benchmarking/top-performing-countries/south-korea-overview/

Coxon, S. V. (2010). FIRST LEGO league, the sport of the mind. *Teaching for High Potential*, 6–8.
Coxon, S. V. (2012a). The malleability of spatial ability under treatment of a FIRST LEGO league simulation. *Journal for the Education of the Gifted*, *35*, 291–316.
Coxon, S. V. (2012b). Innovative allies: Spatial and creative abilities. *Gifted Child Today*, *35*, 277–284.
Davis, G. A., & Rimm, S. B. (1998). *Education of the gifted and talented* (4th ed.). Needham Heights, MA: Allyn.
Eow, Y. L., Ali, W., Mahmud, R., & Baki, R. (2010). Computer game development and appreciative learning approach in enhancing students' creative potential. *Computers*, *54*, 146–161.
Erez, R. (2004). Freedom and creativity: An approach to science education for excellent students and its realization in the Israel arts and science academy's curriculum. *Journal of Secondary Gifted Education*, *15*, 133–140.
Flanagan, J. C. (1979). Findings from project TALENT. *Educational Forum*, *43*, 489–490.
Flynn, J. R. (1984). The mean IQ of Americans-Massive gains 1932 to 1978. *Psychological Bulletin*, *95*, 29–51.
Flynn, J. R. (2007). *What is intelligence? Beyond the Flynn effect*. New York, NY: Cambridge University Press.
Geeter, D. D., Golder, J. E., & Nordin, T. A. (2002). *Creating engineers for the future*. Proceedings of the 2002 American Society for Engineering Education Annual Conference, Montreal, Canada.
Guilford, J. P. (1956). Structure of intellect. *Psychological Bulletin*, *53*, 267–293.
Guilford, J. P. (1959). *Personality*. New York, NY: McGraw-Hill.
Guilford, J. P. (1986). *Creative talents: Their nature, uses, and development*. Buffalo, NY: Bearly Limited.
Kim, K. H. (2006). Can we trust creativity tests? A review of the torrance tests of creative thinking (TTCT). *Creativity Research Journal*, *18*, 3–14.
Kim, K. H. (2008). Meta-analyses of the relationship of creative achievement to both IQ and divergent thinking test scores. *Journal of Creative Behavior*, *42*, 106–130.
Kim, K. H. (2011). The creativity crisis: The decrease in creative thinking scores on the torrance tests of creative thinking. *Creativity Research Journal*, *23*, 1–11.
Liben, L. S. (2009). Giftedness during childhood: The spatial-graphic domain. In F. D. Horowitz, R. F. Subotnik, & D. J. Matthews (Eds.), *The development of giftedness and talent across the life span* (pp. 59–74). Washington, DC: American Psychological Association.
Lohman, D. (1993). *Spatial ability and g*. Paper presented at the Spearman seminar, University of Plymouth, Plymouth, England. Retrieved from http://faculty.education.uiowa.edu/dlohman/pdf/Spatial_Ability_and_G.pdf
Melchior, A., Cutter, T., & Cohen, F. (2004). *Evaluation of FIRST LEGO league*. Waltham, MA: Center for Youth and Communities, Brandeis University. Retrieved from http://www.usfirst.org/who/content.aspx?id=46
Melchior, A., Cohen, F., Cutter, T., & Leavitt, T. (2005). *More than robots: An evaluation of the first robotics competition participant and institutional impacts*. Waltham, MA: Center for Youth and Communities, Brandeis University. Retrieved from http://www.usfirst.org/who/content.aspx?id=46
National Science Board. (2010). *Preparing the next generation of STEM innovators: Identifying and developing our nation's human capital*. Arlington, VA: National Science Foundation.
Newton, L. D. (August, 2010). Creativity in science and science education: A response to Ghassib. *Gifted and Talented International*, *25*(1), 105–109.
Novak, J. D. (2005). Results and implications of a 12-year longitudinal study of science concept learning. *Research in Science Education*, *35*, 23–40.
Schroeder, C. M., Scott, T. P., Tolson, H., Huang, T.-Y., & Lee, Y.-H. (2007). A meta-analysis of national research: Effects of teaching strategies on student achievement in science in the United States. *Journal of Research in Science Teaching*, *44*, 1436–1460.
Shawareb, A. (2011). The effects of computer use on creative thinking among kindergarten children in Jordan. *Journal of Instructional Psychology*, *38*, 213–220.
Sternberg, R. (1984). *Beyong IQ: A triarchic theory of human intelligence*. Cambridge, England: Cambridge University Press.

Torrance, E. P. (1966). *The torrance tests of creative thinking - Norms-technical manual research edition – Verbal tests, forms A and B – Figural tests, forms A and B*. Princeton, NJ: Personnel Press.

Torrance. E. P. (2000). *Research review of the torrance tests of creative thinking: Figural forms A and B*. Bensenville, IL: Scholastic Testing Service.

Torrance, E. P. (2002). *The manifesto: A guide to developing a creative career*. Westport, CT: Ablex Publishing.

Treffinger, D. J., Isaksen, S. G., & Dorval, K. B. (2006). *Creative problem solving: An introduction* (4th ed.). Waco, TX: Prufrock.

US FIRST. (2012). *FIRST at a glance*. Retrieved from http://www.usfirst.org/aboutus/first-at-a-glance

Wai, J., Lubinski, D., & Benbow, C. P. (2009). Spatial ability for STEM domains: Aligning over 50 years of cumulative psychological knowledge solidifies its importance. *Journal of Educational Psychology*, *101*, 817–835.

Kyung Hee Kim
School of Education
William and Mary
USA

Steve V. Coxon
School of Education
Maryville University of St. Louis
USA

MARY LIGHTBODY AND LISA MARY HUELSKAMP

20. ATTRACTING DYNAMOS

How Problem Based Science Opens Doors and Creates Opportunities

INTRODUCTION

As teachers of science across the United States implement revised state standards, and in many cases the Next Generation National Science Standards, in their classrooms, the nature of science teaching and learning has been slowly changing. In the most enlightened classrooms, students are offered opportunities to engage in open inquiry investigations including posing their own questions and testing their hypotheses, as well as collecting data and offering conclusions and explanations based on evidence. The best inquiries also address real world problems or authentic situations, and in an ideal science classroom no limits are placed on student thinking or creativity. In other science classrooms, a more traditional curriculum is employed, involving experiments that provide step-by-step directions and are confirmatory in nature. In still others the lack of science materials and equipment limit opportunities for hands-on investigations; instead students are expected to learn science by reading about science from textbooks or weekly science newspapers. Worst of all, in early childhood classrooms, the pressure to improve student reading and mathematics test scores and achievement has reduced the amount of time devoted to teaching science to nearly zero.

Simultaneously, outside the classroom challenges to life on earth continue to mount. While the day-to-day existence for many in the United States is comfortable, this is not true for everyone. There are homeless in our cities, and our prisons are full and overflowing. There are thousands of men and women who are un- and under-employed, yet manufacturing jobs are moving overseas where labor and production costs are lower, perhaps because environmental protection laws are lacking or poorly enforced. Human population growth across the globe challenges our ability to provide clean water and adequate food to all. Wars displace civilians who are forced into overcrowded refugee camps, which have inadequate facilities and sanitation systems, leading to outbreaks of highly infectious diseases. Our heavy reliance on fossil fuels for transportation and energy correlates to escalating levels of carbon dioxide in the atmosphere and global climate change. Rising global temperatures are shrinking glaciers and polar ice caps and raising levels of the oceans. Problems such as these abound and solutions are desperately needed.

M. K. Demetrikopoulos & J. L. Pecore (Eds.), Interplay of Creativity and Giftedness in Science, 367–380.

Our best chance to solve the problems may well rest on the shoulders of the next generation. It therefore behooves us to encourage the best and brightest in that generation to continue to enroll in science and mathematics classes, to become interested in looking for solutions through the creative application of scientific principles and logical thinking, and therefore to push the boundaries of what we know and understand.

While all children have the right to learn something new every day, and all students should be able to achieve their full potential as they grow and develop and learn, *all* children, *including the gifted* should be offered every opportunity to excel. As Slavin indicated years ago in a Point-Counterpoint article (1990), the question is not whether we should structure our classrooms and learning experiences to meet the needs of all students, but how we should do so. What follows are our suggestions and research findings on how problem based learning offers the most potential, especially when combined with differentiated teaching and learning opportunities, to meet the needs of our most talented, gifted, and creative students of science.

This chapter describes opportunities that are presented to elementary, middle, and high school students through problem based learning (PBL) and other high-end experiences, explores the factors that contribute to or hinder student success in these experiences, and evaluates the impact of participation in these experiences on gifted students' subsequent interest in continuing to enroll in higher level science classes and to consider science related career pathways.

SETTING THE STAGE

Human ingenuity, creative genius, and science offer the promise of solutions for many of our global problems, but we will need to attract and retain the best and brightest minds to the study of science if we are to find workable and affordable solutions in time. Attracting gifted students to scientific career choices would facilitate a better future, so we should be asking ourselves, as teachers and educators, how can this best be accomplished? When interviewed about how they got started in their field, scientists often reveal that their initial interest and fascination with their particular field of study started very early, frequently by the age of ten or twelve, and often through experiences in school with science teachers who were knowledgeable and passionate about science, and enthusiastic about teaching science by providing opportunities for students to *do* science (Archer et al., 2010). Teachers with these characteristics can provide enough encouragement and support for gifted students in their classes to prevent boredom and lack of intellectual challenges from sabotaging the gifted students' interest in school and their enthusiasm for science in particular. Sparking the interest by opening doors to the possibilities and the excitement of new discoveries can make a difference for gifted students especially, so teachers with students who have been identified for their creativity, aptitude for science, and superior cognitive abilities should look for opportunities to challenge and stretch the imagination of their students.

Research studies that have examined why even our most gifted students lose interest in science can provide interesting insights that may improve our ability to attract and retain highly gifted individuals in scientific fields. In an early study, Kahle, Matyas, and Cho (1985) provided a foundation to our understanding about factors that contributed to gender differences in pursuit of science careers. Their work confirmed that the number of mathematics courses students took in high school, the level of achievement students attained in their science classes, and the students' attitudes towards science played a significant role in subsequent career decisions. Further, their study determined while boys were far more likely to have additional extracurricular activities in science outside of school than girls, efforts to increase the opportunities for girls to participate in similar activities dramatically increased girls' attitudes towards science, which also increased the number of science and mathematics courses they took. Fifteen years later Aschbacher, Li, and Roth (2010) found that high school students who experienced success in their science classes, and who received support and encouragement from people who were important in their lives either at home or at school, were far more likely to persist in their studies and to appreciate science, recognize career opportunities, and enjoy doing and learning science than those who did not have these experiences. Jacobs (2005) also identified the importance of mentors who share clear information about what job opportunities are possible with advanced degrees, and acknowledged that students' feelings of competence and interest in science are important factors in choosing science as a career option. These studies, and others, support the conclusion that school districts across the country which recruit and employ knowledgeable and highly engaging science and mathematics teachers, who then engage and challenge their students without being threatened by the sheer brainpower of the most gifted and creative students, will graduate and send on to colleges and universities highly capable students who are prepared to take advantage of the opportunities higher education offers.

WHAT IT TAKES

Teachers need certain skills to be able to accomplish this. Pedagogical content knowledge can be as important as content knowledge and experience (Abell et al., 2008; Appleton, 2008; Settlage, 2013), and experts in a field think about their field differently than novices. Students come to classrooms with preconceptions about how the world works, and formative assessments that teachers use to help students uncover and confront their misconceptions are critical to the students' ability to modify what they thought they knew and to replace their naïve ideas with more scientifically correct understanding (Gomez-Zweip, 2008; Pringle, 2006). Gifted students are well served when they have options and opportunities to challenge themselves through curriculum compacting and acceleration, choice of assignments based on interest, enrichment opportunities such as creative problem solving, developing independent research skills, and using curriculum which features "advanced content, high-level process and product work, and interdisciplinary concept development and understanding" (VanTassel-Baska, 2007, p. 350).

Branford (2000) demonstrated that the rapid rate at which new information and knowledge is discovered and developed makes it less important for students to be able to remember and repeat facts as it is for them to be able to retrieve information and use it to frame and ask more questions that will lead to additional discoveries. The Next Generation Science Standards and a growing body of research support argumentation in science, through which students process information more thoroughly than in the past. Students are now asked to construct scientific explanations, by conducting experiments, collecting and analyzing data, drawing conclusions to answer their initial questions and evaluate their hypotheses, support their conclusions with evidence, and explain their reasoning (McNeil & Krajcik, 2008). Science teachers who have never experienced an open inquiry investigation to learn something new are not likely to feel comfortable providing opportunities for students to learn through even a guided inquiry experience (Capps & Crawford, 2013; Andersen, 2007; Blanchard et al., 2009). Furthermore, differentiation allows teachers to consider their students' strengths, interests, and level of conceptual understanding to learn, and professional development can increase a teacher's willingness to implement the process of differentiation regardless of the grade level of their students (Dixon, 2014; Lightbody, 2004).

Unfortunately, there is no national model for professional development for teachers that could ensure that all teachers of science are well prepared to meet the needs of the gifted students in their classrooms. No standard curriculum is in place to help them provide suitable challenges for their gifted students, although there are several models and a few outstanding curricular options available such as those described in VanTassel-Baska's 2007 comparative content analysis of the characteristics and features of fourteen curricular models. However, without materials or resources from the district, teachers are left to their own devices. Planning time during the school day is never sufficient to create thoughtfully differentiated lessons that provide challenge to high achieving and gifted students. Teachers may never have been taught nor required to prepare different assignments for their gifted students, although they do have to modify their lessons for other special needs students (for whom services are mandated by law). Pressure to do so is increasing at this time, however, because states and public school districts are now tracking the academic gains achieved by categories of students in their schools, including the gifted.

Procedures designed to identify whether various groups of students are making adequate yearly progress are in place, and the results are publicized. Nonetheless, gifted students in district after district are falling short of the gains expected of them. Regression to the mean can explain away only some of the results that are emerging in the analysis, but not all. Many teachers admit using heterogeneous grouping with their students, hoping this will allow all students to learn through collaborative discussions of the content. Many teachers assume that the gifted students will learn the content on their own without any special accommodation or attention. In fact, gifted students do have learning needs that have been ignored, and over time even the most able student can become an underachiever, a trouble maker, or a drop out. Learning to provide more appropriate challenge and support for gifted students requires professional development, and it requires access to quality programs and curriculum.

Professional Development for Teachers

Teachers who want to address the needs of gifted students and help them retain their interest in science can begin by focusing their own professional development on curriculum (what is taught), instruction (how it is taught), assessment (how learning is measured), and the learning environment (including the social culture of the classroom, the physical setting and arrangement). Teachers who do this, and focus on continuous improvement in these four areas by participating in a sequence of professional development opportunities can significantly improve what happens in their classrooms, and can have a positive impact on other teachers in their schools (Loucks-Horsley et al., 2010). The author's prior findings on the amount of professional development support teachers of science needed to be able to differentiate their lessons to meet the needs of gifted students demonstrated that teachers who appreciated the value of inquiry to teach science and had experience teaching inquiry science lessons were able to modify their lesson plans relatively easily (Lightbody, 2004). By planning in advance for students who already had a solid understanding of the science concepts, teachers were able to encourage their gifted and high achieving students to solve open ended problems, to develop a deeper understanding of the science content, and to make connections across science disciplines to related concepts. At the same time, teachers also addressed the needs of students who were struggling, held misconceptions, or lacked prior knowledge of the content under study. Teachers who felt the need to remain in total control of the classroom, who wanted students to remain seated and quiet throughout class, who wanted everyone to be working on the same lesson, and who infrequently offered students opportunities to work together in small groups were less able to differentiate because changes in the classroom environment that typically occur during a differentiated lesson pushed them out of their comfort zone (Lightbody, 2004).

Unfortunately for the gifted students in classrooms where instruction is one size fits all, the lack of challenge has a significant negative impact (Gallagher et al., 1997; Mau, 2003). The slow pace of instruction, repetition of information that the students already know, the inability to move on to new material, few opportunities to develop a personal interest, and an emphasis in the instruction on memorizing facts rather than using information to solve new problems (Coleman et al., 1997) dulls the minds of the gifted, and causes them to either find creative ways to disrupt the classroom or tune out. Contrast this with a classroom in which students are encouraged to identify something of interest to them, set goals, and work independently or with a few other students in the class to achieve those goals.

Curricular Options for Students: Problem Based Learning (PBL)

To help science teachers meet the needs of creative and gifted students in their classes, programs which challenge gifted students and allow them to use their creativity and to experience success need to be identified, and teachers of science need to be encouraged to implement these programs in their schools. Problem based learning can be implemented by the teacher in a variety of ways:

Table 1. Examples of problem based learning options

	Option	*Examples*
1	Modify existing curriculum through differentiation	– Work long division problems using Roman Numerals or – Engage in a highly challenging variation of an experiment in which students control the question and the procedures, and present and defend their findings. – Select a learning challenge from a set of choices as an alternative to the instruction done with the whole class
2	Have students identify a local problem that interests and concerns them	– Find a solution to the invasion of non-natives (http://www.invasive.org) – Address the issue of cars speeding past the school and in the neighborhoods
3	Implement new curriculum	– Try a VanTassel-Baska science unit – Implement Parallel Curriculum
4	Take advantage of new opportunities made available over the Internet	– Start with One Hour of Coding (http://code.org) – Encourage students to create a new app and sell it through the App Store or iTunes – Try the science questions at brilliant.org
5	Engage students in engaging and high quality educational programs	– Science Olympiad (http://soinc.org) – Intel Science Talent Search (Intel STS) – Odyssey of the Mind (Odyssey)
6	Use professional resources available through the National Science Teachers Association (NSTA) (http://nsta.org)	– ExploraVision (http://www.exploravision.org)
7	Participate in programs that integrate science with engineering and or mathematics	– First Robotics (http://www.usfirst.org) – Future City (http://futurecity.org) – Siemens Competition in Math, Science and Technology (Siemens)

In the best PBL experiences, students learn significant content as they work collaboratively to solve the problem. They identify factors that may contribute to the problem, learn about those factors and attempt to propose solutions that will mitigate the problem at the very least, and solve it in an elegant and efficient manner at best. The problems are uniquely suited to the needs of gifted students and offer opportunities for the creativity of the students to emerge and develop, as there are few limits in place on process or product, and there are no artificially low ceilings, glass or otherwise. Teachers who initiate a PBL experience can fully integrate state and national science standards with their mathematics and language arts standards, making the experience a rich learning experience for their gifted students. Teachers are wise to communicate clearly with district personnel, parents, and community members who might be willing to serve as volunteers or mentors, and to line up all necessary collaborators and materials before introducing the problem to the students.

Future City: An Example of Problem Based Learning (PBL)

Given the diversity of the people on the Earth, and their cultural belief systems, the most gifted among us, including those with superior cognitive abilities, those with skills in communication, and those with creative abilities, need to be involved in decision making, development, and implementation of solutions to real world problems. Therefore problem based learning provides a microcosm in which these students and their teachers can explore the boundaries and practice defining the problem, proposing solutions, and making decisions on a smaller scale. In the author's study of the impact of Future City (a problem based learning situation with a computer simulation) on middle school teachers and their students, problem based learning was defined as an instructional strategy that is student-centered, collaborative in nature, open-ended, relevant to the learner and allows for transfer to real-world situations and settings; teachers serve as coaches or facilitators (Huelskamp, 2009). Problem based learning can be described as an instructional strategy in which the teacher or the students identify a real world problem that is messy and ill defined, and use it as a platform for powerful learning.

The literature review outlined in Huelskamp's 2009 dissertation continues to inform work with gifted students in science education. Research from cognitive science on brain development suggests that PBL is developmentally appropriate for students of all ages, but, even though PBL originated in medical schools and is very appropriate for post-secondary students, the fit is especially appropriate for middle school students in general, and for gifted students in particular, given dendrite development at this age and how the brain works. PBL provides students with contextual learning through inquiry and problem solving, and ensures that gifted students experience cognitive challenge and allows them to process and internalize knowledge gained to good effect and application.

Given that PBL delivers the curriculum by combining process, logical thinking skills, creativity, and team-building skills, with product, the acquisition of knowledge and skills, PBL allows students to address real-world problems, not hypothetical case studies, which might have concise, one-path or one-answer outcomes. Thomas (2000) proposed that through the progression of struggling with actual problems, students learn both content and critical thinking skills.

Research into the impact of participating in the Future City Competition from the perspective of the teachers and the students illustrates how one specific problem based learning experience provided a powerful professional development experience for teachers and a challenging growth experience for students. Future City fits into the category of PBL and offers significant opportunities to capture the attention of students, to help them identify areas of science and learning that interest them, and to see themselves working towards future careers in science.

The competition, which is national in reach, is a partnership between the National Society of Professional Engineers and several national corporations. The Future City competition started in 1992, and has been conducted every year since that time, with a growing number of schools across the United States now participating. Since the first year, it is estimated that over 40,000 students in seventh and eight grade have participated in the competition, many of them having participated both years of their eligibility. The national competition is held each year during National Engineer's Week, although participating teams do not have to enter the competition to take advantage of the program. The mission of the Future City Competition is to promote an interest in math, science, technology and engineering through hands-on real-world applications, and the organizers would rather make the program available to as many students and teachers as possible than insist everyone who participates must also be part of the competition.

Future City consists of four parts, and is taken on by teams of students, male and female, who are in 7th or 8th grade. Students start by using SimCity software to imagine, design, create, and manage their own futuristic city. Working with an educator and an engineer mentor from the community, students plan and build a virtual city, which fills with people and grows over time. The software, produced by Maxis, allows students to confront issues in transportation systems, waterways, and electrical production, as well as appreciate the importance of urban planning, maintaining the city infrastructure, and controlling growth. As the students manipulate the program and time passes, social ramifications become apparent, and students confront issues of economics, crime, pollution, quality of education, shortages of goods and services, health issues, taxes, etc. This part of the program is the most intense, but students are given a limited amount of time for this part of the program, and eventually address the remaining parts.

In the second part of the competition, students research and write a well-reasoned and logical essay presenting a solution to an engineering problem, which varies every year but is always academically demanding and complex. Students have to teach themselves enough science content to understand the problem, project the situation into the future, and then apply engineering principles to create as elegant and simple a solution as possible. For some students, this is the least favorite part of the competition; for others it is the most significant and challenging part (IEEE Topline report, 2004). The third task is to build a tabletop scale model of a portion of their city, using recycled materials and within a strict budget. The model must be built within dimensional limits, as those teams in the competition who win at their school, district, and state level competitions are invited to attend the national competition in Washington, D.C, and must take the model with them, through doorways and onto elevators as required by the location. The final part of the competition is for the students to present their ideas to a panel of judges. Judges ask each member of the team to take part in the presentation, to address certain aspects of their city and what they learned, and to answer questions that the judges pose. At the state competitions, an initial judging process will select a limited number of top teams, and a final round of judging in front of a large audience is held to select the ultimate state winner. The competition is intense, and students spend countless hours beyond the school day and on top of their normal school work to complete all sections of the competition. The IEEE report (2004) estimated that students spend on average just over 100 hours during the Future City competition time window.

Giving students roles to play in the context of real-life problems that need solutions is very powerful (Foreman, 2004), so the Sim City software and the Future City program presents precisely the messy, ill-defined problems that make PBL so powerful. In PBL environments, students act as professionals and confront problems with little guidance, a low level of information, and a need to determine the best solution possible by a set deadline. This provides an opportunity for students in the classroom to model the way adult professionals approach problem solving in the workforce (Checkley, K., Glasgow, 1996; Jones, Rasmussen, & Moffitt, 1996)

Driver, Guesne, and Tiberghien (1985) and Weil (1989) state that knowledge stems from challenge and that, in order to have learning take place, current schemas of understanding must be tested as described below:

> Current schema → problems/questions → schema challenged → investigations/observations → solutions/answers → new schema formed → learning → new knowledge.

EVALUATING THE IMPACT OF THE FUTURE CITY EXPERIENCE

To evaluate the impact of the Future City competition on teachers and gifted students, Huelskamp (2009) designed and completed a study in one state during the 2008–2009 academic year. Following a review by a panel of experts and a field test, a questionnaire was given to all teachers in the designated state who had enrolled and competed in the Future City program on the state level that year, as well as all teachers competing at the national competition in 2009. In addition to demographics and background questions, the teachers were asked to self-report on the impact of problem based learning with computer simulation on the frequency of inquiry-based teaching strategies, on the importance of technology education, on the significance of integrating the science disciplines, and on their understanding of their middle level students. Via sampling of the participants, 15 interviews were conducted after the questionnaire. The dataset (n = 101) for this study was generated by a high response rate (91.1% average) from participants, and the robust Cronbach's Alpha reliability coefficient (.912) provided a high degree of confidence in questionnaire results. When the data analysis was complete, some conclusions emerged about the impact of Future City on teachers and students.

A high percentage of the teachers reported that students were highly engaged and took more ownership of their learning. The teachers also reported that problem based learning with computer simulation, such as that provided through the Future City program, allowed them to integrate the curriculum and to help the students make connections, that their students saw the relevance of real scientific knowledge, and that their students developed a far better understanding of the real world than they had achieved through any other instruction. In terms of their teaching, more of the participating teachers who were gifted intervention specialists reported using an inquiry approach in their instruction daily or weekly, compared to classroom teachers, who reported less frequent use of inquiry in their teaching. Teachers who had taken more courses using technology indicated greater comfort with the software, even when glitches, Internet access issues, or computer malfunctions interrupted their students' work. The teachers felt that the SimCity software provided a surprisingly accurate model of issues in city life, and presented challenging intellectual problems that the students had to solve.

Teachers who were interviewed provided additional insight into the value of the program for their students, including increased student ownership of the learning, additional opportunities for diverse students to work together collaboratively, and having fewer constraints on the teaching and learning than more traditional formats and instruction. The teachers also reported that the PBL experience improved their ability to reach "all types of learners" (66.6%) and, specifically, found they were better able to individualize the content for each learner (46.7%) and maximize skill sets (40.0%). These teachers felt strongly that participating in PBL allowed them to better differentiate their instruction, and to understand and reach their students more effectively. Teachers mentioned the students were able to work at their own pace and levels (40.0%), yet were challenged with fewer behavior issues. Given the needs of middle school students, the pressures that gifted students in particular may feel during the turbulent years of adolescence, participating in PBL seemed to provide sufficient gratification for the students, and boosted their self-confidence in their ability to work at the highest levels of mental challenge.

In addition, the Future City organization itself has also been interested in learning more about students' aspirations and career goals. The National Engineers Week Future City Competition, in partnership with the IEEE, commissioned a brief survey among a sample of teachers and participating students to get a better understanding of just who participates in the competition, and how their involvement benefits them

(IEEE, 2004). Their findings were based on a small sample, and should be viewed as informative although not rigorous. Teachers gave high ratings on the quality of the program, and indicated that participating in the program helped develop their students' interpersonal and communication skills, and their organizational skills and ability to co-operate with others. A large majority of students felt that the competition helped by improving their teamwork skills, and was a valuable learning experience. Many students also report a high likelihood of recommending the competition to a sibling or a friend. Further, significantly more seventh grade students report that the competition helped with improving their problem solving skills than the eighth grade students. 88% of the 7th graders and 81% of the 8th graders indicated that the Future City competition had helped them think creatively. When asked about their career goals, the students indicated that they were more interested in science, math, technology, and engineering than other areas of study. Most teachers felt that the competition accomplishes the mission established for the Future City: to promote an interest in math, science, technology and engineering through hands-on real-world applications.

CONCLUSION

The research in this study focused on effects associated with problem based learning with computer simulation in a middle school setting and showed an increased use of promising practices among teachers, as well as a greater understanding of the nature of middle level learners. Problem based learning with computer simulation influenced the participants' overall teaching as the participants had a high level of agreement to promising practices and used them frequently. There are implications that introducing educators to a structured PBL with computer simulation teaching strategy does positively influence middle school teachers. If this can be generalized to other populations, and students in a wider age span, then PBL offers an opportunity for teachers to capture the attention and interest of their students with sufficient strength that the students will be inspired to continue to take advanced science and mathematics courses through high school and college, and to seek careers in scientific fields or engineering, where problems are regularly identified and solutions are developed.

The conceptual model can be used to support and strengthen both in-service teacher education and pre-service instruction in teacher preparation programs. Teachers who implement promising practices and understand the needs of their learners may not only lower professional attrition, but ultimately enable their students to be successful in the international marketplace.

Meeting the right teacher, participating in enough brain tingling activities to sustain and stoke the passion a gifted student develops for science through the perils of adolescence, and, in the case of female students, past the additional challenges of navigating a male dominant society, is part luck and part determination. Getting into the right place and the right time, allowing the young, unfettered, and blinder-less eyes to pursue unique and creative ideas to their logical conclusion, and having good choices for mentors can make all the difference. Many of the leading scientists in the world today, making breakthrough discoveries, proposing new solutions and new ideas, have been mentored by Nobel laureates, and have worked in the very best and most advanced scientific labs (Rothwell, 2002). These opportunities did not fall in their laps because the students were gifted and talented, they happened because the gifted student was able to add to the body of knowledge, to test the limits of her own brain, to ask and answer an intriguing and challenging question. And this can only happen if we find and nurture the dynamos who can lead us to a better tomorrow.

Where are these dynamos now? They are the kindergarteners with bright eyes who are always asking "why." They are the pesky fourth graders who produce a product with a new technology that is so creative that adults marvel at the intuition and insights that produced the idea. They are the unique individuals in our middle school classrooms who ask the questions we don't know how to answer, or challenge an explanation we provided for the class as illogical and wrong thinking. They are the high school students who shake hands with the President of the United States at the Intel Science Talent Search for their outstanding science research. They may even be the college dropouts who launch dynamic new companies from their basement or garage that make innovative products or provide services that serve people across the globe.

If these dynamos are to be successful in later years tackling the global problems that threaten our continued existence on Earth, they will need a high level of content knowledge in the sciences, some prior experience identifying and solving problems, and the ability to combine their understanding of science with other fields of knowledge (Schmidt, 2011). Future City, and programs like it, or other options proposed in this chapter may well hold the key for these dynamos.

Science educators should pay attention, and both capture and support the dynamos as they move through our schools and become our futures. We need to find them early, meet their needs, give them the courage and the determination to continue to follow their interests in the face of criticism and provide them with meaningful academic challenges, appropriate mentoring from trusted family members and talented scientists, and applaud their efforts from the very beginning.

REFERENCES

Abell, S. K., Park Rogers, M. A., Hanuscin, D. L., Lee, M. H., & Gagnon, M. J. (2008). Preparing the next generation of science teachers: A model for developing PCK for teaching science teachers. *Journal of Science Teacher Education*, *20*(1), 77–93.

Appleton, K. (2008) Developing science pedagogical content knowledge through mentoring elementary teachers. *Journal of Science Teacher Education*, *19*(6), 523–545.

Archer, L., DeWitt, J., Osborne, J., Dillon, J., Willis, B., & Wong, B. (2010). "Doing" science versus "being" a scientist: Examining 10/11-year-old schoolchildren's constructions of science through the lens of identity. *Science Education*, *94*(4), 617–639.

Aschbacher, P. R., Li, E., & Roth, E. J. (2010). Is science me? High school students' identities, participation and aspirations in science, engineering, and medicine. *Journal of Research in Science Teaching, 47*(5), 564–582. doi:10.1002/tea.20353

Blanchard, M. R., Southerland, S. A., & Granger, E. M. (2009). No silver bullet for inquiry: Making sense of teacher change following an inquiry-based research experience for teachers. *Science Education, 93*(2), 322–360.

Bransford, J. D., Brown, A. L., & Cocking, R. R. (Eds.). (2000). *How people learn: Brain, mind, experience, and school* (2nd ed.). Washington, DC: National Academy Press.

Capps, D. K., & Crawford, B. A. (2013). Inquiry-based instruction and teaching about nature of science: Are they happening? *Journal of Science Teacher Education, 24*(3), 497–526.

Checkley, K. (1997). *Problem-based learning: The search for solutions to life's messy problems.* In the newsletter of the Association for Supervision and Curriculum Development (ASCD), Summer issue. Alexandria, VA: ASCD.

Coleman, M. R., Gallagher, J., & Howard, J. (1997). *Blending middle school philosophy and the education of gifted students: Five case studies.* Chapel Hill, NC: University of North Carolina, Gifted Education Studies Program.

Driver, R., Guesne, E., & Tiberghien, A. (Eds.) (2002). *Children's ideas in science.* Milton Keynes, England: Open University Press. (Original work published in1985)

Gallagher, J., Harradine, C. C., & Coleman, R. M. (1997). Challenge or boredom? Gifted students' views on their schooling. *Roeper Review, 19*(3), 132–136.

Gomez-Zweip, S. (2008). Elementary teachers' understanding of students' science misconceptions: Implications for practice and teacher education. *Journal of Science Teacher Education, 19*(5), 437–454.

Huelskamp, L. M. (2009). *The impact of problem-based learning with computer simulation on middle level educators' instructional practices and understanding of the nature of middle level learners* (Unpublished dissertation). Retrieved from ProQuest Dissertations and Theses database, (UMI No. 3367883).

Institute of Electrical and Electronics Engineers (IEEE). (2004). *Future city competition top-line summary.* Report prepared by the Response Center and made available by Pender M. McCarter, Program Manager through personal communication.

Jacobs, J. E. (2005) Twenty-five years of research on gender and ethnic differences in math and science career choices: What have we learned? *New Directions for Child & Adolescent Development, 110*, 85–94.

Jones, B. F., Rasmussen, C., & Moffitt, M. (1996). *Real-life problem solving: A collaborative approach to interdisciplinary learning.* Washington, DC: American Psychological Association.

Lightbody, M. (2004). *Onsite professional development: Using differentiation to support instruction in middle school science* (Unpublished dissertation). Retrieved from ProQuest Dissertations and Theses database, (UMI No. 3141744).

Loucks-Horsley, Stiles, K., Mundry S. E., Love, N. B., & Hewson, P. W. (2010). *Designing professional development for teachers of science and mathematics* (3rd ed.). Thousand Oaks, CA: Corwin.

Mau, W-C. (2003). Factors that influence persistence in science and engineering career aspirations. *Career Development Quarterly, 51*(3), 234–243.

McNeil, K. L., & Krajcik, J. (2008). Scientific explanations: Characterizing and evaluating the effects of teachers' instructional practices on student learning. *Journal of Research in Science Teaching, 45*(1), 53–78.

Pringle, R. M. (2006). Preservice teachers' exploration of children's alternatie conceptions: Cornerstone for planning to teach science. *Journal of Science Teacher Education, 17*(3), 291–307.

Rothwell, N. (2002). *Who wants to be a scientist: Choosing science as a career.* Cambridge, England: Cambridge University Press.

Schmidt, A. L. (2011). Creativity in science: Tensions between perception and practice. *Creative Education, 2*(5), 435–445. doi:10.4236/ce.2011.25063

Settlage, J. (2013). On acknowledging PCK's shortcomings. *Journal of Science Teacher Education, 24*(1), 1–12.
Slavin, R. E. (1990). Point-counterpoint: Ability grouping, cooperative learning, and the gifted. *Journal for the Education of the Gifted, 14*(1), 3–8. doi: 10.1177/016235329001400102
Thomas, J. W. (2000). *A review of research on project-based learning*. San Rafael, CA: The Autodesk Foundation. Retrieved March 1, 2004, from http://www.k12reform.org/foundation/pbl/research/
VanTassel-Baska, J., & Brown, E. F. (2007). Toward best practice: An analysis of the efficacy of curriculum models in gifted education. *Gifted Child Quarterly, 51*(4), 342–358.
Weil, S. W., & McGill, I., (Eds.). (1989). *Making sense of experiential learning: Diversity in theory and practice*. Milton Keynes, England: Open University Press.

Mary Lightbody
School of Teaching and Learning
The Ohio State University Newark
USA

Lisa Mary Huelskamp
Westerville City Schools
USA

CARRIE RAINWATER AND NANCY WITTNER

21. DEVELOPING A REBEL WITH A CAUSE THROUGH CREATIVE RISK-TAKING IN GIFTED STUDENTS

INTRODUCTION

Gifted students exhibit insatiable curiosity and a keen ability to make connections. These are characteristics of scientists as well. However, great scientists need the ability to take risks, to think outside the box, and to generate and test hypotheses without being stymied by the fear of failure. Creative scientists accept that scientific research may include a series of failures as theories are tested, reviewed and revised. In order to evoke their creativity, gifted students need to learn how to be risk-takers. Unfortunately, the standard science classroom is an environment of memorization and correct answers, and is not an environment conducive to risk-taking. Such an environment drives students toward perfectionism and a low tolerance for failure. To create the next generation of top scientists, educators must explicitly model and teach gifted students a willingness to embrace novelty in a supportive collaborative environment. Students must learn to expand their current ideas through a process of observation and revision so that they might not only learn to question the world around them, but also create innovative solutions for the problems they encounter.

Gifted students are challenged to mirror what scientists do in the real world through questioning, exploration, observation, discussion, revision and analysis. Questioning and experimentation produce data that drive future questions. Training students in this classic Scientific Method approach provides a solid foundation in facts and laboratory procedures. However, some scientists can lack the crucial element of creativity or a willingness to try something new (Sasso, 2009). This model is designed to challenge even the most gifted thinkers to view failure, not as an end point, but as an opportunity to refine their thinking. As scientists they must not merely learn facts, but solve problems.

This chapter explores model science classroom elements that include: the big idea, questioning, demonstration, making connections, providing evidence, visualization and synthesis. This model runs on a two to three day cycle, using an interactive notebook approach. Socratic discussions and the process of revision run fluidly throughout the model framework.

M. K. Demetrikopoulos & J. L. Pecore (Eds.), Interplay of Creativity and Giftedness in Science, 381–398.

CREATIVITY, SCIENCE AND GIFTEDNESS: IT'S ALL ABOUT RISK-TAKING

In J. K. Rowling's Harry Potter book series, Hermione Granger was the most gifted wizard in her class (Rowling, 1997). Like many gifted students, she was knowledgeable. She had read all the required texts before the term of Hogwarts School of Witchcraft even began. Eager to share her knowledge, her hand shot instantly into the air to answer every question asked in class. Curious and capable, she sought a solution to every problem in class and in life. But aptitude and book learning were not enough to elevate her above the ranks of other talented witches and wizards. Hermione needed to learn a skill not on the class syllabus—risk-taking. Under the tutelage of her brave and daring friends, Harry Potter and Ron Weasley, Hermione learned to step out of her comfort zone to battle trolls, dark magic and her own fears and doubts. Real magic was not rote repetition of an incantation or following a potions recipe in a spell book. Real magic was not an off-the-shelf enterprise. Only by risking failure and trying something that had never been done before did Hermione develop the creativity that elevated her to the brightest witch of her era.

In science classrooms, many gifted students are knowledgeable. Curious and capable, they ask questions and seek answers to the problems set before them. The missing element for many gifted students is the ability and willingness to take risks since risk-taking is what sets apart the best and the brightest scientists in their fields from those that are merely competent. As psychologist Dean Keith Simonton said, "You can't be creative unless you come up with something that hasn't been done before" (Kersting, 2003, p. 40). In order to develop the next generation of Hermiones—individuals who will create paradigm-shifting answers to the scientific problems our world faces—those science teachers currently in the field must explicitly teach our best and brightest students *how* to take risks and *why* they must be taken.

Risk-Taking is Difficult for Many Gifted Students

Learned behavior. The stereotype that gifted students are not creative is espoused by teachers so often that it may have become a self-fulfilling prophecy. While many gifted students are risk-avoidant, it has not been determined if this is a learned behavior or an inherent characteristic. According to Dr. Linda Silverman (1999), a psychologist specializing in the gifted population, avoiding risk is only natural.

> From their earliest years, they [gifted students] have been able to avoid failure and act in a manner that will assure success in their endeavors. They have succeeded in the past, so they expect to be successful in the future, no matter how difficult the challenge. Since they are accustomed to success, and relatively unfamiliar with failure, some gifted children become quite failure-avoidant. (Silverman, 1999, p. 10)

Yet maybe it is not that gifted students are unable to take risks. Perhaps the more chilling notion is that teachers have taught gifted students NOT to take risks by the way they measure success in our classrooms. When teachers continually reward the right answer to questions instead of divergent thinking, they unconsciously condition patterns of behavior in gifted students that may not be in the students' best interests. Silverman also noted that gifted girls tend to not guess unless they are sure that they are correct and that they are often socialized to hide their giftedness (Silverman, 1999).

Gifted students face all the insecurities and self-doubts of all adolescents—only their worries are often magnified. Gifted students often do not fit in with their chronological age peers, and some are socially awkward or on the social fringe of pop-culture-driven middle and high schools (Silverman, 1999). However, even if risk-taking is difficult for gifted students, if teachers set creativity as a measure of success, gifted students will strive to meet these expectations as confidently as they strive to acquire straight A's. Risk-taking may merely lie dormant in gifted students, awaiting the encouragement of teachers and mentors.

Perfectionism. Several studies report that the majority of gifted students exhibit tendencies of perfectionism (Silverman, 1999). To compound this attribute, introverts, who represent over half the gifted population (Silverman, 1999), are inherently cautious. When perfectionism and cautiousness combine, creativity is the casualty in the classroom and the lab. Teachers must ask why perfectionism is so pervasive in gifted students and then find ways to relieve that tension. One reason might be that this characteristic is used to alleviate the boredom that these students face during the drudgery of the repetitive tasks that are found in the average classroom. "*Since I am not being challenged by the material, I will challenge myself to do this dull work perfectly. No room for anything less than 100%. I must make a perfect score.*"

Recently a girl was noticed writing furiously in a small notebook during a class observation. As a gifted student, her classroom behavior was classic early Hermione—first to answer every question, never in trouble, first to finish every task. Was this a student diligently recording the information being provided? Was she outlining the material to study for a future test? Oh, no. Upon peeking over her shoulder to observe her writing, "*I am bored. I am bored. I am bored*" (personal communication) was found in tiny neat letters filling every inch of the lined paper. Bored?? Oh, yes, but still intent on having perfect papers, perfect tests, perfect homework assignments, and being seen as the perfect, always attentive student!

Introverts and underachievers. Unfortunately, some gifted students have decided that the only way to win at this game in our classrooms is not to play. As gifted consultant Lisa Natcharian (2010) notes "Underachievement is often a result of perfectionism, when a child wants so much to be perfect, but is afraid to fail, and so decides not to try at all" (para. 11).

An underachieving gifted student frustrates many excellent teachers, but perhaps the solution lies not in student behavior but in the behavior of teachers. Teachers need to explicitly teach gifted students how to take risks and allow them to practice this approach repeatedly as they would any other vital skill. "If we can show them that learning is a process that by definition involves mistakes, they can begin to focus more on that process, rather than on a perfect product" (Natcharian, 2010, para. 14).

This is especially true in the Science classroom. In the world of Science, failure, mistakes, and dead-ends are often part of the process; indeed, without being willing to encounter failure, few new discoveries would come to light. For the gifted scientist, marrying perfectionism with a dash of risk-taking just may elevate student research to ground-breaking levels, since perfectionism can be a strength if it produces a willingness to pursue one's goals in the face of obstacles, setbacks and failures. An example of this confluence is Sara Seager, a planetary scientist at the Massachusetts Institute of Technology and the first discoverer of an exoplanet atmosphere. She continued her research in the nascent and then controversial field of exoplanets even though this commitment meant she repeatedly failed to obtain faculty positions (Sasso, 2009).

Teacher as Mentor: Nurturing Creative Risk-Taking in Gifted Students

In order for students to become creative risk-takers teachers must teach differently. For years the educational system in many western countries including the United States has used the factory model, treating students like cars on an assembly line. Teachers assemble students piece-by-piece, year-by-year until they have the completed product at the end—a scientist with a university degree ready to take his or her place in a research lab. To produce a different product, a creative risk-taker, teachers need to begin with themselves. Developing creative risk-taking scientists requires a personalized and individualized model of education. As Sir Ken Robinson (2010) said in his 2010 TED talk, "Education must be transformed into an organic, agriculturally based model where students are grown—not assembled."

Presently teachers are trained to educate the mind, filling our students' brains with facts. To cultivate risk-taking in our students, teachers need to become mentors, teaching students not only with their minds but also with their eyes, ears, and hearts. Nurturing risk-taking requires that teachers see students as individuals and personalize education as much as possible.

> Human resources are like natural resources, they are often buried deep. You have to go looking for them. They are not just lying around on the surface. You have to create the circumstances where they show themselves. (Robinson, 2010)

Neuroscience research has demonstrated that students process information using different modalities including visual, auditory or kinesthetic (Fleming,1992). Classroom assignments designed to teach risk taking may include all three components of learning styles in order to maximize the probability of student success. Just as parents hold their toddler's hands when their child rises up to take those first halting steps, teachers need to provide support for fledgling risk takers. Providing opportunities for students to take risks using varied learning styles may provide a support system while the process of risk-taking is learned and habituated.

Risk-taking is scary. Mentor teachers listen to students' fears and concerns. Their job is not to make the process of risk-taking easier, but simply to acknowledge that risk-taking can be hard and/or uncomfortable. Students need opportunities to share their concerns with their mentors and with each other. One approach includes a reflection component where students discuss or write what they learned about a topic. Mentor teachers build into lesson plans small group discussions or blog entries with a questions stem designed to capture student reactions to risk-taking. This helps gifted students think through the worst-case scenarios. Perfectionist gifted students easily discern the myriad ways a project can go wrong. Mentor teachers use listening, not to minimize the difficulties inherent in risk-taking, but to remind students that fear is part of the process.

A 5th grade teacher recently shared her frustration with a gifted student. The girl was categorically refusing to make a short presentation in front of the class. Threatening the student with a zero for the assignment had not reduced her denial; neither had an offer to make the presentation in front of a smaller group. Had this student's concerns been solicited and addressed at the beginning of the project, the teacher possibly could have coached her to take this risk. But at this point, fear had effectively paralyzed this student. The creative scientist keeps going despite his/her apprehensions. Just as a coach trains an athlete to use the adrenalin of arousal to spur peak performance, a mentor teacher does not teach students to sublimate their concerns about risk taking but to channel them. This transforms negative arousal into a powerful drive to follow their scientific curiosity.

Creative scientists take risks because they are passionate about their research. The driving need to know why or how to answer their personal curiosity enables them to take risks that other scientists, lacking that passion, do not. Many science teachers bring a passion for their subject into the classroom. Mentor teachers seek to inspire students to find their own passion—to follow their hearts. Mentor teachers notice what sparks their students' curiosity, even if it is outside the scope of the curriculum. While the demands of the curriculum may not allow time for individual learning projects during the school day, mentor teachers can kindle the passion that drives risk-taking by encouraging students to explore their questions beyond the classroom. Teachers have a unique opportunity to revolutionize education from within. By acting as mentors to provide individualized academic and emotional support for gifted students, teachers can create classroom environments where risk-taking can take root and flourish in every student.

REBELS WITH A CAUSE: RISK-TAKING IS CRUCIAL FOR TOP-LEVEL SCIENCE

Ground-breaking, paradigm-shifting, extending the frontiers of knowledge. These are terms used to describe the scientific research that changes the face of history. Nobel Prizes are awarded to scientists who take risks, scientists who view their fields with new eyes—asking questions or pursuing research avenues other scientists have not. Real science requires risk. To develop creativity in gifted students, first teach them why they need to become risk-takers.

Rebel, outcast, revolutionary. Not many students equate these terms with the practice of science, but they should. That is part of the problem. In the process of teaching science, teachers and textbooks give the impression that the greatest scientific minds of history were recognized as sages in their own times—just as they are today. Most students recognize the names Galileo, Copernicus, and Darwin and possibly even their scientific achievements but they do not perceive them as risk-takers. The stiff portraits in the textbooks belie these scientists' inherently rebellious risk-taking. Teachers need to flesh out the scientific advances that earned these men a place in the annals of science and a featured spot in their textbooks and elaborate upon the amount of risk their discoveries entailed. Perception is reality. Students who aspire to be professional athletes realize that top-tier players have certain physical and psychological attributes that set them apart from average players. We need to explicitly teach gifted students that great scientists are audacious. For example, if gifted students learn that Galileo was a rebel who endured censure and house arrest, (Linder, 2002) their perception of the character traits he needed to be successful will expand beyond just being smart. Our goal is for gifted students to realize that it is not enough to be smart. It is not enough to be knowledgeable. It's not enough to get into the right school. They must be risk-takers as well.

DEVELOPING RISK-TAKING IN GIFTED STUDENTS

Changing How We Teach Science

How we teach Science matters. Lesson presentation and teacher expectations for our gifted students absolutely impact their behaviors. Currently, teachers and textbooks present the study of Science as a list of facts—largely static and unchanging. Students learn the vocabulary, the laws, the properties and the systems by rote methods, and even our classroom experiments have predetermined outcomes. Because of the way it is taught, gifted students quickly conclude that Science is a set of questions with one correct answer for each. History and Science are presented in much the same way—names, dates, formulas and definitions. No wonder waggish students refer to these classes as the dead and the done.

While students must acquire a thorough foundation in the facts and formulas of Science, a crucial distinction is lost when scientific knowledge is only presented as a series of facts to be memorized. Unconsciously, students internalize an unfortunate message. All the important questions in Science have been answered. I have nothing to add as a scientist because there is nothing new left to discover. Science does not involve risk-taking because we have already discovered everything important or interesting.

Presentation matters. Teachers need to explicitly model Science as an ever-evolving discipline where exciting new discoveries are being made every day. While students might be aware of ongoing advances in medicine and certainly they are cognizant of new technology, often students do not equate medicine and invention with Science with a capital 'S'. Science, as presented in today's classroom, is merely the chapters covered in the textbooks—the 'ologies', biology, geology and cosmology. Rescripting Science as a fluid and dynamic endeavor both worth pursuing and requiring a dash of risk can be accomplished without major changes to the Science curriculum. In fact, with the correct presentation of the material, only tweaks—a change of scientific attitude—will be needed.

Given that fact, what can a teacher do to create awareness of Science as an exciting, ever-changing subject? Consider these ideas! First, new discoveries in all areas of science are published every day. Every morning *Science Daily*, publishes an internet-based compendium of late-breaking scientific news. A recent morning's perusal netted the following: a novel use for a key enzyme in cancer therapy, previously unknown courtship rituals in fish, self-assembling nanocubes, and an explanation for weird lunar soil behavior. And these are just samples from hundreds of papers published every day by working scientists. Any teacher can subtly shift students' perception of Science from a rehashing of the previously discovered to an exciting moving target, full of questions that still need to be answered, perhaps even answered by themselves by establishing a Using a New Science Discovery of the Day teaching segment.

Theory Busters is another inquiry activity to develop risk-taking that is based on the Discovery Channel's show entitled Myth Busters. Today's refuted theories, quaint and sometimes mildly amusing, were yesterday's prevailing consensus. For example, in the 1930s, cigarette smoking was considered healthy and restorative. Today, cigarette packages come with graphic labels warning of the dangers of smoking. Presenting science in the context of this is what scientists know now shifts the world of science from a dogma of facts to a flow of changing information. A Theory Busters module can highlight an aspect of science as important as the scientific method, the recognition that scientific theories are sometimes falsified, or at least, updated when new data renders them outmoded.

For example, an earth and space unit could highlight famous scientific theories that turned out to be wrong from *TopTenz.net*. In the nineteenth century, scientist Urbain Jean Joseph Le Verrier thought he had discovered a new planet located between Mercury and the Sun, which he named Vulcan. According to Le Verrier, only the existence of a planet in that location would explain the odd fluctuations in Mercury's orbit. Decades later Einstein explained Mercury's orbital peculiarities with his theory of general relativity which did not require the theoretical planet Vulcan and so the idea that there was a hidden planet fell out of favor and was eventually dropped. In Chemistry, the Phlogiston Theory sought to explain combustion and rusting of metals by means of a special element called phlogiston that was released during burning, thus causing a reduction in weight. Unfortunately for its leading proponent, Johan Joachim Becher, it was later determined that some metals actually gain weight after burning—thus oxidation replaced phlogiston in the science textbooks. Even Einstein was capable of making a mistake. He referred to his theory of the Static Universe—that the size of the universe was an unchanging constant—as his biggest blunder. Scientists still do not have consensus as to whether the universe is ultimately expanding or contracting. Hopefully, gifted students will learn from and internalize the message that it is OK to go out on a limb, to take a chance in science, because even Einstein, possibly the greatest scientist of our time, made a very public mistake.

Gifted students, along with young people everywhere, consider anything that happened before they were born as ancient history. The discoveries of Johannes Kepler, Louis Pasteur and Edwin Hubble—although each of these men lived and worked in different centuries—coexist neatly in students' minds as an equivalent long ago. A more immediate way to revise students' perception of science from a static discipline into an ever-evolving one is with a time-line of new scientific knowledge gained within their own lifetimes. A Science in My Lifetime segment would begin with key ideas discovered or theorized about a particular scientific topic in the year of the student's birth. The next segments of the time-line would include data gained each year up to the present day. In the final segment, labeled Today, current, up-to-date information would be listed. Students would be encouraged to add to this section of the time-line as they regularly monitor science news sources for recently published research on the topic being studied.

For example the Science in My Lifetime for an Earth and Space unit would note that 10 years ago astronomers discovered 11 new moons orbiting Jupiter, making it not only the largest planet in our solar system but the one with the most moons- a total of 39 (Hartman, 2002). Other entries might include the first possibly earthlike extrasolar planet that was discovered in 2007—warm enough to have liquid water on its surface (Plait, 2007). In 2008, NASA's Phoenix's Mars Lander collected water ice, the first evidence of this main ingredient for life (Bryner, 2008). The timeline—a concrete visual representation of the mutating nature of scientific knowledge—explicitly points out to gifted students that there are new discoveries to be made in science and that they just might be the scientists who make them.

Practicing Risk-Taking—We Are What We Do

It is not enough for gifted students to internalize the message that Science is an exciting field with important discoveries made literally every day. To aspire to be an audacious scientist, a trail-blazing explorer, gifted students must learn to overcome possibly their biggest challenge. *They need to become risk-takers.* For as Sir Ken Robinson states in his popular TED talk on creativity, "If you're not prepared to be wrong you'll never come up with anything original" (Robinson, 2006).

Since risk-taking is essential to ground-breaking science, teachers must start early to create a pattern of behavior that ingrains the process of risk-taking in our gifted students just as they teach them the Scientific Method. Aristotle wrote "We are what we repeatedly do. Excellence therefore, is not an act but a habit."

A professional basketball player does not decide at age 20 to pick up a ball, shoot a few hoops and, voila, sign a multimillion-dollar contract with the Chicago Bulls. No, a professional athlete generally begins at a young age practicing his skills at a local recreation center or at school, day after day, year after year until he attracts the attention of an agent or scout. The key ingredient is practice, the repetition of a set of movements over and over until it becomes engrained in what is popularly termed 'muscle memory'. To develop top-tier scientists, science teachers need to recreate this system of repetition, providing gifted students the opportunity to learn how to take risks, and then to practice this attribute until it is internalized—until it becomes part and parcel of who they are as science students and who they will become as adults. Fortunately, risk-taking can be built into the existing science curriculum through a series of small tasks.

A first step would involve encouraging gifted students to become comfortable with making guesses. Most gifted students are comfortable raising their hands in the classroom, because they usually know the right answer. However if the answer is outside their compendium of knowledge, many gifted students will not readily venture a guess. In the Happy Scientist Photo of the Day activity students, must practice both their observation skills and their willingness to be wrong. Robert Krampf, the self-titled Happy Scientist, publishes a daily blog with The Science Photo of the Day. Recent selections included a photo of raindrops with the question, do you get more wet by walking or running through the rain? Another entry posted a strange object spotted hanging from a cypress tree. Students were challenged to scan the photograph for clues to answer the accompanying question (Krampf, 2012). This creates an opportunity for gifted students to practice the first baby steps of risk-taking, as they must hazard a guess, and back it up with evidence from their background knowledge and close examination of the picture. Students then lay claim to their theories by committing their statements to writing in their science notebooks and even discussing it with others. This public extrapolation without certainty is uncharted territory for many gifted students.

In today's science classrooms, teachers spend a great deal of time exposing students to facts and developing what is commonly termed critical thinking. Handel (2011) describes critical thinking as the ability to formulate a correct answer through analysis and dissection of a problem, but points out that while schools provide opportunities for students to seek specific identified answers, they do not provide opportunities for students to engage in divergent and creative thinking.

Sir Ken Robinson, a creativity expert agrees: "Creativity is as important in education as literacy, and we should treat it with the same status" (Robinson, 2006).

By incorporating simple yet powerful divergent thinking tasks into classroom routine, gifted students will practice risk-taking by thinking outside the box. Since these open-ended activities are graded based upon participation, not upon outcomes, gifted students are able to practice these risk-taking drills in a supportive environment. A classic exercise in divergent thinking is the Alternative Uses Task. A science classroom is filled with tools and equipment. Familiarizing the students with their correct use is an important classroom procedure. To practice risk-taking, challenge the students to brainstorm as many diverse uses as they can come up with for a test tube or pipette or any other piece of equipment. Encourage them to go beyond the realm of ordinary ideas by striving to list at least 20 alternative uses. Then ask students to work with a partner or small group, sharing their ideas and collaborating with each other in order to expand their lists. By publicly claiming their ideas, even though some of them are offbeat, students experience a greater degree of risk-taking. Including this short exercise regularly when new equipment is introduced or before it is put away incorporates a layer of divergent thinking into any science classroom.

Another classroom activity that encourages risk-taking practice can be introduced through an open-ended divergent thinking exercise—Connected Note Taking. Described as a stream of consciousness style of structured note-taking, this exercise builds gifted students' tolerance for risk-taking by requiring them to capture not only the main ideas and details in their notes but also to involve themselves in two additional divergent thinking tasks. In Connected Note Taking, students divide a blank sheet of paper into three sections or boxes. In the first box, students write the vocabulary definition or key fact gleaned from the class discussion. In Box Two, students must create their own unique illustration to depict the concept written in Box One. To ensure that this exercise builds their creativity, do not allow students to merely copy textbook depictions. This activity is surprisingly difficult for many of my gifted students. They may dislike coming up with their own illustrations when an approved image is available. Since their pictures are required be substantially different from those in the textbook, students are encouraged to close their eyes and visualize their concepts first by making a movie in their minds. In Box Three, students jot down words, memories or any other personal connections that come to mind about the topic. The element of risk is most evident in their personal connections to the key idea. One gifted student's connection to the concept latitude and longitude on maps was that she remembers longitude because she has long hair that goes up and down and longitude has the word long in it.

To further encourage risk-taking, students share their notes with a partner or small group. At first some gifted students are uncomfortable sharing their personal connections with others since they often have to explain how their association or memory is connected to the key fact. To ensure that students practice risk-taking safely, it is key that all students in the classroom are taught to collaborate respectfully, without hurtful comments or laughter.

A perennial part of the science curriculum is the end-of-chapter or unit review that includes divergent thinking and the practice of risk-taking by using Mind Mapping as a review technique. Mind mapping was originally developed by Tony Buzan as a creative way to organize information using key words and icons branching from a central theme depicted in a circle in the center of a blank sheet of paper. Main ideas radiate outward from the central theme circle with a line and a single key word. Usually each idea is represented with a different color. Subtopics or details branch from the main ideas, like tributaries flowing from a river (Margulies, 1991). Mind Mapping is another example of incorporating risk taking by forcing students to make a choice when there is no one right answer. While students will generally select the same key facts from a unit, the words and icons each student selects to encapsulate the main ideas and detail will be different.

These activities may be only small forays into the realm of risk-taking, but first steps are important. As students practice these simple classroom activities they will create new habits and new patterns of behavior that will serve them well as they become comfortable taking larger constructive risks in classroom discussion, science fair projects and university level research.

IMPLEMENTATION: A PRACTITIONER'S NOTES

All of the strategies discussed, while having their origins in a reading classroom, can and should be applied in every science classroom.

Imagination is more important than Knowledge. (Albert Einstein, 1931)

Perhaps one of the most imaginative people of our times, Albert Einstein valued creativity. For Einstein, it was the key to developing his thought experiments. He believed ideas could be understood in many ways. The exact steps of the thought experiment have not been documented, but at its root is curiosity. The biggest challenge for teachers of any student and any subject is to get the student to be actively engaged. This is referred to as creating efficacy for learning. This can be very challenging in a world of smart phones and social media. Children and adults are bombarded with images, videos, sounds and messages. Processing content has become fast-paced—so much so that modern day scientists filter through massive amounts of totally useless and conflicting information in order to arrive at what is seemingly at best a possibility. Most students do not have the patience to formulate ideas in such an environment. At the core of Science is problem-solving. The models most commonly used for teaching new knowledge in science are varied, but typically involve introduction of new materials through notes and interactive labs that help simulate new knowledge and application through inference. Student's work with and store new information and are later tested in isolation from any real application of their newly learned facts. Most standardized testing deals with problem solving through a complex cause and effect relationship.

The If–Then model

In this model, gifted students are challenged to create new learning pathways through questioning, exploration, observation, discussion, revision and analysis. This model mirrors what scientists do in the real world. Questioning and experimentation lead to relative data that drive the development of future questions. Training students in this way creates habits of mind that result in the ability to solve difficult problems. Our gifted students have always had the ability to learn knowledge, but they sometimes lack the crucial element of creativity, which is a willingness to try something new. As scientists they must not merely learn facts, but solve problems.

Thinking has two components, imagination lies at one end of the model and logic lies at the opposite extreme. Creativity requires a little of both, to be creative implies to create. The goal is to inspire new ideas from old truths. You start with basic knowledge and then you apply that to new applications. An example of that would be as follows. Fiber is required for the proper digestion of nutrients into the bloodstream, however, the longer the food stays in the stomach the more likely the nutrients will be available for dissolution. This concept can be applied to other situations such as how plants obtain their nutrients. This type of learning requires some basic knowledge in order to make the connection necessary to improve a method. Now this is just a thought, an analogy, of how one system may relate to another. However, this is the first step to experimentation. Science requires basic knowledge and then it requires curiosity. Accepting ideas only at face value, does not allow the creative state to develop, which could be defined as looking for answers and then seeking different applications for the same analogy of a system. So in teaching students, there should be an expectation to do more than teach basic facts. The expectation should be to teach basic problem-solving and emphasize the importance of connecting these ideas to new applications. So how does learning progress from basic knowledge to application, which is the basis of creativity? One way to answer this question would be through rethinking the way teachers build their lessons. Science instruction must be taught through the scientific method. This process starts with a question, researching possible solutions, being challenged to make connections between this information and new applications, providing evidence to support the hypothesis, visualizing outcomes, and synthesizing through experimentation and questioning. This process describes a possible explanation of the traditional thought experiment—*If* this is true, *then*, this will happen.

Einstein (1919) published an essay entitled the, Induction and Deduction in Physics, where he wrote;

> The most simple picture one can form about the creation of empirical science is along the lines of an inductive method. Individual facts are selected and grouped together such that their lawful connection becomes apparent. By grouping these laws together, one can achieve other more general laws until an more or less uniform system for the available individual facts has been established—such however, that the intellect, looking backwards, could arrive at the individual facts reversely in a merely mental way. (p. 1)

The method of Science is logic; much like a detective makes observations and collects evidence to arrive at a conclusion based on factual data. But the detective also uses other tools—intuition is arguably as important, and in many instances it determines the outcome of the case. Following the right lead is crucial.

What it looks like. It begins with a notebook. All scientists must be good at writing down their observations. The notebook is nothing new. Interactive notebooks have been used effectively for many years. The only twist is the way they are viewed by scientists. The right side is for If, and the left side is for Then. The If side is for exploring and the Then side is for explaining. This will be the basis for our collection of data. Student participants should always be operating out of a If (why) or Then (how) state of mind.

The interactive notebook is set up with a table of contents, explanation of strategies and a rubric on the final page. The right side is for teacher-directed learning and the left side is for student-directed exploration. Each day in class the student will have an input and an output page. If the student requires more than one page for input or output, then the student can glue additional pages to the bottom of the page where more room is needed.

There are many models for the Interactive Notebook (INB) all over the web; the most common version used is closely modeled after the AVID notebook. The model you choose is unimportant, however consistency within the classroom is crucial. The students must know where things go; this routine is established through constant modeling. Assessing notebooks weekly is an important part of the experience. The students do a self-assessment. They have a quick reference checklist, which is generated by the instructor from the running table of contents. They are assessed on accuracy, completion, creativity and style. This self-assessment holds them accountable, and it keeps them on track. It could be said that this, literally keeps us on the same page.

Big idea scientific concept. Each science classroom has a set curriculum—big ideas that must be covered. This is the basis for teaching. So, it is suggested that the lesson starts with the singular facts that Einstein wrote about. The Big idea is the cornerstone of the unit. Most Big Ideas are covered for several units, and facts are taught in isolation. For example, the earth's motion, the earth and its moon, the inner planets and the outer planets may be covered in a unit about systems in space. These concepts would be introduced as separate chapters of a unit, singular facts, disconnected. However there are basic laws at play between all these systems. The laws of motion and gravitational pull do not change, they are the same and only the mass of the objects in a system changes its unique path. Exploration requires some basic facts. What do we know? After engaging in a discussion of these basic facts, the first challenge is to try to make sense of them. How are these similar and are they different. What patterns developing between these objects can be seen? All of these things orbit something else, perhaps. Yet, how can this be explained within the hierarchy of their orbital relationships? This may be one question posed. During the exploration phase, questioning and modeling are imperative. There must be access to various materials in order for the students to explore. Their observations would be recorded. In the explanation phase, the students must make a claim, such as, the closer things are to the sun, the more solid the surface. In this stage, the student would be challenged to make a claim, discuss it using a Socratic seminar and support it with evidence. By creating an environment where debate is encouraged, gifted students can thrive.

Questions student generated. The *If-Then* model is about the grouping and regrouping of ideas and discovering through discovery. Let's break it down even further. Perhaps the most difficult lesson to teach in science is Earth and Space. This is a very abstract concept for many students since there is not any real way to experience the magnitude of the universe without creating awkward models that do not give the true depth that the students need to grasp the vastness. A random list of facts is presented including basic facts about the planets themselves and their location in the solar system. Articles are read about their discovery, video clips are supplied and basic simulations are done as part of a group interaction. How can this process be imagined differently in order to inspire creativity? Start with discovery. There is a great simulation found on the Internet called, The Toilet Paper Universe. Kids love that toilet paper is used. In this simulation a scale is chosen to map out a mini solar system constructed by unrolling the toilet paper according to the incremental scale in order to view distances between objects in space. This visual simulation would be the start of the unit, afterwards posing questions based on what is seen and observed. Rather than saying, Wow! Look kids. Look at the distances; pose the question, what relationships exist between these objects? As the students begin to explore this question they uncover truths, such as the inner and outer planets are similar and different in certain specific ways. Distance from the sun creates certain characteristics in objects. They begin to predict behavior and composition of objects based on the location in our solar system. That is to say, the students discover the facts through exploration instead of note taking. They are participating in a three dimensional experience to come up with a one-dimensional idea. The students are taking singular observations and regrouping them into ideas that can lead to new discoveries. The students become the explorers. They form a kindred relationship with the scientists and share in their delight.

Making connections. For creativity to occur, an experimental environment where the groundwork for discovery is modeled through exploration by design has to be the focus. This is nothing new in education and many teachers teach through hands-on activities yet, advocating an advanced thought process is uncommon. Novel ideas are part of the discovery process and for each unit of study, the student is encouraged to produce at least one novel idea. They can accomplish this by making a connection between learning and observation to create a new idea. These ideas may not be entirely novel, and may just be new to them. Discovery has a fluid nature and is constantly changing as new thinking reshapes prior ideas and understanding. When an environment is created based on the concept of fluidity, students begin to feel safe to create new thinking and to take risks.

This challenge of a puzzle or problem is critical for gifted students. As part of this model, the student helps to develop challenges through questioning, looking at the facts, and trying to discover the connections between them, to develop these habits of mind. In this model the teacher is part observer, part facilitator. The teacher provokes thought through questioning, but patiently waits for the student to bring forth answers. The teacher plays the role of questioner; the student plays the role of explorer. In many ways patience is the most important virtue for the teacher, the ability to hold back the knowledge and allow it to be discovered somewhat naturally.

The connectivity piece is most important in that connections to new ideas allow new discoveries to take place. Creativity is born from making connections between ideas and bridging a gap between traditional applications. In this phase of thinking, teachers ask students to create an application for the new knowledge learned. Students compare two systems and create a hypothesis based on the new information. The statement itself need not be completely correct but it needs to be internally consistent and logical. An example might be the law of conservation states that matter cannot be created or destroyed then it can be said that the weight of a person would fluctuate as meals are eaten due to the transfer of matter. Now, it is understood that this is not entirely correct in that work is required to eat the food which expends energy which is produced from the conversion of matter to energy. However, this hypothesis could be tested, and the student could be encouraged to find out why this hypothesis does not hold true. The connection component is the synthesis of the idea; this is the proof in the pudding.

Providing evidence. During the research component of the students' experience they must look for evidence to support their hypothesis, belief or idea. They can bring in articles, journals, noted observations or excerpts from texts. They must make a claim and support it. This is an excerpt from the Common Core for implementation in 8th grade science. Over the course of the next two years, all classrooms will be implementing this national curriculum. It is important that teachers begin to incorporate the components listed below into their planning in order to make the transitions smoother.

Write arguments to support claims with clear reasons and relevant evidence.

a. Introduce claim(s), acknowledge and distinguish the claim(s) from alternate or opposing claims, and organize the reasons and evidence logically.
b. Support claim(s) with logical reasoning and relevant evidence, using accurate, credible sources and demonstrating an understanding of the topic or text.
c. Use words, phrases, and clauses to create cohesion and clarify the relationships among claim(s), counterclaims, reasons, and evidence.
d. Establish and maintain a formal style.
e. Provide a concluding statement or section that follows from and supports the argument presented.

During this stage students reflect on new ideas by supporting or refuting their understanding of what is known. Students can participate in debates, such as Socratic seminars, philosophical chairs and four corners. This will allow students to actively engage in the content, while formulating ideas. So writing becomes an important aspect of the output or explanation model it the INB. The students must make predictions and argue their relevance through research of multiple resources.

Visualization. Being able to visualize is very important in the creative process since you must first process your understanding of that concept or idea. It is necessary for students to process in multiple ways in order to truly know or find new meaning. When faced with a difficult concept, students could first be asked to visualize the process or law behind what is taking place. They may not understand all the complexities, but they can explain what they see and encouraging students to illustrate their new learning will support later recall. Illustrations allow students to communicate what they are not only seeing but also what they are thinking. Students can compare these illustrations and ask questions of one another in order to develop new conversations.

For example, there is a demonstration in class of a gentleman walking across a bed of hot coals. Students could be asked to illustrate what they believe may be happening scientifically to allow for this to happen without causing harm to the participant. Some will draw a picture of a magician; some will draw a picture of religious icons, believing this phenomenon is due to a transient state. Others will illustrate with question marks, offering no explanation. Once all the students have put forth an explanation, they will begin to peer-review the illustrations. What commonalities can be seen? Many students would believe this phenomenon to be mystical. Now, the instructor can interject, this is based on a very basic scientific principle. At this point, take suggestions from the students. As a group, choose four possible solutions, and even if the correct answer is not represented, the students will then place themselves in the corner of the room that represents their choice. The class will then begin to discuss their theories through this four-corner debate, to support or refute these hypotheses. The conductivity of coals is very low, because this is evident, if the subject walks very quickly, then the coals will not be able to transfer energy through heat. The instructor should not participate in the discussion; they are merely going to watch the discussion run its course, until the proper response is generated. The teacher may pose questions such as, how would the speed of motion effect the outcome? What surfaces are most commonly used for this activity? These questions will serve as a model of digging deeper. Allowing the students to consider more than what they see, they are encouraged to give meaning to what they see. If the students do not arrive at an acceptable answer in our ten-minute activity, then the question remains on our questions board, until someone brings in a viable solution. This encourages exploration. The illustration component helps students to make connections to new material through visual representations, which we know helps in the process of learning and creativity.

Synthesis. This is the point where all learning is pieced together to create new understanding of a concept. Think of it as the reflective component, where learners demonstrate their understanding of a topic based on new ideas. The synthesis part of the lesson is best evidenced by a paragraph written by the students that discusses their understanding of a concept and the most important components of that topic. Students could create visual collages or multimedia web pages that link old thoughts to new understandings. This component is generally in a project based format that is student driven by design.

CONCLUSION

Gifted students have the potential to become top-level creative scientists who will take on the challenges of the 21st century. In order to reach their full creative abilities, gifted students must be willing to step out of their comfort zone and learn how to take risks. Classroom teachers can nurture these nascent risk-taking skills through mentoring, explicit instruction and ongoing practice. Using higher order critical thinking skills, drawn from reading comprehension strategies, science teachers can develop habits of mind that support creative risk-taking. Finally, implementing an interactive notebook in daily classroom instruction models and refines this process of observation, inquiry and renewal. This model is designed to challenge and support gifted students as they explore the exciting field of science and begin their own journey toward scientific genius.

REFERENCES

Bryner, J. (2008, December 23). Top 5 amazing astronomy discoveries in 2008. *Space and NASA News.* Retrieved July 21, 2012, from http://www.space.com/6247-top-5-amazing-astronomy-discoveries-2008.html

Calle, C. (2005). Essential facts about Albert Einstein's life. In *Einstein for dummies*. New York, NY: Wiley Publishing.
Curious about Astronomy? Extrasolar Planets. (2011, December 16). *Curious about astronomy? Ask an astronomer*. Retrieved June 28, 2012, from http://curious.astro.cornell.edu/extrasolarplanets.php
Einstein, A. (1919). Induction and deduction in physics. *Berliner Tageblatt*, Beiblatt.
Einstein, A. (1931). *Cosmic religion: With other opinions and aphorisms*. New York, NY: Covici-Friede.
Fleming, N. (1992). Not another inventory, rather a catalyst for reflection. *To Improve the Academy*, *11*, 137–142.
Hartman, H. (2002). Roundup of recent science discoveries, 2002. *Infoplease*. Retrieved July 21, 2012, from http://www.infoplease.com/ipa/A0905090.html
Kersting, K. (2003, November). What exactly is creativity?: Psychologists continue their quest to better understand creativity. *American Psychological Association 10, 40*. Retrieved August 7, 2012, from http://www.apa.org/monitor/nov03/creativity.aspx
Krampf, R. (n.d.). Science videos and experiments by Robert Krampf, the happy scientist. *Robert Krampf, The Happy Scientist*. Retrieved July 1, 2012, from http://thehappyscientist.com/
Linder, D. (2002). The trial of Galileo: An account. *UMKC school of law*. Retrieved July 27, 2012, from http://law2.umkc.edu/faculty/projects/ftrials/galileo/galileoaccount.html
Margulies, N. (1991). *Mapping inner space: Learning and teaching mind mapping*. Tucson, AZ: Zephyr Press.
Natcharian, L. (2010, July 15). Real learning: Meet the perfectionists [blog]. *masslive.com*. Retrieved June 10, 2012, from http://blog.masslive.com/ real_learning/2010/07/meet_the_perfectionists.html
National Governors Association Center for Best Practices. (2010). *Common core state standards*. Retrieved July 29, 2012, from Common Core State Standards website: http://www.corestandards.org/the-standards
Plait, P. (2007, April 24). Huge news: First possibly earthlike extrasolar planet found! *Discover Magazine*. Retrieved June 28, 2012, from http://blogs.discovermagazine.com/badastronomy/2007/04/24/huge-news-first-possibly-earthlike-extrasolar-planet-found/
Robinson, K. (2006). Ken Robinson says schools kill creativity. *TED: Ideas worth spreading*. Retrieved June 19, 2012, from http://www.ted.com/talks/ken_robinson_says_schools_kill_creativity.html
Robinson, K. (2010). Sir Ken Robinson: Bring on the learning revolution! *TED: Ideas worth spreading*. Retrieved June 19, 2012, from http://www.ted.com/talks/sir_ken_robinson_bring_on_the_revolution.html
Rowling, J. K. (1997). *Harry Potter and the sorcerer's stone*. New York, NY: Scholastic Press.
Sasso, A. (2009, October 9). Audacity, part 2: A blueprint for audacious science. *Science Careers, from the Journal Science*. Retrieved June 11, 2012, from http://sciencecareers.sciencemag.org/career_magazine/previous_issues/articles/2009_10_09/caredit.a0900120
Silverman, L. (1999). Perfectionism: The crucible of giftedness. *Advanced Development Journal*, *8*, 47–61.
Top 10 Most Famous Scientific Theories (That Turned Out to Be Wrong). (n.d.). *TopTenz.net*. Retrieved July 1, 2012, from http://www.toptenz.net/top-10-most-famous-scientific-theories-that-turned-out-to-be-wrong.php

Carrie Rainwater
Manatee County Schools
USA

Nancy Wittner
Manatee County Schools
USA

Lightning Source UK Ltd.
Milton Keynes UK
UKOW06f1106080216

267939UK00001B/11/P

9 789463 001618